On Flexibility

On Flexibility

**RECOVERY FROM TECHNOLOGICAL
AND DOCTRINAL SURPRISE ON THE BATTLEFIELD**

Meir Finkel

Translated by Moshe Tlamim

STANFORD SECURITY STUDIES
An Imprint of Stanford University Press
Stanford, California

Stanford University Press
Stanford, California

English translation © 2011 by the Board of Trustees of the Leland Stanford Junior University. All rights reserved.

On Flexibility was originally published in Hebrew in 2007 under the title *Al Hagmishut* © 2007, Ma'arachot.

No part of this book may be reproduced or transmitted in any form or by any means, electronic or mechanical, including photocopying and recording, or in any information storage or retrieval system without the prior written permission of Stanford University Press.

Special discounts for bulk quantities of Stanford Security Studies are available to corporations, professional associations, and other organizations. For details and discount information, contact the special sales department of Stanford University Press.
Tel: (650) 736-1782, Fax: (650) 736-1784

Printed in the United States of America on acid-free, archival-quality paper

Library of Congress Cataloging-in-Publication Data

Finkel, Meir, 1968- author.
 ['Al ha-gemishut. English]
 On flexibility : recovery from technological and doctrinal surprise on the battlefield / Meir Finkel ; translated by Moshe Tlamim.
    pages cm
 Includes bibliographical references and index.
 ISBN 978-0-8047-7488-8 (cloth : alk. paper) -- ISBN 978-0-8047-7489-5 (pbk. : alk. paper)
 1. Military readiness. 2. Military doctrine. 3. Military art and science. 4. Surprise (Military science) 5. Military history, Modern--20th century. 6. Technology--Military aspects. I. Tlamim, Moshe, translator. II. Title.
 UA10.F5413 2011
 355.02--dc22

2010045287

Typeset by Bruce Lundquist in 10/14 Minion

To my wife Amira, and children Yael, Alon, and Yiftach
for their patience and support

TABLE OF CONTENTS

Acknowledgments	ix
Introduction	1

PART ONE
THE CHALLENGE OF FORCE PLANNING FACING FUTURE SURPRISES — 19

1. Prediction and Intelligence—The Dominant Approach in Force Planning and Its Failure to Answer the Challenge of Technological and Doctrinal Surprise — 21

PART TWO
FLEXIBILITY-BASED RECOVERY—A THEORETICAL VIEW — 53

2. Conceptual and Doctrinal Flexibility — 55
3. Organizational and Technological Flexibility — 73
4. Cognitive and Command and Control (C2) Flexibility — 98
5. The Mechanism for Lesson Learning and Rapid Dissemination — 111

PART THREE
RECOVERY FROM SURPRISE—A HISTORICAL VIEW — 121

6. The German Recovery From the Surprise of British Chaff — 123
7. The German Recovery From the Soviet T-34 Tank Surprise — 138
8. The Israeli Recovery From the Egyptian Sagger Missile Surprise — 150
9. The Israeli Air Force Recovery From the Arab Anti-Aircraft Missile Surprise — 164
10. The Slow British Recovery From the German Armor and Anti-Tank Tactics — 179

11	The Slow Soviet Recovery From the Surprise of Low-Intensity Conflict in Afghanistan	191
12	The French Failure to Recover From the Surprise of the German Blitzkrieg	205
	Summary and Conclusions	223
	Appendixes to Chapter 1	233
	Notes	249
	Bibliography	301
	Index	321

ACKNOWLEDGMENTS

I am deeply indebted to my Ph.D. thesis supervisor Dr. Avi Kober of Bar-Ilan University. His intellectual and methodological approaches have been veritable guiding lights, and his classical style of mentor-pupil instruction the ideal relationship for me.

I began my research as a student in the thirty-first class of the National Security College and was fortunate to have the willing support of many people. Professor Gavriel Ben Dor, the college's academic advisor, and Dr. Uri Bar-Josef of Haifa University were my M.A. thesis advisors on military intelligence in force planning.

Sincerest thanks are extended to Eado Hecht for his incisive comments on my analysis of historical events, especially the parts on Sagger missiles, the T-34 tank, the blitzkrieg, and German armor in the Western Desert. I am grateful to both Major General (res.) Giora Rom for his generous contribution to the chapter on the way the Israeli air force dealt with surprise in the Yom Kippur War, and Brigadier General (res.) Dr. Dani Asher for the knowledge and experience that he collegially shared with me on the surprise effect of the Sagger missile on Israeli armor forces in the Yom Kippur War.

Thanks also go to Dr. Yossi Hochbaum for enriching the book with historical examples and analyses, and to Dr. Zeev Bonen for his erudite input on military technology.

I wish to extend my heartfelt appreciation to Lieutenant Colonel Hagai Golan without whose abiding support the Hebrew edition of this book would never have been published. Publishing the book in English was not an easy task for me. Many thanks to those who made the dream of translating and publishing the book a reality. Warm words of gratitude are due to Moshe Tlamim who translated the book into English and to U.S. Army Brigadier General H. R. McMaster who encouraged me to publish this book in English. Last, but not least, many thanks to Stanford University Press, beginning with the director Geoffrey H. Burn, who guided me as a friend, and to his professional and dedicated team: Jessica Walsh, assistant editor; Emily Smith, production editor; and Leslie Rubin, copy editor.

On Flexibility

INTRODUCTION

ON OCTOBER 3, 1973, the renowned British military historian Michael Howard presented a Chesney Gold Medal Lecture entitled "Military Science in an Age of Peace." Referring to the innovative use of technology in the battlefield, Howard stated:

> It is this flexibility both in the minds of the Armed Forces and in their organization, that needs above all to be developed in peace time . . . This is the aspect of military science which needs to be studied above all others in the Armed Forces: the capacity to adapt oneself to the utterly unpredictable, the entirely unknown.[1]

Three days later, October 6, the Yom Kippur War broke out with an Egyptian-Syrian, two-pronged surprise attack. The Israeli army was suddenly forced to adapt itself to harsh battlefield realities that it was unprepared for: anti-tank and anti-aircraft layouts specifically designed to neutralize its armor and air force superiority.

This book addresses one of the basic questions in military studies: how do armies cope with technological and doctrinal surprises that render them vulnerable to unexpected weapons systems and/or combat doctrines?

Armed forces must develop the ability to overcome technological and doctrinal surprise in order to prepare themselves for future confrontations. This may be the most urgent challenge facing military forces today. In the past, armies made every effort to reduce being caught by surprise. Nevertheless, history offers many cases in which surprise was the key factor in determining battlefield victory or defeat.

Military research has generally focused on strategic surprise attacks while paying less attention to technological and doctrinal surprise, although the latter's importance is constantly increasing. Traditional research states that the main solution for a surprise attack lies in improving the intelligence layout.

Decisions regarding force planning are still based on intelligence reports of the enemy's specific capabilities and estimates of the future battlefield, especially its technological and doctrinal aspects.

This book proposes an innovative track for dealing with technological and doctrinal surprise: preparing military forces, but with only minimal dependence on predictions of the future battlefield and information on the enemy's capabilities. The book presents a force planning process that enables armies to cope with the uncertainties of future wars by employing optimal flexibility and adaptability.

The underlying assumption is that the continuous effort to meet the challenge of technological and doctrinal surprise often fails because of various factors that increase the likelihood of surprise occurring, the main one being *a force planning concept overly dependent on intelligence reports that tries too hard to predict the future battlefield.* My main argument is that the solution to technological and doctrinal surprise lies not in predicting the nature of the future battlefield or obtaining information about the enemy's preparations for the coming war, *but in the ability to recuperate swiftly from the initial surprise.*

The following research question is intensely scrutinized: given the difficulties created by overreliance on prediction and intelligence in force planning, and its too common failure to avert technological or doctrinal surprises, how have armies managed to cope with surprises once they occurred?

I contend that armies have quickly recovered from technological and doctrinal surprises by using a variety of abilities that come under the general heading of flexibility. Flexibility combines doctrinal, cognitive, command, organizational, and technological elements that, if properly applied, can eliminate most obstacles in the current paradigm that stem from biases caused by: overdependence on a specific concept, group-think, problems inherent in large organizations, relations between intelligence agencies and decision makers, failure to learn from mistakes, and so forth. The study shows that when armies markedly improve their response skills and reaction time to technological and doctrinal surprise, most of the obstacles based on prediction and intelligence solutions become superfluous. The theory of flexibility-based force planning envisions transferring the solution onto the battlefield in real time in order to overcome peacetime obstacles. This theory is built on four strata.

The first stratum is conceptual and doctrinal (Ch. 2). Conceptual and doctrinal flexibility occurs when senior civilian officials and military officers create an organizational atmosphere that encourages lower-ranking commanders to broach *ideas that challenge the official doctrine*. Officers (and enlisted men) who come forth with original ideas augment the number of options, thus enabling the army that has been caught by surprise to modify its doctrine and tactics. (In this chapter, German open-mindedness to the idea of armor maneuver is contrasted with British dogmatism.) A doctrine based on such an approach presents a *balanced view of all forms of war* and reduces the danger of getting stuck in a dogmatic rut. Without the conceptual stratum, the other strata cannot develop.

Two examples of balanced and imbalanced doctrines are given:

Israel's ground forces in the Yom Kippur War exemplify the consequences of an inflexible doctrine. For years Israel's "cult of the offensive" dominated unit training and war games, rendering the army unprepared for waging defensive battles. The need to fight defensively came as a shock to the majority of Israeli commanders. And although the army recovered from the surprise, the "one-dimensional" doctrine stymied its ability to respond quickly and effectively to the surprise of massive anti-tank missiles.

An example of a balanced doctrine that enabled an army to recover from an unexpected situation is Germany's pre-World War II (WW II) doctrine. Like Israel, Germany emphasized the offensive; but in contrast to the Israeli doctrine, the German one did not neglect defensive training and the development of defensive weapons. In late 1941, when the Wehrmacht realized it would have to wage a defensive war in Russia, its "multi-dimensional" doctrine enabled it to recuperate quickly from this surprising reality.

The second stratum is organizational and technological (Ch. 3). Flexibility in these fields is obtained by: a *balance* among basic military capabilities (attack and defense, firepower and maneuvering, assault and logistics layouts). At the unit and weapon levels, *organizational diversity* is based on the realization that "super weapons," no matter how dazzling their potential, eventually will be confronted with countermeasures and will have to be supplemented with other weapons. When dealing with a major operational challenge, *redundancy* is of utmost importance. Israel's development and deployment of its bridging equipment on the Suez Canal during the Yom Kippur War illustrates this. *Technological versatility and changeability* add another stratum of flexibility to combat units, best exemplified by Germany's use of the 88-mm anti-aircraft gun against Allied tanks in WW II.

The third stratum, flexibility in command and cognitive skills (Ch. 4), is currently considered of supreme importance in modern military organizations, notwithstanding the inordinate difficulty of its implementation. *Mental flexibility* is an acquired cognitive trait of commanders who have learned and operated in an environment that encourages questioning and creativity. In the volatile conditions of the battlefield, it enables a commander to adapt quickly and keep his wits. *Flexible command* expects junior commanders to take the initiative. The wide berth of action should enable them to generate original solutions in surprise situations and receive their superiors' backing. Conceptual and doctrinal flexibility is essential for the development of this stratum; otherwise conditions will not exist that cultivate mental elasticity and decentralized command and control (C2) methods.

German commanders displayed an outstanding capacity for improvisation. Two brief case studies illustrate German C2 decentralization: Rommel's use of 88-mm anti-aircraft guns in Arras (northern France) during the British counterattack in 1940 and Manteuffel's response to the Soviet introduction of Stalin tanks in the Battle of Târgul-Frumos (Romania) in May 1944. The Soviet centralized, rigid C2 system is presented as the source of recovery delay.

The fourth stratum (Ch. 5) is the mechanism that facilitates *fast learning* and *rapid circulation of lessons* so that the entire military system is updated on surprises and informed of their solutions. This stratum takes into account the need to link past, present, and future, and to rely on communications measures that permit a swift flow of information. Britain's failure to utilize all available data on Japanese Zero fighters prior to the battle of Singapore, and American shortsightedness in not implementing the operational experience of the Flying Tigers in China before the Japanese attack on Formosa, are classic examples of what happens when the fourth element is ignored. The arms industry is another area that can provide swift feedback enabling recovery from technological surprise. Close working relations between the military and arms industry can counter surprises by modifying existing equipment even while the battle is still in progress.

In conclusion, these strata constitute flexibility. The order in which they are presented goes from the general to the particular. Concept and doctrine predicate C2 method, organization, and weapons systems. The strata are mutually dependent. Unless uncertainty is recognized as a major problem, the other strata designed to cope with uncertainty will remain undeveloped. In the current state of research, the first, second, and fourth strata receive minimal attention in comparison to the third. The main innovation of this book is the integration of all four strata into a unified theory.

Part Two discusses the four strata and adds brief historical examples to substantiate the arguments. Flexibility or its lack is in many cases the result of the military culture. Various aspects of military culture that encourage or hamper flexibility will be analyzed and presented throughout the theoretical and historical parts.

THE CURRENT STATE OF RESEARCH

Studies on technological and doctrinal surprise and attempts to deal with it can be divided into three parts.

The first part—studies on technological and doctrinal surprise—has been unsatisfactory to date. As a result, most of the material is derived from works that focus on theory, historical analysis of strategic surprise, and the intelligence systems intended to cope with it. The authors are university professors, members of the intelligence community, many of them Israel Defense Forces (IDF) officers (who often provide their own views of surprise in the Yom Kippur War and general conclusions on how to cope with surprise).

Studies on technological and doctrinal surprise as a research subject are few and far between. Michael Handel noted this in an article published in 1987 when he tried to clarify its definition, types, conditions and best times for employment, repercussions, and place in future wars. "Yet while strategic surprise has been studied extensively as a strategic and intelligence problem, technological surprise has received only scant attention in the open literature."[2] The few works that intelligence analysts have published focus on the definition and description of the field rather than on solutions to the problem, or concentrate on improving intelligence work as a solution. Richard Betts discussed the influence of surprise (that resulted from technical and doctrinal innovation) on creating a strategic surprise and briefly reviewed the typology of these surprises.[3] Thomas G. Mahnken's *Uncovering Ways of War* deals with doctrinal and technological surprise and the challenge posed to intelligence organizations to identify military innovations in the interwar periods.[4] An article in the IDF monthly *Ma'arachot* by Eado Hecht, an Israeli intelligence expert, deals with the growing status of technological deception.[5] George Heilmeier, the Director of the Defense Advanced Research Projects Agency (DARPA), analyzed the importance of confronting technological surprise and offered some cogent points for coping with it by integrating intelligence and flexibility.[6]

Another related topic is the impact of technology on combat. This field has long been researched from both the theoretical and historical perspectives, but it does not concentrate specifically on technological and doctrinal surprise. The writers come from academic as well as military backgrounds in doctrinal research. The following reasons may explain why technological and doctrinal surprise has not been studied intensively:

Strategic surprise is of greater interest because of its role in historical analysis and its general engrossment (Operation Barbarossa, Pearl Harbor, the Yom Kippur War).

When analyzing a combat environment, trying to isolate the influence and uniqueness of technological and doctrinal surprise often proves an elusive task.

For security reasons military establishments are loath to discuss their reactions to such surprises.

The second type of professional literature deals with the way flexibility provides a solution to technological and doctrinal surprise (see Part Two on flexibility theory). Unfortunately, professional literature has dealt neither profoundly nor comprehensively with this subject.

Military thinkers have paid relatively little attention to the use of flexibility as a solution to the uncertainty challenge. In fact, excluding the British military theorist B. H. Liddell Hart, all academic discourse has been limited to the general framework of battlefield uncertainty and has concentrated on the optimal command method for dealing with change. The term *flexibility* rarely appears in military literature except as a synonym for related concepts such as mobility, adaptability, and so forth.

Specialists in air war and logistics deal with the matter in greater depth; theoreticians of the future battlefield treat the subject only superficially; and the military doctrines discuss mainly mental and command flexibility.

Students of force planning discuss flexibility, but rarely elaborate on its outcome. Their concern is with flexibility in development and acquisition so that the most effective, up-to-date equipment will be employed on the battlefield. One searches in vain for details on the composition of weapon systems or unit structure that are supposed to provide battlefield flexibility.

Experts in the field of military organization generally stress the importance of combined-arms warfare, in which units from different branches participate, but they do not elaborate on the composition and balance needed to create flexibility.

My book presents the findings of political scientists (who deal with flexibility-related factors that contribute to military success and effectiveness), academic scholars, members of the RAND Corporation, and Israeli engineers in the arms industry (who have studied the feasibility of military innovation and the conditions under which it can be implemented).

In summary, organizational and technological flexibility and the related information flow are discussed in military research, but until now have not been analyzed in detail, nor has a comprehensive theoretical framework of flexibility been proposed.

The empirical literature presented in Part Three is divided into three relatively clear categories. The first consists of operational reports and the lessons of units that have been caught in surprise situations. This part consists mainly of raw material, void of scientific analysis. The second category contains works on technological responses to surprise—from a broad perspective (e.g., radar technology) to specific weapons systems. The authors of this literature are academic scholars and designers of weapons systems who endeavor to improve development and acquisition processes. The third category is literature that describes warfare from a personal, non-academic viewpoint, reflecting the way individuals and military units coped with surprise.

In the absence of a theory that offers a flexibility-based solution for overcoming battlefield surprise, the empirical literature deals almost exclusively with surprise itself and not with recovery. Thus, most of this part is based on two types of literature. The first consists of stories about military units and combatants. Books in this category are not the fruit of academic research, but they provide the facts that enable a story to be woven about the way battlefield surprise can be dealt with. From the other type of literature, which describes technical and technological layouts, we can learn how lessons were learned, and how weapons systems were developed in response to surprise.

In conclusion, the basic problem is that to date all of the studies on battlefield surprise have been lacking a theory that integrates the various types of flexibility into a comprehensive solution, especially for technological and doctrinal surprise. Empirical literature contains reports on force employment; but unless backed by a viable theory, it fails to link cases where flexibility worked, with a meaningful modus operandi that an army can adopt to meet the challenge of battlefield surprise.

THE CONTRIBUTION OF THE PRESENT STUDY TO THE FIELD OF MILITARY SURPRISE

This book is a part of ongoing academic and pragmatic efforts to deal with military surprise. As stated, the book's theory relates to three basic areas:

First, it introduces a detailed analysis of technological and doctrinal surprise. Second, it offers a comprehensive theoretical and historical approach to intensive battlefield "under fire" innovations, which in this case are meant to counter the enemy's interwar or wartime innovations. This is different from other works that deal with interwar innovation, such as Williamson Murray's and Allan Millett's *Military Innovation in the Interwar Period*,[7] or interwar innovation as a reaction to the other side's innovation as Kimberly Marten Zisk's *Engaging the Enemy*,[8] or long-term wartime innovations as described in Stephen Rosen's *Winning the Next War*.[9] Third, the book presents the universal principles of force planning. Although many studies examine the principles of war, this work is one of the very few that attempts to lay down the principles of *preparing* for war.

THE RESEARCH METHOD

The book is a deductive inquiry into the proposed theory based on a number of events in twentieth-century military history.

Two criteria were used to determine the extent of recovery: the time needed to recover (which is relatively easy to measure) and the assessment of battlefield effectiveness (which is the more difficult to gauge because of the difficulty in any analysis of military events to separate the effects of various factors on the outcome of combat). This difficulty notwithstanding, whenever the size of the forces and their weapons systems are known, the components can be isolated and conclusions reached regarding the relative importance of doctrine, C2 method, and so forth. Examples of this are the studies carried out by the American military historian T. N. Dupuy on the German army's successes and the Israeli historian Martin van Creveld on the effectiveness of the German, British, and American armed forces in WW II.[10] My book analyzes the available data on rival armies, which, I argue, is sufficient to identify with relative certainty the factors that contributed to recovery. The effectiveness of the response is based on a graded criteria scale: the best solution results in complete recovery and generates a new problem for the enemy; the second level of solution neutralizes the problem without causing the enemy an operational counter-challenge; the third solution minimizes the amount of damage inflicted by the surprise; and the lowest level of effectiveness is a failure to recover from surprise.

The independent variable—flexibility—is measured according to the number of strata involved in each case study. Thus, C2 flexibility, unless accompanied by other flexibility elements, constitutes low-level flexibility. A military organization where flexibility in command, organization, and "lesson learning" are significant is defined as highly flexible. In addition to the quantitative measurement of the elements employed, great importance is attached to the extent to which the basic stratum—conceptual and doctrinal flexibility—is employed.

The test cases were chosen according to four criteria. The first is the solution's effectiveness. In some cases one party was caught by surprise, but after it recovered, an operational problem was created for the other party; in other cases, recovery failed altogether. The second criterion is the number of flexible or inflexible elements discernable in the recovery. Each example illustrates several elements. The third criterion is surprise at various levels of war: from the technical-tactical level (the German reaction to the Soviet T-34 tank) to the strategic level (the German response to the challenge of British chaff). It is important to show the effect of surprise at a variety of levels of war to demonstrate that the thesis is valid for all manifestations of warfare, despite the commonly held—but mistaken—opinion that technological and doctrinal surprise is limited to the tactical level and, therefore, is relatively unimportant. A distinction must be made between the level at which the surprise occurs and the level at which its consequences have the greatest impact (see Ch. 1). The fourth criterion is the various reasons for surprise. Thus in some cases, the enemy intentionally planned the surprise, whereas in others, it was unintentional and occurred because of the victim's overconfidence or failure to understand the enemy.

In the empirical part of the book, the examples of successful recuperation are limited to the German and the Israeli armies; both share a common basis, especially regarding C2 and commanders' mental flexibility. Specific armies were not chosen a priori or at random, but through the process of elimination based on the abovementioned criteria (especially the first criterion—solution effectiveness).

The American scholar Williamson Murray arrives at the same conclusion and argues that although recent studies have shown that throughout the twentieth century, there were many cases in which military doctrine made innovation possible in the course of a war, this was true mainly in WW II for Germany, whose armed forces exhibited considerable ability to respond to battlefield events by adapting and innovating within their doctrinal framework. Murray regards this as a unique example, with the possible exception of the Israeli response in 1973.[11] On the other hand, failure due to lack of flexibility can be found in many armed forces. The reader, however, should keep in mind that the military organizations discussed in the book have undergone significant change since the time of the events described. Debriefings and lesson-learning mechanisms in the United States Army and Marine Corps are examples of the transformation that the American armed forces have undergone since WW II.[12]

The parameters for determining the degree of recuperation from surprise are defined for the purpose of analyzing the events. The time dimension is the easiest parameter to measure. Many armies have come up with solutions to technological and doctrinal surprises that they implemented in a different war from the one in which the surprise occurred or, in the case of drawn-out conflicts like WW I and WW II, after a number of years. My study does not deal with this kind of situation. All of the solutions discussed in the book were implemented during the war itself, from the moment the surprise was sprung to the months following it (in cases of an extended war); and the time dimension is classified according to immediate (hours or days), short-term (days to weeks), or long-term (weeks to months) solutions. The exception is the Russians' slow recovery from guerrilla warfare in Afghanistan, where the time span was measured in years.

Seven test cases are meticulously analyzed. The first four illustrate recovery based on flexibility; the last three demonstrate recovery failure because of inflexibility. The cases are presented in descending order according to the application of flexibility.

The first case (Ch. 6) deals with the Luftwaffe's recovery after Britain's surprise use of chaff (metal foil released in the air to obstruct radar detection). Recovery commenced with a preliminary tactical response just days after the British bombed Hamburg in late July 1943 and lasted until the Allies' final attempts to defeat Germany by bombing Berlin in March 1944. It included a combination of all four elements of flexibility. The German doctrine stressed the uncertainty factor on the battlefield and as a result, it encouraged a form of C2 that emphasized initiative and independence. Added to this were: cognitive flexibility of officers and soldiers that was realized in the form of two new, night interception tactics (the "Wild Boar" and the "Tame Boar"); a variety of means capable of serving as the basis of improvised solutions (such as the Naxos radar); the ceaseless effort to improve and learn from mistakes; and the close ties between the Luftwaffe and the arms industry. In this case, the Germans quashed a strategic operation whose purpose was to bring the war to a quick end. This case meets the criterion of total recovery plus the boomerang effect—that is, it posed a new problem for the enemy. It also constitutes a classic example of technological surprise with repercussions at the strategic level.

The second case (Ch. 7) concerns the German ground forces' recovery from the surprise introduction of Soviet T-34 tanks—from their first appearance on the battlefield at the start of Operation Barbarossa in the summer of 1941 until the Germans introduced their new Panther and Tiger tanks in August–November 1942. The T-34 was superior to the German tanks in firepower, survivability, and maneuverability. Its entry into the battle zone caught the German commanders by surprise. The effectiveness of German recuperation significantly reduced the damage caused by the Russian surprise so that the impact of the blow was only at the tactical level. In this case, surprise was the result of German conceptual failure rather than the Soviet intention to spring a surprise.

Recovery occurred through the combination of conceptual and doctrinal flexibility (see Ch. 2), cognitive and C2 flexibility (see Ch. 4), organizational flexibility, and fast learning. The Wehrmacht Panzer divisions were probably the most diverse military formations of their time, integrating armor, artillery, infantry, engineers, and anti-aircraft units—together with close air support. When confronted with the superior T-34 Russian tank in the early days of Operation Barbarossa, the Wehrmacht used its branch integration for fast recovery. For example, German armored units had infantrymen fasten anti-tank mines to the chassis of Russian tanks; and anti-aircraft and artillery flat trajectory fire was used against the T-34s, whose heavy armor plating had been practically impenetrable to German tank and anti-tank guns.

The German army also illustrates what modern military thinking refers to as "learning organization," that is, an emphasis on post-action reports and unit training, even in wartime, according to the most recently learned lessons. The product of close military-industry cooperation was the replacement of the Panzer III's short 50-mm gun with a longer gun of the same caliber so that the shells could penetrate the Russian armor. The long-term (one year) response to the T-34 was the design and production of Panther and Tiger tanks.

The third case (Ch. 8) discusses how IDF ground units recovered from the Egyptians' massive use of anti-tank weapons in Sinai in the Yom Kippur War of 1973. Although nineteen prewar intelligence publications had noted the danger of Sagger missiles, and a number of Israeli tanks had been hit by them in the Golan Heights a few months before the war, the IDF ignored the enormity of the threat. The shock was felt mainly in the armored corps, whose commanders and soldiers had been reared on the glory of the Six-Day War's lightning victory.

Despite Israel's obsession with the "cult of the offensive" (see Ch. 2), what eventually saved the day was cognitive and C2 flexibility based on the Israeli civilian culture's extensive use of improvisation. Numerous cases of tactical improvisation occurred on the battlefield. Organization was another area of flexibility that came to expression. Branch uniformity at the divisional level partially accounted for Israel's tribulations in the first days of the war. Lacking sizable artillery and infantry support, Israeli armor divisions were at a loss to respond effectively. When artillery and infantry units finally entered the battle zone, they provided invaluable assistance in recovering from the surprise. The IDF's recuperation almost neutralized the initial damage and had a major impact on the tactical and operational levels. In this case, Israel's conceptual limitation had been the cause of the surprise, not Egypt's attempt to gain a technological or doctrinal surprise.

The fourth case (Ch. 9) deals with the Israeli Air Force (IAF) recovery from the unsuccessful October 7, 1973 attack against the Arab surface-to-air missile layout. The IAF was caught short because of an intelligence omission regarding the latest Soviet SAM-6s (surface-to-air missiles) and the IAF's failure to implement its combat doctrine, which was predicated on near-perfect conditions, such as a preemptive strike (the Israeli government refused to authorize it) and clear skies (the Golan Heights were overcast on the morning of October 7). The inability of the air force to assist the ground forces had dire operational—even strategic—consequences. Air squadrons developed flexible responses based on their commanders' initiatives and the squadrons' inherent cognitive flexibility and fast-learning ability. But the IAF command reacted more slowly to the surprise. It took two weeks before it formally changed its fighting methods. Close cooperation between the military and the arms industry also helped. In this case, recovery minimized the damage wrought by the surprise.

Three additional cases demonstrate with considerable plausibility that inflexibility was at the root of the military's inability to overcome technological and doctrinal surprises.

The first case (Ch. 10) analyzes British slowness in developing an effective response to German anti-tank warfare in the Western Desert between 1941 and 1942. The delay had repercussions at the operational and strategic levels. In this case, surprise was not the result of a German decision, but occurred because the British failed to comprehend the nature of the battlefield, namely the lethal combination of a concealed anti-tank (50- and 88-mm) gun layout and tank maneuvers that tricked the British into thinking that the latter were of primary importance. The absence of British flexibility can be ascribed to conceptual, organizational, and technological deficiencies.

A classic example of British low-level cognitive flexibility is the 3.7-inch aircraft gun. This weapon—the potential equivalent to the German 88-mm—was not exploited as an anti-tank gun because of the British fixed mindset and centralized C2 that restricted the junior commanders' freedom of action. Low-level organizational diversity was another shortcoming. The British regimental system frowned upon close cooperation between infantry, armor, and artillery. The absence of anti-personal explosive ammunition in British tanks was also a factor in their slow recovery. It took an entire year, from the time of the German invasion of North Africa to the appearance of American Grant tanks on the battlefield, for the British to upgrade their ability to deal with German anti-tank guns (and even then, improvement was only minimal).

The second case (Ch. 11) involves the Soviets' lack of flexibility and delayed response to the doctrinal surprise of the guerrilla warfare waged by the Afghan Mujahideen, a type of fighting that the Soviets were totally unprepared for. The Soviets began adapting to these conditions in the first years of conflict (1979–1984), but because Afghanistan was a protracted, low-intensity conflict, it is difficult to determine precisely the extent of recuperation even though the adjustment had considerable influence on the level of Soviet success. This appears to be a case of "self-surprise." The low level of Soviet flexibility may be explained by doctrinaire dogmatism that was blind to the differences between high-intensity conflict against a conventional enemy and low-intensity conflict against guerrilla forces. The Soviets emphasized the operational level in warfare but lost sight of the importance of tactical operations. This may have been applicable to the vast open plains of central Europe, but not for the mountainous terrain and guerrilla warfare of Afghanistan. The Soviets' "Afghani concept" evolved slowly, going from division-size operations (armored-column attacks preceded by massive artillery bombardments) to battalion-size, airborne assaults accompanied by new methods of employing armor and artillery in mountainous terrain. The change involved experimenting in unit reorganization in the course of almost three years of combat. The intrinsic low-level cognitive flexibility, which stemmed from the Soviet system of centralized C2, was a key factor in the inchmeal rate of adaptation and especially in the resistance to decentralizing authority so that junior commanders could function more freely.

The third case (Ch. 12) discusses the French failure to cope with the German blitzkrieg in May 1940. The surprise came not from the German tanks but from the inability of France to confront the German concept of dynamic, fluid operations. The French were stunned by the German forces' lightning speed and their ability to fight unremittingly, without bringing the artillery forward (which, according to the French doctrine, was essential in preparing for attack). Low flexibility on the part of the French was a key element in the ensuing strategic fiasco. Here, too, the Germans did not intend to surprise the enemy. Indeed, the French were caught off-guard, at least partially, because of their insistence on adhering to erroneous concepts and C2 methods.

In the interwar period, French doctrinal dogmatism went from the "cult of the offensive" to the "cult of the defensive." Having been bled white in WW I by futile attacks against concentrated firepower, the French operational concept emphasized the defense and when attacking, advancing slowly under cover of artillery fire. This modus operandi—the methodical battle—stressed strict operational phases, tight control, and obedience. Over the years, French commanders lost the aptitude to improvise; so, when faced with a blitzkrieg, they were at a loss how to respond.

French dogmatism also inhibited original thinking among those officers who warned that an enemy armored attack could seriously upset France's defensive preparations. Despite the urgency of this issue, articles on it were denied publication in military journals. Charles de Gaulle, the main advocate of a mechanized, armored army, was harshly criticized and his promotion held in abeyance. The refusal to even discuss the possibility of rapid armor maneuvers was a key factor in the French army's failure to respond effectively to Germany's invasion in May 1940.

The absence of case studies of low-intensity conflict (except for the war in Afghanistan in the 1980s) is not because this type of conflict is unimportant but because in these instances, the need for flexibility is less severe due to the reduced influence of battlefield surprise on the total result of the confronta-

Table I.1 Historical events demonstrating successful recuperation from surprise

Surprise, extent of recuperation, level of warfare	Speed in devising solution; main strata at each stage		
	Immediate (hours to days)	Short-term (days to weeks)	Long-term (weeks to months)
British chaff ("Window") in WW II. *Surprised party*: Germany *Extent of recuperation*: Full, creating a problem for the enemy *Level of warfare*: Strategic	*Cognitive, command*: Changes in bomber interception techniques and tactics	*Technological*: Introduction of steep-angle-firing guns; attacking bomber blind spots	*Technological*: Development of "Nexus" receiver, homing in on British airborne radar; development of "Lichtenstein SN2" radar, unaffected by chaff
Soviet T-34 tank in WW II. *Surprised party*: Germany *Extent of recuperation*: Damage minimization *Level of warfare*: Tactical	*Conceptual, command, organizational*: Use of diverse means; development of innovative combat techniques	*Technological*: Versatility in weapons systems (replacing the main gun in the Panzer MK III tank)	*Technological*: Military-industrial coordination—swift development of Tiger and Panther tanks
Egyptian anti-tank warfare in the Yom Kippur War *Surprised party*: Israel *Extent of recuperation*: Between minimization and neutralization of damage	*Cognitive, command*: Development of combat techniques in tank battalions	*Organizational*: Rebuilding of unit diversity	
Arab anti-aircraft warfare in the Yom Kippur War *Surprised party*: Israel *Extent of recuperation*: Damage minimized *Level of warfare*: Operational and strategic	*Cognitive, command, technological*: Development of combat techniques in squadrons; reduction of aircraft thermal signature	*Cognitive, command, technological*: Changes in air attack doctrine; military-industrial cooperation; deciphering SA-6 electronic data by RAFAEL by the end of the war	

tion and because of the relatively longer time frame that the surprised party has to recuperate.

The cases in the book date from WW II and later, which is not to say that technological and doctrinal surprises did not occur earlier. The choice of events was dictated by the fact that the study deals with modern warfare, where technology is recognized as a major element of military power.

The analysis of the German response to British chaff is based mostly on secondary sources (though also on primary sources, such as the accounts of German pilots). The analysis of the German response to the Soviet T-34 tanks is based mainly on primary sources dealing with lessons learned at the tactical level, chiefly in the form of English translations of operational diaries and unit reports on lessons learned. The analysis of the IDF's response to the surprise use of anti-tank and anti-aircraft weapons in the Yom Kippur War is based on unclassified primary sources (mainly lessons learned from the fighting and unit battle descriptions). This source material comes from IDF libraries (the armored corps, IAF, the Defense Ministry's R&D Administration) and interviews.

The material for the British response to German anti-tank warfare in the Western Desert comes from primary sources (memoirs by soldiers and commanders) and secondary sources (analyses of the fighting). The chapter on Soviet warfare in Afghanistan is based mainly on secondary sources that analyzed the nature of the fighting. The French failure to cope with the blitzkrieg has been examined using battle accounts and an inquiry into France's prewar doctrine.

THE ORGANIZATION OF THE BOOK

The book consists of three main parts.

Part One contains one chapter and appendixes. It defines and analyzes technological and doctrinal surprise and the process that should produce the solution, namely, military force planning. Then it presents the traditional paradigm for solving the uncertainty dilemma: prediction and intelligence. It then discusses the reasons for the failure of the traditional paradigm and argues that technological and doctrinal surprise is constantly on the rise and constitutes the main challenge to force planning, and that any intelligence attempt to predict the nature of the future battlefield and discover the enemy's intentions will be only partially successful at best.

Part Two discusses flexibility as the general solution for uncertainty. It analyzes the issue through the eyes of military theorists and looks at the command structures, military organizations, and the military technologies and doctrines of various armed forces. It describes in detail the various components of flexibility—conceptual and doctrinal, organizational and technological, cognitive and command, and the military system's mechanisms for lesson learning and information flow (Chs. 2–5). The cognitive and command element is more familiarly known as mission-oriented command. Since this kind of flexibility has been treated exhaustively, it is discussed only when relevant to technological and doctrinal surprise. In developing the book's theory, brief historical illustrations are presented to clarify abstruse points.

Part Three furnishes an in-depth analysis of the seven abovementioned historical test cases. After defining the nature of the surprise that the forces encountered, a discussion ensues on the level of recovery, its effects at the various levels of warfare, and the elements of flexibility that made it possible: the basic warfighting concept of armies that either succeeded or failed in the confrontation with surprise; the combat doctrine derived from this concept; and the concept's connection to the command system, force organization and structure, weapons system development, and the circulation of information to other units. Each case appears in a separate chapter.

The last part, Summary and Conclusions, offers a number of suggestions for theory implementation.

Part One
THE CHALLENGE OF FORCE PLANNING FACING FUTURE SURPRISES

1 PREDICTION AND INTELLIGENCE
The Dominant Approach in Force Planning and Its Failure to Answer the Challenge of Technological and Doctrinal Surprise

MILITARY FORCE PLANNING: PRAXIS AND RESEARCH

The goal of military force planning is to enable the "planner" to deal with security threats in the best possible way. According to Avi Kober, an Israeli expert in defense matters, "[t]he theory of force planning is a complex set of principles that directs force organization, structure and arming so that it can wage war successfully according to parameters laid down in the security doctrine. There is no universal doctrine of military force planning."[1]

Various military organizations use different terms and definitions in the force planning process.[2] Since technological and doctrinal surprise challenges force planning on all levels and at every stage of war, the definition of military force planning should be as comprehensive as possible, that is, it should include all aspects of military force development, beginning with the development of the concept of force employment(as a combat doctrine), through the planning of organization, armament, equipping, education, training, and human resources management, to the implementation of the plans and adapting them to changes. Military force planning deals with the strategic, operational, and tactical levels of war.

Given the sensitivity to surprise of warfare doctrine, this book treats the doctrine as an integral part of force planning. All of the abovementioned elements support the development of abilities and expertise in six main areas: maneuvering, fire, intelligence, command and control (C2), logistics, and force protection.

The Sensitivity of Force Planning to Uncertainty

Force planning is directly influenced by the security doctrine, which itself is derived from the national security doctrine, the latter being based on geopolitical, social, economic, ethical, and ideological factors. For example, Israel's national security doctrine is predicated, inter alia, on limited geographical depth, a plethora of enemies and confrontation fronts, a small but high-quality, technologically developed society, and limited economic resources that preclude a protracted war. These factors led to the development of a warfare doctrine that strives for deterrence, strategic warning based on intelligence gathering, a lightning decision in enemy territory, and force planning based on a limited conscript army, a large, well-trained reserve force, compulsory conscription that puts manpower to the best use, reliance on an extensive intelligence layout for strategic warning, and offensive branches such as armor and air.[3]

Force planning at the operational and tactical levels is strongly influenced by technological development and scenarios that depict the probable fighting method against defined enemies in specific combat arenas. Uncertainty is an inherent feature in force planning, derived from the need to predict future influences on each of the abovementioned factors and on the army's ability to achieve superiority on the battlefield.

Thus, while force planning consists of relatively invariable factors, such as geopolitics and population size (or quality), variable factors are also present, so that the uncertainty factor is of supreme importance. In addition, it should be recalled that force planning is carried out on a number of levels. At the strategic level, it is based on relatively invariable factors and deals with the general structure of the army: size, the composition of the conscript army and reserves, command structure (according to the number of fronts), and type of confrontation (high or low profile, conventional or non-conventional). Structure generally remains the same and has practically no need of intelligence input. Force planning at the operational and tactical levels, however, involves the creation of weapons systems superior to the enemy's and the development of the military capabilities of its units (organization, C2, and so forth) that surpass the enemy's parallel units. The success of force planning at these levels depends to a large degree on knowledge of the enemy's capabilities; otherwise, it is difficult to ascertain whether an advantage has been attained. This chapter discusses in detail the basic uncertainty at the heart of force planning.

The planning phase of force planning—the essence of the process—has two main features. The first is realism, based on a rationalistic link between force structure and security requirements; strategic and technological trends; and a compromise between what is desirable and attainable (given the limited resources) and between the military branches that compete for a slice of the pie. The second feature is reality-based flexibility, whose goal is updating and modifying planning decisions according to changes.[4]

Both characteristics illustrate that force planning for the next war is a field that contains a large element of prediction about the nature of the future battlefield. This is also why it suffers from uncertainty-related problems. "External friction" comes from changes that are not dependent on the force "builder," but from changes in technology, geopolitical conditions,[5] interaction with the enemy,[6] and so forth.

Regarding the enemy, the uncertainty factor may expand because of changes in Side B's force planning in response to Side A's, a pattern that becomes a vicious cycle as in the case of the arms race. Yehezkel Dror claims that "one of the common failures in military force planning is the inherent assumption that while we develop our force, the enemy's force planning remains uninfluenced by our activity."[7] Changes and modifications in force planning are introduced not only after a battlefield clash, but also on a regular basis in times of quiet. Thomas Schelling, an expert in international relations, studied the action-reaction cycle in the American-Soviet arms race and noted that each side was capable of spurring its rival to boost weapons production:

> Thus, by the end of the decade, we (Americans) may be reacting to Soviet decisions early in the decade; and vice versa. The Soviets should have realized in 1957 that their military requirements in the middle 1960s would be, to an appreciable extent, a result of their own military programs and military public relations in the late 1950s.[8]

The source of the "internal friction" lies in problems intrinsic to force planning, such as the development process of military technology,[9] the difficulty involved in assessing the effect of the weapons and combat doctrine in a full-scale confrontation,[10] the tension between innovation and conservatism,[11] and the need to depend on long-term, often unreliable economic support.[12] Another area of internal uncertainty lies in the leaders of state themselves or, to be more exact, government policy on military operations and the level of risk the nation's captains are prepared to assume at the moment of truth.

UNCERTAINTY AND SURPRISE IN BATTLE

Uncertainty is one of the most basic elements in war and is inherent in every combat situation, frequently taking the form of surprise. Surprise on the battlefield can stem from the enemy's intention or an assessment failure by the victim even when the enemy did not intend to spring a surprise. Sometimes surprise occurs because of a failure in executing the plans and without the enemy's resistance, or it may occur because of unplanned success.

In his book *Surprise Attack*, Efraim Kam divides the surprise-causing factors into four types:[13] the attack itself or more commonly known as the surprise attack, timing, place, and method and means of applying force (type of attack). Kam includes technological and doctrinal innovations and their application in the last category. He identifies two types of technological surprise: surprise due to unawareness (for whatever reason) of innovations (e.g., the Japanese torpedo at Pearl Harbor) and surprise due to ignorance and misunderstanding of the impact of a known technological or doctrinal factor. The innovative use may be expressed in quantity and/or the manner of implementation that catches the victim by surprise. For example, Egypt's method of using the Sagger anti-tank missile in the Yom Kippur War or Israel's "Operation Focus" that destroyed the Egyptian air force in the opening hours of the Six-Day War. This factor had little to do with the enemy's intentions; instead, it depended almost entirely on its capabilities.

In Barton Whaley's classic work, *Stratagem: Deception and Surprise in War*, five elements or modes of surprise are categorized: intention, time, place, strength, and style. The last category incorporates doctrinal and technological surprise.[14]

Surprise in strength relates to the order of battle (ORBAT) employed by the enemy. The Germans' force concentration in the Battle of the Ardennes Salient in December 1944 is an example of this kind of surprise. Surprise in combat strength relies to a greater extent on an assessment of the enemy's intentions and to a lesser extent on its ability.

The type of surprise least influenced by identification of enemy intentions is technological and doctrinal surprise. However, the likelihood of this type of surprise occurring is quite high (see below).

Is battlefield surprise an ordinary occurrence requiring a basic, systematic approach to the problem, or is it a rare and unique phenomenon? A brief look at military history and the doctrines of the world's armies shows that surprise may be, paradoxically, the most consistent element on the battlefield because it lies at the core of combat activity and is the epitome of the art of warfare. As a war principle, surprise is employed by all armies. The British military theoretician Richard Simpkin asserts, "Perhaps the one military matter over which there is no dissent is the value of surprise."[15] The Israeli scholar Yehoshafat Harkabi is of the same opinion.[16] Evidence of the universality of surprise is its presence in the works of military theoreticians throughout military history[17] and in military doctrines the world over.[18] Robert Leonhard, an American officer who has written about war principles in the information age, rejects most of the traditional principles but insists that the principle of surprise will always remain valid.[19]

The universality of surprise has led many armies to devise an entire doctrine on the art of deception, dissemblance, and stratagem in order to achieve surprise at various levels of war.[20] Deception combines active operations with passive activities (such as concealment and camouflage). Even when basic deceptive moves are not employed, concealment alone can cause surprise. Ronald Sherwin and Barton Whaley made a statistical analysis of ninety-three cases of strategic attack in the Western world (between 1914 and 1973) and found that when deception was applied, there was a strong likelihood that surprise resulted; but even without deception, surprise was attained in many cases.[21] An expert on the psychology of military intelligence, Richards Heuer, claims that an enemy who is aware of the various cognitive and conceptual biases lying at the psychological base of deception, "holds most of the cards. . . . Perceptual tendencies and cognitive biases strongly favor the deceiver as long as the goal of deception is to reinforce a victim's preconceptions, which is by far the most common form of deception."[22]

Deception, like intelligence in general, works against the enemy's capabilities and intentions. Since the book's main focus is technological and doctrinal surprise, it deals only with the enemy's capabilities. Handel divides the deception of capabilities into two categories:

A. Concealing one's ability in order to trick the enemy into underestimating the real strength of the side perpetrating the deception (example: Germany before 1933, the Soviet Union before 1941, and Israel before 1967).

B. Enhancing one's ability by dissemblance, selective exposure of equipment, weapons, and so forth to create deterrence (examples: Hitler and Mussolini in the 1930s, the Soviet "bomber gap," and the "ballistic missile gap").[23]

According to Hecht, technological deception is concerned only with misleading the portrayal of technological capability (he removes the quantitative element from deception of capabilities). This is done for three purposes: to mislead the enemy regarding the technological capability of the deceiver; to mislead the enemy regarding his own technological capability; and to convince the enemy that the deceiver possesses technology that it does not really have.[24]

Deception of the fighting method can occur as a last-minute change. This is not done as a ruse but for other operational needs, though the side perpetrating it is aware of the information gap being created in the enemy and exploits it accordingly. Handel summarizes the use of deception by stating:

> Since no effective measures to counter or identify deception have yet been devised, the unavoidable conclusion is that deception—even if it fails to achieve its original goals—almost never fails and will therefore always favor the deceiver, the initiating party . . . Rationality dictates that a move which involves little cost and negligible risk of failure should never be left out of one's repertoire.[25]

Since deception is a vital element that must be taken into account in force planning, surprise too should be assumed to be an expected occurrence. The significance of the universal imperative to "spring a surprise," and the corollary that deception is always beneficial, means that the probability of technological and doctrinal surprise happening is high and that surprise is indeed a permanent, constant, major element in war.

Another common feature of the probability of surprise is the friction in war and interwar periods. The friction between battlefield enemies or in the interwar periods when force planning is in progress creates paradoxical situations related to surprise: surprise stemming from an unexpected and unplanned "exemplary" success in the use of weapons or combat doctrine; the inability to cope with surprise even after it occurs because of its concealment; a surprise reaction by the enemy after earlier failure on the battlefield has prompted it to launch a surprise in the next engagement.

Surprise stemming from "too great" a success can happen when one side misjudges the potential of its own technology or combat doctrine. In this case, the fortuitous effect of the surprise is left unexploited. This occurred in November 1917 in the Battle of Cambrai (northern France) when the British failed to follow through in the overwhelming success of their first-ever massive tank attack. In a single day of combat, an unprecedented breach was attained: ten kilometers wide and eight kilometers deep, exceeding that of the entire Third Battle of Ypres (also known as the Battle of Passchendael), which raged for four consecutive months (August–November 1917). Caught off guard by the dimensions of the breakthrough, the British High Command felt it lacked sufficient reserves to exploit the surprise. The Germans counterattacked and reclaimed what had been lost earlier to the British tanks.[26] A similar surprise occurred when the Germans employed gas in the Second Battle of Ypres (April 1915).

Surprise can also take place following "too great" a success by Side A in the first stage of a war that provokes Side B to respond unexpectedly. In his book *Strategy: The Logic of War and Peace*, Edward Luttwak claims that one of the basic characteristics of strategy is the paradoxical connection between the devastating effect of armaments or combat doctrine in one conflict and their abysmal showing in the next. A common phenomenon of an arms race is that every weapon development spurs the development of a counter-weapon. Less common is the realization that the more surprising and lethal the weapon, the faster and more concerted an effort will be made by the enemy to neutralize it. Because of this paradox, surprise in one war or battle leads to a counter-surprise in the next war or battle. Surprise will increase if the side that first sprang the surprise makes an all-out effort to concentrate in the next engagement on the factor that produced the first success.[27]

In conclusion, with surprise as a war principle, deception as a tool for achieving surprise, and friction on the battlefield and during the force planning process, it seems that force planning must be based on the assumption that surprises will always occur.

THE CHALLENGE—TECHNOLOGICAL AND DOCTRINAL SURPRISE

Handel defines technological surprise as "unilateral advantage gained by the introduction of a *new* weapon (or by the use of a known weapon in an innovative way) in *war* against an adversary who is either *unaware* of its existence or *not ready* with effective counter-measures, the development of which requires *time*."[28] According to Handel, true technological surprise includes the integration of new weapons (though a new doctrine of applying known weapons can also produce surprise).[29]

This definition, with its emphasis on a number of issues, suits the present book's requirements. Since the proof of surprise is its result, it makes no difference whether it occurs because of the introduction of a new weapon or a new doctrine. Regarding force planning, the book describes the magnitude of the difficulty in identifying the impact of a familiar weapon used in a surprise-creating, warfighting doctrine. Handel claims that only in rare cases is the victim unaware of the cause of the surprise. The victim usually knows what "he's up against" but is unprepared for dealing with it because of inadequate countermeasures. For our study, there is no significant difference between the categories. Handel defines "time" as the period required to develop countermeasures. This is a key criterion for the effectiveness of technological surprise. If surprise faces immediate countermeasures, its impact on the war will probably be slight. As far as research is concerned, this factor is of minor significance since some surprises are so lethal that countermeasures cannot be used in the course of the war unless they are already in advanced stages of development.

Doctrinal surprise, like military doctrine, can be divided according to levels of war: strategic, operational, or tactical. The first usually appears under the title of "doctrine," while the last comes under the heading of "combat doctrine." As in all matters related to the levels of war, a problem arises when defining the level. For example, Betts sees the blitzkrieg as the clearest case of doctrinal surprise,[30] whereas Shimon Naveh, an expert on the operational level of war, argues that the blitzkrieg was not even a combat doctrine.[31]

To the best of my knowledge, Betts is the only scholar who uses the term doctrinal surprise, though he does not define it. He claims that "surprises in doctrinal applications of weaponry involve long-term misjudgment."[32] While most security scholars have observed the link between technology and doctrine, few deal with them in the context of surprise.

In the following chapters, the term doctrinal surprise refers to three situations: a warfighting concept that catches the enemy by surprise; familiar weapons used in an unexpected way; and an innovative battle technique.

The technological-doctrinal link is bidirectional. Dromi notes that the "doctrine appends technology to its needs, and presents it with challenges and demands in weapons development. This system (doctrine-technology-weapons) is based on a constant dataflow which is expressed in directives and feedback."[33] Sometimes technological changes foster and enable doctrinal changes, for example, the American AirLand Battle (ALB) doctrine and NATO's Follow on Forces Attack (FOFA) in the late 1970s and early 1980s. In these cases, technological advances in precise target acquisition and precision fire enabled the development of the abovementioned doctrines.

Despite the connection between technology and combat doctrine, doctrinal surprise can occur without any technological change. Examples of this are WW I infiltration tactics developed by the German general Oskar von Hutier and the German submarine "wolf packs" of WW II. Two well-known cases should clarify the definition and the difference between strategic doctrine surprise and tactical-operational (combat) doctrinal surprise: Israel's strategic doctrine prior to the Yom Kippur War was based on swiftly halting the enemy and then defeating him in a strategic attack deep in his territory.

The operational-tactical combat doctrine called for an offensive carried out solely by large armored formations.

German strategic doctrine before WW II envisioned lighting victory by an offensive that would spare Germany the need to fight simultaneously on two fronts. The operational-tactical doctrine, known as the blitzkrieg, was based on the concentration of a critical mass of combined arms units, including tactical air support, operated with a large degree of initiative for a surprise penetration deep into enemy territory.

Doctrinal surprise for our purposes is *the application of a combat doctrine in a way that the victim cannot disrupt it in the course of an engagement.*

A concise definition of technological and doctrinal surprise is: *the use of weapons and combat doctrine that the victim does not anticipate and cannot obstruct with countermeasures during an engagement.* (For a discussion on definitions and demarcations of military surprise in the technological-doctrinal context, see Appendix A.)

Technological and Doctrinal Surprise—
The Solution Inherent in the Force Planning Process

The basic question in force planning is which factors determine battlefield success or failure. The question is riddled with controversy. According to Harkabi, "victory factors" in war include quantity, quality (troops, officers, and morale), technological superiority, organization and force implementation, chance, luck, and the attributes of the commanding officer.[34] Of these, only chance, luck, and troop morale are outside the realm of force planning. The commander's character is problematic and, as will be seen shortly, some of the approaches to the definition of force planning include commanders' education. Clausewitz, for example, put great store in the tenacious, original-thinking commander and in the sum total of forces concentrated for the decisive move. Most historians believe that use of a general staff and decentralized command played a key role in Germany's victory in the Franco-Prussian War and that the blitzkrieg combat doctrine played a major role in Germany's victory over France in 1940.

In his study on the optimal conditions for achieving military decision, Kober lists those that are outside the military system and force planning and others that are not even connected to decision in the battlefield. The main conditions, according to Kober, are superpower backing and political leadership that is determined to attain victory. Other conditions directly related to force planning are the ratio between firepower and maneuvering, the military branches and dominant weapons systems, and an attrition ratio that is better than the force ratio—all of which are based on the effectiveness of the fighting forces.[35]

As the "technology factor" for battlefield victory grows in importance (see below), it follows that the ability to counter the enemy's capability to implement a technological and doctrinal surprise that could lead to victory is also gaining in importance. This is not to say that other factors contributing to the victory are on the wane; rather, the status of technology in achieving military supremacy is on the rise.

What is the critical stage in developing the ability to cope with technological and doctrinal surprise? A correlation can be made between types of surprise (and their manner of implementation), time phases that influence surprise, the focus of military activity, and the type of intelligence required to identify them (see Table 1.1).

Table 1.1 illustrates why research on the way armies cope with technological and doctrinal surprise should concentrate on force planning. Naturally, in the case of a prolonged war, the different types of military activity become intertwined; nevertheless, the solution to the technological and doctrinal surprises that the enemy developed during the war remains: force planning. WW II provides a wide range of examples: electronic warfare, submarine warfare, tank warfare. In each of these warfares, new countermeasures and fighting methods were developed and employed during the war to neutralize the enemy surprises.

Table 1.1 The correlation between types of surprise (and the ways they are expressed), time phase that influences surprise, the focus of military activity, and the type of intelligence required to identify them:

Type of surprise	Way surprise is expressed	Time phase that influences surprise	Focus of military activity	Type of intelligence required to identify surprise
Method of force employment	Technological and doctrinal	Peace, pre-war	Force planning	Technological and doctrinal
Occurrence of attack	Surprise attack (strategic surprise)	Opening stage of the war	Mobilization and deployment of the force	Warning, strategic
Time, place, and intensity of force employment	Tactical and operational	During the war	Employment of force	Operational, tactical

To reiterate, technological and doctrinal solutions that are developed in the course of force planning are not the only answer to technological and doctrinal surprise. The Soviets, for example, eventually overcame the German blitzkrieg during Operation Barbarossa by exploiting their quantitative advantage and their country's strategic depth. Another solution is the troops' esprit de corps that gives them the fighting mettle to overpower a superior enemy. These are possible solutions, but they come at a high cost in human life, and the more that war depends on technology, the more their battlefield effectiveness subsides. In the empirical part of the book, the case studies illustrate the importance of the flexibility element as part of the complex answer to surprise.

Dealing With Uncertainty in War Through Force Planning—Basic Approaches

Various approaches exist for dealing with uncertainty in wartime. Some try to provide a solution for a number of uncertainty factors; others focus on a single factor.

Clausewitz regarded uncertainty in battle as the main problem of war. From his writings we can recognize/identify three solutions to the conundrum and a hint to a fourth. In this section a link is made between the solutions to battlefield uncertainty and the basic approaches to the solution of the problem when already in the stage of force planning. Clausewitz claimed that the solution to uncertainty depended on two factors. The first is mass; the second is the genius commander who directs this mass at the critical point. Clausewitz believed that a solution involving mass came from lessons gained from experience. This view dominated military theory until WW II and through the Cold War when various armies put a different emphasis on it. Today Clausewitz's solution is considered anachronistic (see below).

The second solution to uncertainty emphasizes Clausewitz's second key element: a concentrated attack to achieve a quick victory. This approach to force planning tries to overcome uncertainty by winning a swift victory that reduces uncertainty by forcing the fighting method on the victim and reducing its ability to use its fighting methods. This is the accepted solution in a number of armies and lies at the heart of IDF thinking. However, this solution is becoming increasingly limited, especially because the importance of the battlefield decision has declined in recent decades.

The third theoretical solution—the one currently most accepted—for dealing with uncertainty by force planning is based on predicting the characteristics of the future battlefield and relying on intelligence. It will be remembered that Clausewitz did not concern himself with force planning and rejected intelligence, which he consigned to the list of factors that heighten uncertainty. The reasons for the rising status of prediction and intelligence today is discussed later; at this point, suffice it to say that it is a relatively new solution that began to gain prominence in the WW II era.

The fourth solution—flexibility—is the main subject of this book. While Clausewitz makes no explicit mention of flexibility, there seems to be a hint of it in his emphasis on the military commander's need to think creatively. Historically, the demand for command flexibility, independence of subordinates, and flexibility in planning developed from the thinking of Helmuth von Moltke, the German chief of staff (1857–1887) and Clausewitz's exponent, and was introduced into the German army as a warfighting doctrine that promoted flexibility in thinking. The German army made supreme use of this doctrine in WW II when it vied with Soviet, British, and American combat doctrines that placed more of an emphasis on mass and firepower and less on creative thinking and flexibility. After WW II, Western armies and the Red Army came to realize that the solution to uncertainty did not lie in the quantity of men and material. This paralleled the accelerated development of military intelligence and the partial adoption of a command system that stressed flexibility—similar to the German method of mission command. Currently, a comprehensive flexibility-based solution is in the initial stage of development and is basically what this book is all about. These are not dichotomous solutions incapable of integration but basic trends that can be identified in a historical perspective.

The solution for uncertainty	Clausewitz (end of 18th century–19th century)	WW I	WW II	Early 21st century
Mass				
Quick decision				
Intelligence				
Flexibility				

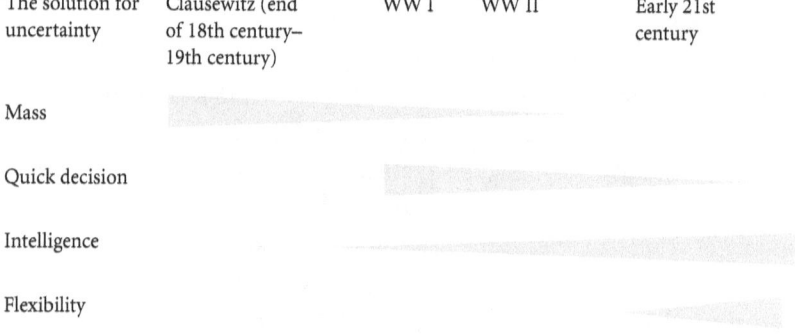

Figure 1.1 Status of uncertainty solutions

The Status of the Mass (Numerical Preponderance) Solution to the Uncertainty Problem

The term *mass* is used for the numerical preponderance of soldiers, equipment, weapons, or firepower, whose loss can be sustained as a result of uncertainty and surprise. It is not a specific solution to technological and doctrinal surprise but a general solution. The mass of personnel and material is designed to solve the question of uncertainty by minimizing every type of surprise through unlimited strength to withstand loss, as in the case of the Soviet reaction to Operation Barbarossa. This type of flexibility, I believe, is in decline. Clausewitz wrote, "An impartial student of war must admit that superior numbers are becoming more decisive with each passing day. The principle of bringing the maximum possible strength to the decisive engagement must therefore rank rather higher than it did in the past."[36]

Liddell Hart viewed Clausewitz's thinking as the source of the catastrophes of WW I.[37] Liddell Hart wrote that the new standard for measuring strength is firepower and mobility—not "numerical superiority." He criticized the "worship of numbers" but admitted in 1936 that "in spite of all the experience which shows that numbers have little real meaning, the governments and generals of Europe still, as a whole, continue to count their forces in term of 'numerical men.'"[38] Evidence of this can be seen in the American[39] and British[40] tendency to trust in material and firepower supremacy in WW II.

Leonhard analyzed the validity of the "mass" principle of war in the age of information and concluded that much of the prestige surrounding this principle stems from its ability to overcome doubt, ignorance, and uncertainty. He noted that there is still a strong propensity in the American army to look to numbers to overcome uncertainty even when it has a complete picture of the battlefield. According to Leonhard, American military education stresses the "numerical solution." Be that as it may, he argues that because of the development of intelligence layouts and precision weapons, mass fighting must switch to precision fighting.[41] This being the case, the solution proposed by modern force planning to the uncertainty problem is mass precision fire.

Can this type of force planning be defined as defensive or offensive? Apparently not. Its high potential to win in defensive engagements notwithstanding, the "mass" solution can be combined with an offensive approach (as the Soviet Union and Warsaw Pact forces did, using their quantitative superiority).[42]

Can the uncertainty conundrum be solved by calculating the size of the mass still relevant? In the past the "human mass" solution was applied by countries such as Russia and China. In the 1950s, Chinese quantity was expressed in a form then called "human waves," that is, a massive number of troops assaulting in wave after wave against enemy firepower. This concept failed in the Korean War and led to a change from a solution based on human quantity to one that gave priority to modernization and technology.[43] Luttwak claims that the growing sensitivity to human life has annulled the validity of this type of solution.[44]

Today the mass approach must be altered from the use of relatively "cheap" human masses to the mass employment of sophisticated weapons. For the foreseeable future, the United States remains the only superpower capable of implementing this type of solution to the uncertainty dilemma,[45] which means its theoretical value for force planning is practically worthless.

The Status of the "Quick Decision by a Preventative Strike" Solution

The second basic solution for reducing battlefield uncertainty is by attaining a quick victory. The standard definition for battlefield decision is the neutralization of the enemy's fighting capability, as well as its ability to employ essential weapons systems, launch a technological and doctrinal surprise, and so forth. Aggressiveness and swift decision create certainty for the aggressor by forcing his plan on the enemy and neutralizing the enemy's chances of realizing its fighting strength. If a war is quickly consummated, it diminishes the need to deal with any form of surprise (including technological-doctrinal surprise) that the enemy is capable of employing. Moreover, it reduces the uncertainty involved in changes that Side A would have to make if the fighting were prolonged. "Generally, however, professional soldiers appear to believe that striking the first blow is beneficial because, at least initially, it reduces the attacker's necessity to improvise and the defender's ability to improvise."[46] On the other hand, defense demands the ability to improvise when confronted by enemy initiatives; an ability that armies try to avoid employing.[47]

Barry Posen's study, *The Sources of Military Doctrine,* tries to lay down rules for the development of military doctrines. The author writes that according to organizational theory, armies develop doctrines because of their organizational needs, which is why armies generally prefer an offensive doctrine since it reduces the element of uncertainty.[48]

According to Kober, one of the most sought-after action paradigms for attaining battlefield decision is first-strike delivery[49] since it drastically reduces the element of uncertainty in the rest of the war. Sometimes a first strike is intended to reduce uncertainty while the enemy's force planning is still in progress, as in the case of the German invasion of the Soviet Union in WW II.

Because of Germany's and Israel's demographic and economic inferiority vis-à-vis their enemies—the Soviet Union and the Arab states, respectively—the uncertainty problem could not be solved with the mass solution that characterized their enemies' force structures. In these cases, uncertainty meant dictating erosive battles by the enemy or being forced into a war of attrition. This was solved by employing offensive operations as decisively as possible. Kober acknowledges that since the 1956 Sinai Campaign, "the desire to deliver a first strike has become a key element, although not stated so explicitly, in Israel's security and victory concept."[50] Israel launched a first strike in 1956, 1967, and considered doing it in 1973.

How is the force built for achieving victory with these paradigms? It must have an offensive doctrine that emphasizes speed, initiative, deception, and determination, and is based on offensive weapons, such as tanks and planes, whose mobility enables the rapid concentration of force and movement to the enemy's flanks and depth, all this as part of an indirect approach. An excellent example is the IDF's force structure, which is based on tanks and planes as the agents of decision. Discussing IDF force planning before the Yom Kippur War, Israel's Major General Adan acknowledged that "the IDF increased its strength according to the doctrine that advocated carrying the war into enemy territory and striving for a quick decision. The offensive shock was the preferred element in force planning. Armor and combat aircraft were practically doubled, and the corps of engineers strengthened, equipped, and trained . . ."[51] Although the IDF still retains the decisiveness concept, there has been a noticeable decline in recent decades in the status of the maneuvering element relative to firepower (the air force's status has remained stable and even has risen).

Defensive weapons are generally relegated to the sidelines when a decisive force is being built. For example, German radar was developed before the war, mainly for offensive needs—bomber navigation over enemy territory—not for spotting enemy aircraft entering Germany's defended air space. This miscalculation was later corrected.

Force planning designed for a quick decision usually allocates relatively modest resources to long-range supply and maintenance. Posen notes that the Luftwaffe was set up for short-term wars, not for campaigns of attrition, thus it lacked reserve aircraft and spare parts. It was built as a "shock force"—not a holding force. In 1940 the Luftwaffe was able to carry out intensive operations for only short time spans. Even the relatively brief Battle of France reduced some Luftwaffe units by almost half. Obviously it was not designed for a long-term campaign like the Battle of Britain.[52]

PREDICTION- AND INTELLIGENCE-BASED FORCE PLANNING

This section discusses two elements in the paradigm currently dominant in force planning solutions to battlefield uncertainty: prediction of the future battlefield and force-planning-oriented intelligence, both of which invariably fall short of the mark.

In practice, military organizations generally make scientific-technological predictions of the battlefield within the context of future enemies and fighting environment. It is difficult to separate the prediction of military technologies and warfighting doctrine in a future battlefield from intelligence gathering on a specific enemy's conduct (that is, armies that face a ballistic missile threat look for technologies that are different from those who face short-range rockets). Nevertheless, these elements must be differentiated since they reside in two different spheres. The first (prediction) is based on scientific thinking that attempts to examine the most likely use of future weapons. Theoretically, prediction does not depend on information about the enemy but on an understanding of trends in technological development, and it is usually performed by civilian and military scientists and personnel in the defense industry. The second element is intelligence gathering on whether the enemy possesses weapons based on this technological development, and assuming it does, how they will be employed. This analysis should be carried out by intelligence experts who have a scientific background.

Predicting the Nature of the Future Battlefield

Given the time span involved in predicting the future battlefield, most methods come under the rubric of long-range (strategic) planning. The majority of the methods were developed for business management or resources planning but because of the similar need to predict future trends, they were adopted for military planning. When employed in the military context, they attempt to clear up uncertainty by predicting the nature of the future battlefield, especially from the technological and doctrinal view. Since most of today's prediction methods were developed in recent decades, only in rare instances have states and armies divulged their results. Thus, a historical perspective on the reduction of uncertainty in war is missing, which makes the examination of the success rate of these methods in force planning a formidable task. Be this as it may, prediction methods are based on two well-established elements: the first is the application of lessons from the past; the second is the testing of new weapons and doctrines. Past and present methods suffer from various limitations and difficulties.[53]

Since this book deals with meeting the challenge of technological and doctrinal surprise, it focuses on the ability, or to be more exact, the inability, to predict the use of first-ever technologies or the novel use of existing weapons on the future battlefield. This subject is relatively easy to prove. The examples cannot be directly tied to a particular prediction method, and though in each case the planner had a specific enemy in mind, the examples deal with the prediction of applied military technological capabilities in general, not necessarily in the context of a particular enemy. No definite connection can be proven between the manner of thinking about the future, or a particular planning method and its success or failure, but the conclusion based on the examples is that for every method used, there are severe limitations on predicting the characteristics of the future battlefield.[54]

Handel offers a number of examples to illustrate the challenge of predicting the way military technology will be used in the next war. The first is the adherence of the world's sea powers (United States, Japan, Britain, Germany, and Italy) after WW I to massive use of heavy warships. In the interwar period, it was not clear whether the warship had lost its supremacy to the airplane or submarine. Unable to experiment "in the field," the admirals kept with what was familiar despite attempts, such as that of American General William "Billy" Mitchell (1879–1936), to prove that the warship's day was over. The might of the aircraft and aircraft carrier became evident in WW II. The prewar proportion of warships to aircraft carriers (2:1) was reversed during the war. Another example of misreading the next generation of sea battles in the interwar period was the underestimation of the submarine. None of the belligerents had prepared adequately for combat in this environment. Handel believes that the failure to predict the optimal use of technology stems from the inbuilt hiatus between military technological development and the formation of the operational concept of its implementation.[55]

Other examples of the failure to predict technologies include the cases of Vannevar Bush, the science director of the American Office of Scientific Research and Development during WW II,[56] and the German scientists' inability to develop or foresee the strides made in centimeter-wave radar for bomber navigation and nighttime detection of surfaced submarines.[57] In general, the more the problem in question is technological in nature, the higher the status of the scientists' views. This becomes problematic when individual opinions among decision makers are biased because of the experts' prestige.

Military commanders, too, who are responsible for translating current technology into practical realization on the battlefield, struggle futilely to predict the direction that battlefield technology will take. Examples include German generals who doubted the development of large-scale armor formations prior to WW II[58] and the British admiralty's overdependence on the Anti-Submarine Detection Investigation Committee (ASDIC) as a counter-submarine measure during the same period.[59]

So far, no mention has been made of the contribution of operations research in the study of battlefield prediction. This area applies the scientific method to systems management and organizations. Attempts at improving the British detection-interception system against German air attack began in 1939. Since then, this field has frequently proved itself when dealing with systems already in operation, but its use for predicting the future battlefield is limited.[60]

Another tool in the Sisyphean attempt to overcome uncertainty regarding the operational use of new technologies on the future battlefield is testing and exercises. Here too lies a basic problem. Before WW II, the United States invested enormous energy in developing its aerial bombing capability. Progress in this field, compared to that in the development of fighter aircraft, led to increasingly high success of bomber capability in field exercises. But the erroneous conclusion was that the bombers could penetrate air defenses without the support of long-range fighter aircraft. This contributed to the development of the B-17 and the doctrine of daylight bombing based on the defensive ability (mainly machine guns) of the planes.[61] The concept led to the loss of tens of thousands of American airmen and the accelerated program to develop long-range fighter aircraft to escort the bombers.

Another problem is that reliance on past lessons often turns out to be a source of mistakes in the prediction of the future battlefield and intelligence for force planning.

INTELLIGENCE RESEARCH FOR FORCE PLANNING REQUIREMENTS

The rise in the status of military intelligence began only in recent decades. In his article, "Clausewitz in the Age of Technology," Michael Handel called the section on intelligence "From Friction to Panacea." This heading sums up the transformation that the status of intelligence has undergone since Clausewitz's time. Handel claims that the rise in status is based not only on the need to warn against strategic surprise, but also on "the need to gather information about the development and production of new weapons systems, their effectiveness and performance, their integration into military doctrine, and so on. The independence of intelligence as a military activity was one aspect of the growing professional specialization and differentiation of military organizations created in response to the technological revolution."[62] It took time until the importance of intelligence bureaus was realized. Until WW I, and even up to WW II, armies viewed military intelligence as an area of secondary importance relative to their goals. Despite the awareness of the importance of intelligence in wartime, the decline of intelligence organizations can be observed in the interwar period. Only at the onset of the Cold War were large intelligence agencies established on a permanent basis.[63]

On the psychological need for intelligence information, Handel notes that Napoleon and Hitler invaded Russia while relying on scant intelligence and exceptional audacity and writes, "[t]oday, the high degree of psychological dependence on every last bit of data can induce paralysis (or at least reluctance) in leaders who must decide whether to take action when such information is available."[64] The dependence of force planning on intelligence is not exceptional in this regard. Intelligence for force planning deals also with the characteristics of the enemy's force structure; however, this book concentrates on technological and doctrinal intelligence rather than the enemy's ORBAT. At first glance, this area of intelligence seems easier to apply than strategic intelligence on the enemy's plans and intentions, since it analyzes "dry," technical data rather than trying to figure out the enemy's thinking. Many of the problems in strategic intelligence, which attempts to decipher policy, are not intended to appear here, but, as will shortly be seen, force planning intelligence is unreliable too.

The discussion begins with information itself. A historical analysis of surprise proves that assessment failure is rarely the result of paucity of information, although surprises resulting from it have occurred.[65] The quality and quantity of intelligence data is another issue. Qualitative uncertainty and quantitative uncertainty have discernable differences. The first is the failure to identify the process; the second relates to the difficulty in attributing the degree of probability in the identified processes. Each dimension has a different influence on the ability to identify technological and doctrinal surprise.

Our understanding of problems related to information quality and quantity is based on Roberta Wohlstetter's theory of signals and noise that she borrowed from communications theory. She uses this theory to analyze the period preceding the Pearl Harbor attack.[66] Data quality is determined by two basic criteria: reliability and analytical value. Data reliability is divided into three levels: the first is unreliable data based on a questionable, contradictory source or perceived as being intentionally influenced by the enemy. Most information emanating from human sources belongs to this group. The second level is reliable information that may be manipulated by the enemy (like air reconnaissance photos). Even when third-level, highly reliable information is obtained, free of enemy control, there is no guarantee that it will be interpreted and/or used correctly.[67]

The analytical value of information lies in whether it corroborates or refutes the premise being examined. This depends on whether or not the information contains data that can differentiate signals from noise and on the volume of information relative to the size of the intelligence problem.[68] No matter how sound the information, it still might be incorrectly or only partially understood. Another problem that often surfaces with such information is the obsession with its primacy over and above other information, which, had this not been the case, would have received a more central place. And still another problem is the possibility of a sudden drop in data quality without the experts being aware of the change. Early warning signs are an important part of information analysis and its problems; however, since they are not directly related to this study, they are not discussed further.

The continuous increase in intelligence-gathering capability, due to sophisticated technologies, affords both advantages and disadvantages. Researchers are inundated with an enormous volume of information that is practically impossible to digest, classify, and evaluate. This can result in inflated intelligence organizations and an undesirable unwieldiness. John Ferris and Michael Handel have termed the problem of superfluous information due to the power of communications, command, and control and intelligence (C3I) systems in data gathering and circulation "Type B uncertainty," as opposed to Clausewitzian "Type A uncertainty." They claim that surplus information creates an unmanageable system that hampers the military's ability to evaluate the incoming information at a rate proportionate to its increasing volume.[69] The crux of the problem is that experts obtain available information that is not always relevant. The greater the volume of data, the greater the noise, so paradoxically, more information can cause an intelligence short-circuit because of the reduced quality of analysis and assessment due to the greater noise.[70]

Data analysis is subject to a wide range of problems and mishaps that can be classified, albeit in a somewhat strained fashion, into objective problems and blunders and those resulting from various types of biases.

Objective Problems in Assessing Technological and Doctrinal Abilities

Intelligence assessment is carried out in two key areas: enemy intentions and enemy capabilities. At first glance, assessment of intentions appears the more complex task, but a deeper look provides a different view. Israel's pre-October 1973 assessment of Egyptian intentions was largely based on an underestimation and misinterpretation of Egypt's capabilities, especially its anti-tank and anti-aircraft weapons, and the part they played in the Egyptian attack plan. Be this as it may, prediction of the enemy's choice of modus operandi is very tricky, especially when the decision maker is someone like Hitler. This was the case of the German decision to employ concentrated armor units. A study carried out by A. W. Marshall of the RAND Corporation on Soviet force planning from the end of WW II to the beginning of the 1960s failed to show a close link between military objectives and the decisions made at the highest level and the products of force planning (e.g., weapons). Decisions in large bureaucratic organizations tend to go through hard-to-predict changes during the long-term force planning process.[71]

Intentions assessment in the force planning context is strongly influenced by the way the intelligence analyst perceives the range of factors that the enemy takes into consideration. Sometimes this perception is fundamentally wrong, having assumed norms and values different from those of the analyst's side.[72] Since assessment of intentions in force planning is a relatively small area in the overall assessment of technological and doctrinal surprise, it is not discussed further. The assessment of technological and doctrinal capabilities contains a number of problems that exist even when information is reliable. These include:

> The formidable challenge involved in calculating essential factors, such as the influence of the warfighting, training, and C2 doctrines on the enemy's performance.[73]
>
> The difficulty in drawing conclusions from a known weapons system that deals with the warfighting doctrine derived from it and its influence on future battlefield performance.[74]

The tendency of intelligence organizations to overlook the enemies' emerging innovations that are not part of their own preconceived concepts of warfare.[75]

The difficulty in understanding how the former capabilities of an army will be translated into future ones.[76]

The difficulty in identifying specific operational goals of weapons designed for ad hoc use.[77]

Finally, although technological and doctrinal surprise is the main subject of the book, we cannot overlook the difficulty in determining the size of the enemy's forces and number of weapons.[78]

Adir Pridor, an Israeli expert on operations research who has studied the assessment of military strength, stated in 1985 that whereas the strength of weapons systems is a measurable factor, an accurate measurement theory for military units, formations, or an army's strength has yet to be invented.[79] Twenty years later, Stephan Biddle, dealing with the components of military power, claimed that ". . . no systematic, formal model of force employment had gained general acceptance."[80] Klaus Knorr expressed the same problem in the following terms:

> There are neither theoretical guidance nor empirical apparatus for measuring and comparing, and essentially predicting, the combat strength of the mobilized forces of different states. The only known measurement test which is accurate is the test of battle. Of course, quantitative comparisons of infantry divisions, aircraft wings, naval vessels, missile launchers and military personnel can be made. . . . But the main problem arises from the fact that the presence of qualitative factors makes quantitative comparisons often inconclusive.[81]

Given the cumulation of difficulties, the conclusion seems to be that even if the enemy refrains from deception and stratagem and only conceals his abilities (and sometimes not even that), assessment analysts may still fail to identify "objective" data looking them straight in the eye and flashing warning signals of an impending technological and doctrinal surprise.

Overestimation and Underestimation in the Field of Force Planning Intelligence

Handel has observed two kinds of basic errors in assessments of enemy ability: underestimation due to the analyst's overconfidence in his capabilities and overestimation for a number of reasons, for example, incorrect information because of gaps or deception, a lack of self-confidence due to previous failure, a tendency to be on "the safe side," and sometimes an intended tactic to gain a larger budget than what the politicians allotted.[82]

Underestimation may generate a sense of superiority in one's capability that is void of an empirical basis but that creates a mindset that obscures the ability to see developments in the enemy's capability. A classic example of this is the IDF's mental state after the Six-Day War, which prevented it from observing changes in the weapons and doctrine of the Egyptian and Syrian armies.[83]

An example of an intelligence overestimation that caused a change of force planning was the British prewar assessment of the Luftwaffe's ability to produce long-range bombers. The Germans did begin a project for manufacturing long-range, four-engine bombers, but they abandoned it in favor of bombers for ground support.[84] The British miscalculated the volume of an expected German bomber attack by 80 percent,[85] which resulted in Britain's choosing an air-defense doctrine and discontinuing its deterrent doctrine based on the construction of long-range bombers.[86]

Gideon Hoshen, dealing with technical intelligence for the development of counter-weapons, notes that the source of overestimation is often the incompatibility of weapons to the local theater. An intelligence officer who fails to match the weapons to the state employing them generally miscalculates their rate of absorption, the ability to maintain them, and their performance envelope. This results in preparing an answer exceeding the threat, and the cost will be a surplus outlay of money, time, and effort.[87]

Another failure to assess technological and doctrinal ability is the projection of the analyst's own technological ability or doctrine on the enemy. For example, the French estimated the enemy's ability on the basis of their own ability. Thus, in pre-WW II armor exercises, the French estimated that the Germans had reached conclusions from tests similar to their own, that is, the Germans would use tanks as infantry support and not in armor concentrations.[88]

Mahnken sees the inability of American technical experts to conceive that the Japanese had developed a liquid oxygen-powered torpedo as stemming from the fact that the Americans had failed in their efforts to develop a similar technology.[89] The Japanese Type 93 "long-lance" torpedo with a larger warhead and longer range than the American version wreaked heavy damage on U.S. naval forces.

Mahnken analyzed the ability of American intelligence agencies to identify enemy innovations. He notes that intelligence organizations tend to focus on familiar areas rather than try to discover fields still in the developmental stage.[90] For example, the Americans failed to perceive the Germans' prewar development of airborne and marine landings because they concentrated on intelligence gathering in traditional areas of development: infantry, artillery, armor, and so forth.[91]

May's study of intelligence estimates of enemy capabilities and intentions before the two world wars points to improvements in the interwar period because of a growing awareness of the technological factor on the battlefield, but he generalizes that "all intelligence agencies performed poorly in making long-term projections useful for planning or procurement."[92]

This concise survey has shown that objective problems in analyzing enemy capability, and subjective problems emanating from a wide range of factors, are crucial elements in the elusive task of identifying technological and doctrinal surprise based on intelligence data. The following section describes the main sources of the problem—human nature.

FACTORS RESPONSIBLE FOR THE FAILURE OF THE DOMINANT APPROACH

The roots of the failure of prediction and intelligence for force planning have been studied in detail from several perspectives.

The first is human shortcomings rooted in human psychology. In our context, this refers to the personal characteristics of the military commander and intelligence officer,[93] and their implications on decision making in the force planning process. Human decision making is biased due to environmental influences and mental (heuristic) strategies[94] and biases stemming from prior concepts, beliefs, and expectations. Broad knowledge and a comprehensive system of beliefs, positions, theories, and assumptions lie at the foundation of human cognitive activity involved in filtering out important signals from the background noise and trying to interpret them correctly. The unavoidable need to rely on a concept contains a number of potential biases: remaining stuck on a particular set of concepts; finding it difficult to perceive that the enemy is willing to take great risks; absorbing information into prior beliefs or assumptions; coming under the influence of expectations in the course of information processing; and dealing with ambiguous information. Some of these difficulties emanate from psychological biases and problems.[95]

A second perspective deals with organizational failure. The intelligence analyst and decision maker do not work alone but are part of a small group of individuals who are reciprocally influential and committed to each other, generally sharing assumptions and concepts. This reality has a major impact on the perception of the enemy's conduct and nature of the approaching war. Problems characteristic of small work teams include group think,[96] excessive influence on the part of the leader or specialist, and the danger of group risk taking.

Another type of problem comes from the influence of the characteristics of the military organization on the ability to evaluate intelligence data and use it to reach a decision. Military conservatism plays a major part in dealing with technological and doctrinal surprise.[97] Problems related to the size of military organizations, interbranch coordination and rivalry,[98] concepts of superiority and aggression,[99] or the military organization's culture that set preconceptions can also detract from the ability to identify and comprehend a threat.[100] The size and complexity of military organizations means that even when warned in time, their chances of dealing with technological and doctrinal surprise are small.

A third area is the interaction between the intelligence analyst and decision maker that, in some cases and despite reliable information, results in faulty decision making.[101]

Because of past failures in coping with technological and doctrinal surprise, this problem continues to plague intelligence analysts and decision makers. Many attempts have been made to solve these problems, inter alia, by improving various factors in the planning and predicting process,[102] providing an answer to human shortcomings,[103] revising the methodology of intelligence work[104] and organizational structure,[105] better use of peacetime foreign military relationships (between future rivals),[106] and revamping the interface between intelligence analysts and decision makers and the decision-making process.[107] The difficulty in implementing the proposed improvements is that they all revert to the basic, insoluble problem of human nature. Despite the awareness of the wide range of problems involved in preventing surprise by prediction, intelligence work, and decision making, none of the solutions have been able to guarantee the identification of technological and doctrinal surprise before it occurs.

All of these problems and their solutions have been analyzed in the field of strategic surprise and can also be identified in the case of technological and doctrinal surprise.

This section concludes with the words of Yoel Ben-Porat, the commander of the Central Warning Unit of the Israeli Intelligence Branch in the Yom Kippur War, who believed that the nature of the future battlefield defies prediction. "The problem lies first and foremost on the theoretical level. Prediction as a pseudo-science has failed. Prediction and other types of intelligence assessments are like equations containing only unknowns."[108]

In addition to the traditional failure in coping with this type of surprise, the complexity of the problem increases because of other developments, especially in the field of technology. My contention is that given the technological development in recent years, a greater likelihood exists for technological and doctrinal surprise than for a surprise attack. There are three cogent arguments for this:

The importance of the technological dimension has surpassed the operational dimension in grand strategy.

The differential influence of technology on military layouts, that is, the accelerated development of observation, tracking, and identification, as opposed to the modest improvements in mobility. This has created a situation in which effective strategic warning is far more possible than in the past.

Developments in numerous technological fields, especially since the 1960s, have produced a wide range of weapons that have increased the likelihood of technological and doctrinal surprise.

For a detailed discussion on this issue, see Appendix B.

Technology and Warfighting Doctrine

The gap between technological innovation and its implementation according to doctrinal guidelines is a familiar issue in military history.[109] Handel claims that the faster technology progresses and produces a wide range of weapons, whose effects and mutual influence are uncertain, the greater the discrepancy between military technology and the development of tactical and strategic doctrines that are supposed to employ the weapons within a general concept.[110]

The chances of doctrinal surprise are increased because of the inability to understand exactly when and how a particular weapon, even if the enemy is known to possess it, will be integrated into the warfighting doctrine that will maximize its influence on the outcome of the battle. Given the rapid pace of change and the gap between technological inventions and doctrinal development, it is difficult to predict whether the weapons will have a significant impact on events. Examples of this are: the gaps between the use of gas and tanks in WW I and their integration into a consolidated doctrine and the nuclear deterrence doctrine that was formulated after the detonation of the atom bomb. This is why accelerated technological development in the last decades seems to have increased the ability to catch the enemy by surprise with warfighting doctrines.

The result of the trends has been a rise in the importance of technology in grand strategy and a decrease in importance of the surprise attack, an accelerated development in strategic warning and a relative freeze in the development of mobility, and the increased spectrum of options for technological and doctrinal surprise, which is illustrated in the following graph:

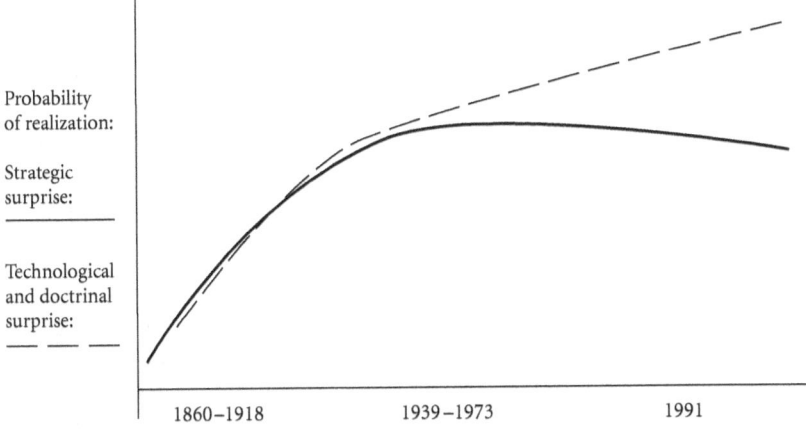

Figure 1.2 Strategic, technological, and doctrinal surprise as a function of technological development

A Test of Sensitivity

The previous sections explained the limitations of the traditional model for dealing with uncertainty through the force planning process. Given the number of intervening factors that reduce the effectiveness of intelligence, technological and doctrinal surprise present force planning with a great challenge; while reliance on intelligence and prediction of the future battlefield as key elements in the solution is problematic.

The two basic approaches to the uncertainty problem in force planning that Israel deliberated in the late 1990s were the "response to gaps" approach and "relative advantages" approach. An examination shows that both are plagued by a prodigious amount of uncertainty. The "response to gaps" approach places great stress on understanding a threat, based on an assessment of the enemy force's structure. This understanding enables force planning to provide an answer to the gaps identified in the army's ability to deal with the enemy. Despite the logic of this approach, it is susceptible to a wide range of problems connected with intelligence work: psychological biases of the individual, conceptual rigidity, objective problems involved in data assessment, risk of lapsing into group think, and so forth. The "relative advantage" approach calls for the development of capabilities that provide relative advantages and then "lures" the enemy to cope with them.[111] This approach, which seems to be less dependent on intelligence, nevertheless requires information on the enemy's strengths and weaknesses. If this data is missing, it will be difficult to focus on the relative advantage. Moreover, even if Side A correctly identifies the relative advantage and builds its strength accordingly, it is still vulnerable to a large array of problems, such as rivalry and coordination within the military organization, conservative thinking, and so forth. Also, the relative advantage approach is liable to create problems stemming from a sense of superiority and pride. Both approaches suffer from uncertainty stemming from technological developments and from the phenomena resulting from battlefield friction and the interwar period friction.

Eliot Cohen identified four approaches to American force planning within the revolution in military affairs (RMA) discourse.[112] The pupils of Admiral Bill Owens are representative of the "system of systems" approach, which is similar to the relative advantage approach and is based on the United States' unparalleled advantage in developing and obtaining military technologies. This approach is predicated on the concept that intelligence and precision fire enable the creation of a single integrated meta-system uniting all the means of intelligence and fire. The strength of the "system of systems" lies in terminating the need to relate to the enemy's capabilities, thus solving the uncertainty problem in one fell swoop. Naturally, at present, this approach is accessible only to the Americans because of the gargantuan price tag attached.

Another approach, which Cohen terms the "uncertain revolutionaries" approach, is also similar to the relative advantage approach and seeks to gain the technological advantages of the United States. Proponents of this approach believe that a drastic change will occur in a future war but are not sure of its direction and are concerned about the certainty that lies at the basis of Owens's "system of systems" approach. They suggest developing a large number of weapons systems, organizations, and warfighting doctrines and choosing the most appropriate ones for dealing with future challenges. This is presumably the best approach for dealing with uncertainty and technological and doctrinal surprise. Its realization entails vast resources, which is why it is currently applicable only to the United States.

Two other groups represent more conservative approaches. The "Gulf War veterans" approach attributes success to the high level of the training of troops in the use of modern technology. This approach sees conventional fighting retaining its primary role in future wars and believes that its continued application depends less on the development of sophisticated technologies than on maintaining highly skilled professional soldiers. The fourth group, which Cohen calls the "skeptics," claims that RMA is a mistaken term since that change has been and always will be gradual. This approach is based on Clausewitz's concept that the human being—not technology—is the dominant factor in battle and casts doubt on Owens's outlook that friction and uncertainty can be eliminated. The "skeptics" approach recommends planning for a wide range of scenarios, without any special emphasis on technological development. The basic concepts of these approaches in trying to overcome the uncertainty factor are similar to the "response to gaps" approach.

Stephen Rosen's book *Winning the Next War* discusses the influence of innovation on winning wars. The author sets forth two solutions to the uncertainty problem in weapons development.[113] The first is called "let the scientist decide," that is, when uncertainty reigns, every possible innovation must be tested. Scientists have the know-how to do this best because they are more independent than military engineers and scientists. This argument is based on the success of civilian scientists in developing weapons in the United States and Britain in WW II. The sensitivity of this approach to uncertainty regarding the operational effectiveness and financial outlay is no lower when scientists are in charge of the development process. One problem inherent in this approach is the limited ability to concentrate the development effort. This is what happened in Germany in WW II.

Another approach postulates that uncertainty is a fact and focuses on two characteristics of flexible development. One is flexible weapons capable of performing a number of operational missions. For example, the aircraft carrier with its various types of planes on board is more expensive than a vessel with cruise missiles but is much more versatile. A second type of flexibility is based on the acquisition of knowledge of various systems by simultaneously developing technologies until their battlefield performance and cost can be precisely estimated. According to this approach, prototypes are developed that can be tested by the military system. This approach prevents units from being armed with weapons that seem promising in their developmental stage but later prove ineffective because of a newer technology or improvement in the enemy's capability.

Bonen suggests different solutions to the problem of weapons and doctrinal development from the linear process of: development; integration into the layout without changes in the structure and warfighting doctrine; absorbing slight changes in the structure and warfighting doctrine into the existing layout; further development; and basic changes in the layout that present maximum use of new weapons. Bonen recommends integrating weapons development into the warfighting doctrine by tactical testing in the field. Actual weapons are not necessary; dummy models can be used. The aim of these experiments is not to test a preconceived approach, but to pursue and examine an assortment of possibilities. In addition to the speed of weapons development, this approach brings unforeseen problems to the surface, preventing expensive modifications in equipment at a later stage.[114]

The application of these approaches contains an inherent answer to the problem of choosing the direction of technological development and a partial answer to the problem of an enemy's dynamic development of its capability. Nevertheless, they suffer from sensitivity to uncertainty because of the difficulty in intelligence gathering and assessment of the enemy capabilities, and the difficulty in comprehending the full impact of a small number of weapon prototypes that have been tested on the operational effectiveness. This means that the entire force planning process, no matter which approach is dominant, is vulnerable to technological and doctrinal surprise.

. . .

The aim of the first part of the book has been to explore whether the answer of force planning to the uncertainty dilemma should be reexamined because of the military's lack of success in dealing with technological and doctrinal surprise. I presented the relationship between force planning and technological-doctrinal surprise as the main challenge in this process. We have seen that the traditional model, based on early detection of surprise by intelligence gathering and the prediction of the future battlefield, suffers from many problems intervening between the development of conditions for surprise and the ability to identify this development, as the basis of force planning for dealing successfully with surprises. These problems are valid for all types of surprise because the main cause for uncertainty—human nature—

lies at the heart of intelligence research and assessment and at the heart of force-planning decisions.

Furthermore, because of the changes in the war phenomenon and the influence of technology on war, the challenge of early identification of technological and doctrinal innovation has become more complex.

The first part of the book presented the problem of technological and doctrinal surprise and the current paradigms for solving it—prediction and intelligence—and their serious limitations. Are military organizations that deal with this problem doomed to failure? The following part offers a different solution—rapid recovery based on flexibility.

Part Two

FLEXIBILITY-BASED RECOVERY—
A THEORETICAL VIEW

As presented in the introduction, flexibility, as a solution to the challenge of technological and doctrinal uncertainty, offers a balanced combination of four strata: conceptual and doctrinal; organizational and technological; command and cognitive; fast learning and rapid circulation of lessons.

The ability of military organizations to change is greatly influenced by their culture. An army's cultural norms are based on and influenced by military tradition, social norms, and ethical standards that have been absorbed into the organization in the form of laws and routine procedures. These laws and routines shape the thinking patterns of an army when changes must be made.[1] The "culture of flexibility" or the opposite form of organizational behavior—dogmatism and resistance to self-criticism or learning from another's experience—is the leitmotif that runs through the four strata.

This part discusses the four strata and adds brief historical examples to clarify the arguments.

2 CONCEPTUAL AND DOCTRINAL FLEXIBILITY

CONCEPTUAL AND DOCTRINAL FLEXIBILITY is the basic stratum in answer to technological and doctrinal surprise. According to Posen, "Military doctrine includes the preferred mode of a group of services, a single service, or a subservice for fighting wars."[1] The British Army elaborates on this: "Doctrine evolves in response to changes in the political and or strategic situation, in light of experience, or as a result of new technology. In turn, it influences the way which policy and plans are developed, how forces are organized and trained, and what equipment is procured."[2]

The "preferred mode for fighting wars" can be gleaned from standard military texts. Doctrinal flexibility can be learned not only by reading the doctrine but also by inquiring into the processes that shaped it and the changes it went through, for example, by studying the nature of the discussion on controversial military topics that challenged the premises of the doctrine.

I contend that conceptual and doctrinal flexibility is the most fundamental stratum in this study. There is no alternative to an approach that encourages critical debate, initiative, and independent thinking and creates the kind of openness that enables commanders to adapt impromptu to unexpected circumstances. If a commander identifies a surprise situation on the battlefield but lacks the authority or intellectual capacity to meet the challenge, then neither flexible weapons nor a unit based on an adaptable organizational structure will be of avail.

Dromi claims that the optimal preparation for the next war requires maintaining a constant interchange ("flow") between doctrine, technology, and weapons systems. Stagnation in the system or one of its components can lead to disaster.[3] A flow of the kind that Dromi envisions can counterbalance the conservatism so typical of military organizations and keep the system's components up-to-date. I argue that any attempt to prepare for a future war by amending doctrine, technology, and weapons systems is of dubious value;

in other words, the product of the flow is of less importance than creating an atmosphere of openness to new ideas and a mindset conducive to dealing with uncertainty. These assets will provide commanders with the mental stamina and conceptual agility for coping with surprises on the battlefield by improvising solutions.

Conceptual and doctrinal flexibility has two fundamental features. The first is tolerance of unconventional ideas, even if they clash with accepted views. This is predicated on:

Tolerance of candid discussion of ideas that challenge the dominant view and permitting the publication of such ideas in articles and books.

Tolerance of people who advocate views other than those of the official line.

Willingness to accept ideas external to the army, (e.g., from foreign armies).

Discussing the possibility of implementing military innovations during war, Michael McNerney argues that an organizational environment is needed that promotes decentralization and creative thinking and recruits people with organizational and developmental skills.[4] I believe that more is required. There must be an environment that intrinsically encourages ideas and opinions to be bandied about, contested, and mulled over so that an army can work out solutions for surprise situations. Any mechanism that allows this must also have a system of checks and balances, that is, mutually independent organizations that can develop their ideas quite independently and compete for resources within the framework of force planning . A doctrine written by thinkers from one military branch should be reviewed by members of other branches as a matter of procedure.

The second feature is a multidimensional warfare doctrine, even though armies generally have a clear strategic preference for offensive or defensive combat. This means that the doctrine in question should give equal importance to all forms of war (offensive, defensive, advance, and withdrawal),[5] even if the doctrine is the brainchild of a conceptual background that attaches more weight to the offense or defense. The undesirable condition of a single-form doctrine is a type of dogmatism that is not imposed by the military establishment per se; rather, it stems from a form of self-inducement or group-think that permeates the entire system. In this case, dogmatism is the result of enervated thinking on key issues.

There are a number of ways for determining the multifaceted nature of a doctrine:

The manner in which different forms of war are treated in doctrinal literature. This element, however, has its limitations. While all forms of combat may be treated equally in theory, in actual training the balance is often lost.

The nature of training prior to the war in which the army was caught by surprise. Was equal weight given to all forms of combat or did one form receive priority (for example, urban warfare or mountain fighting)?

The development of weapons systems. Were they developed according to a biased doctrine or did they provide solutions for situations outside the doctrine's focus?

The eminent historian Liddell Hart has stated that "history shows that armies are liable to become obsessed with the offensive idea to the neglect of the defensive. It also shows that an army which is disproportionately offensive-minded, may, when suddenly thrown on the defensive, prove incapable of a type of action for which it has not been prepared."[6]

By the same token, an army obsessed by the cult of the defense is no different from one that is trained solely for offensive operations.

A doctrine can influence a wide range of military skills. Employing historical analysis to discover a direct link between doctrinal flexibility and success in dealing with technological and doctrinal surprise can be a frustrating task. This chapter looks at military doctrines and their approach to sudden shifts on the battlefield, but it does not concentrate on technological or doctrinal change. Later, in the analysis of historical test cases, the connection between doctrinal flexibility and the successful handling of technological and doctrinal surprise is examined in detail.

At the outset, an important point of methodology must be made: none of the following cases deals with the question of whether the basic rationale of a particular doctrine was proven correct in actual battlefield situations (e.g., was an attack or a defensive deployment the right one?), but instead whether the structural features of the doctrine facilitated or obstructed flexibility.

DOGMATISM VERSUS OPENNESS TO NEW IDEAS

The following examples demonstrate how receptivity toward ideas that challenge the reigning view (by adopting unconventional methods and weapons and by publishing articles and promoting people who call the view into question) has affected an army's ability to respond flexibly to changing conditions on the battlefield. The empirical part of the book describes in detail, inter alia, the British army's failure to recover, due to conceptual and doctrinal dogmatism, from the surprise of German armored and anti-tank warfare in the Western Desert; and the extreme case of reluctance to consider views that are not in lockstep with the official line: the French attitude toward the possibility of a German breakthrough along the Maginot Line prior to WW II.

Tolerance of Those Who Hold Different Views: The Attitude Toward Supporters of Armored Warfare in Great Britain and Germany

During the interwar debates in Britain and Germany over armored warfare, different military concepts were developed. Regardless of the outcome of the debates, a different attitude toward innovation and its supporters can be seen. I wish to emphasize the way the debate was managed rather than the debate itself.

Murray discusses British difficulty in learning the lessons of WW I and in introducing innovations. "The evidence suggests a general disparagement of study in the army's cultural system."[7] Part of the blame is placed on the succession of chiefs of the Imperial General Staff (excluding Lord Milne 1926–1933), who were basically unwilling to countenance critical analysis of the British army's performance in WW I. In his *On the Psychology of Military Incompetence*, Norman Dixon wrote that:

> Successive Chiefs of the Imperial General Staff between 1918 and 1939, with the support of other senior officers, did not exert themselves to mechanize the army. Some were actively obstructionist. Against these reactionary elements stood a handful of progressive army officers and a few like-minded civilians. The progressives, who had assimilated the incontrovertible evidence from the preceding war with Germany and were only too well aware of Hitler's preparations for the next, made their views known through books, essays and lectures, and by word of mouth. These moves were countered by the military establishment in two ways. First, they resisted the dissemination of progressive literature; second, they did their best to curtail the careers of those who questioned their own obsolete ideas.[8]

Dixon recalls that Fuller's career advancement was put on hold (though he eventually attained the rank of major general) because of his ideas on armored warfare. Other British tank supporters found their careers stymied for strange and sundry reasons: Charles Broad, who wrote the 1929 British field manual on the use of armored and mechanized units, was "too hot-headed"; Frederick Pile, who published articles on armor during the 1930s, was "too dapper"; and Percy Hobart, commander of the first regular British armored brigade in 1934, was in the midst of a divorce.[9] In his memoirs Liddell Hart recalled, "If a soldier advocates a new idea of real importance he builds up such a wall of obstruction—compounded of resentment, suspicion and inertia—that the idea only succeeds at the sacrifice of himself: as the wall finally yields to the pressure on the new idea it falls and crushes him."[10] Rommel's views on British armored warfare: "The British had remained conservative and their responsible quarters had almost entirely rejected the doctrine of mechanized warfare that had been so brilliantly set forth by some of their countrymen."[11] Guderian admitted that one of Britain's armor theoreticians, Percy Hobart, strongly influenced him. Hobart contributed prodigiously to the development of tactics and communications in armored warfare and tried to test the theory of strategic deep penetration. "But most senior soldiers viewed the method with doubt and disapproval, and the Chief of the Imperial General Staff, Sir Archibald Montgomery-Massingberd, put a curb on the continuation of such practice in subsequent years."[12] In effect, "In 1934, Hobart refused to participate in the exercises of the experimental armored division because of all the barriers placed in the way of its possible success."[13] Even after the outbreak of war, when the value of armor was recognized, the head of the British General Staff Alan Brooke prevented anyone who had been involved with the experimental tank regiment from having a say in the development of armored warfare theory and ascertained that any officer who served with proponents of armored innovation was kept from leading combat units at the division level and above. Although Hobart was returned to active duty—only on Churchill's insistence—he never received a command[14] but was allowed to train armored units in Egypt, which it should be noted, failed abysmally against the Germans. Why was the British army so conservative? Bond lays part of the blame, at least for the British difficulties, with innovation and innovators on the regimental system, which "although undoubtedly possessed many virtues ... lay at the root of the malaise of the interwar Army in the effects it had of narrowing the horizons and obstructing reform."[15] This is an example of a military organization that narrowed/lessened its own future flexibility by putting aside an optional way of thinking and fighting.

Nevertheless, despite the problems in the development of British armor in the prewar years, those who championed it played a major role in shaping its modus operandi. Many of the snags that British armor faced in the Western Desert (see Ch. 10) were due to their basic misconceptions.

On the other hand, the German army's attitude toward the master theorist of tank warfare, Heinz Wilhelm Guderian, was quite the opposite, although here too some domestic opposition existed (especially on the part of senior commanders). General Ritter von Thoma, whose contribution to the development of Germany's prewar armored force was second only to Guderian's, once told Liddell Hart:

> It may surprise you to hear that the development of armoured forces met with much resistance from the higher generals of the German army, as it did at yours. The older ones were afraid of developing such forces fast—because they themselves did not understand the technique of armoured warfare, and were uncomfortable with such new instruments. At the best they were interested, but dubious and cautious.[16]

Guderian claimed that General Ludwig von Beck, the German chief of the general staff between 1933 and 1938, was the main obstacle blocking new ideas. Beck did not reject Guderian's theories but was hesitant about their application. Guderian's *Achtung Panzer!* was published in 1937 in order to win military and public support for armored warfare[17] and as a counterweight to the chief of staff's skepticism.

Gat states that unlike the picture that Guderian tries to portray, Beck supported armored warfare, but as head of the general staff he had to take a broader view of the army's development.[18] According to the scholar of German tactics Rudolf Steiger: "After the 'tank-less period' from 1919 to 1934, the real struggle began in the army of the German Reich over operational doctrine of armour. Guderian and his staff encountered great resistance to their new ideas."[19] Murray claims that the development of armor was met with skepticism by the top brass. General Stulpnagel told Guderian that his ideas were utopian and that large armored units were preposterous.[20]

According to John Mearsheimer, after the victory in Poland, a coterie of senior officers admitted that the role assigned to the armored divisions had been a mistake. They should not have been used to break through the enemy's defenses but to exploit the successful penetration by infantry divisions. Before the battle for France, the most ardent advocate of cutting back the role of armored divisions was General Georg von Sodenstern, who had replaced Erich von Manstein as chief of staff of the Wehrmacht's Army Group A. His views were supported by Gerd von Rundstedt and Gunther Blumentritt of the same army group. But their arguments were rejected by Chief of Staff Franz Halder whom Guderian and Manstein had convinced that armor units should be concentrated when making a breakthrough.[21]

Matthew Cooper states that "cliques" in the traditional branches, headed by artillery officers in key positions, promoted their own people, while the backers of armored warfare were shunted aside. The official military manuals were off limits to armor theorists, who were forced to express their ideas in military journals. The high command engaged in "Byzantine intrigue" in order to consign Guderian and other advocates of armored warfare to minor positions with minimal authority. In 1939 Guderian almost received command of a reservist infantry corps.[22] He remained a controversial figure throughout the war, being relieved of the position of army group commander following a dispute with Hitler. In the spring of 1943, after almost fifteen months of inactivity, he was appointed inspector general of the armored branch and later chief of the general staff.

The issue here is not German innovation with armored warfare but Guderian's story, which may be illustrative of a military organization that accepts disagreement and controversy since they broaden the range of thinking, and, at the appropriate time, provide a suitable response to changes on the battlefield.

In this context Dupuy wrote that to avoid a conceptual rigidity, the German General Staff recruited "fresh, brilliant minds" and took steps to avoid tendencies toward conservatism and rote thinking.[23] Corum writes that "German officers had free rein to publish their ideas on military affairs and doctrine and to critique the ideas of other officers—even to criticize established doctrine."[24] Murray believes that ". . . the Germans proved surprisingly willing to tolerate outspoken officers, Guderian being a prime example," who, at one time or another, had clashed with every senior officer in the land army, without any appreciable effect on his promotion, until the December 1941 incident after which Hitler relieved him of command.[25] Thus, self-criticism, the willingness to hold discussions, and tolerance toward innovators was part of the German military culture.

Containing disagreement and controversy over innovations is not easily achieved in a military organization, and, as Rosen and the American military scholar Douglas Macgregor have noted, on occasion they should be backed by civilian intervention.[26]

Learning From Foreign Armies

The German army meticulously studied foreign armed forces. Beginning in 1925 German intelligence published a bi-weekly review of technological and theoretical developments in armored warfare across the globe. The publication contained translated articles from American, British, French, Austrian, Polish, and Soviet literature.[27] A British training manual published in 1927 served as the German armored forces' main textbook until 1933.[28] The German military attaché in England observed field maneuvers and sent back regular reports on British military development between 1933 and 1937.[29]

As part of the secret German-Soviet cooperation, several German officers were sent to the Soviet armor school. Between 1929 and 1933 a German armor school operated in Kazan, Russia.[30] In this way the Germans gained reliable information on the number of British armored vehicles the Soviets had acquired in their effort to build up their armor force.[31] The British Carden Loyd tank became available to the Germans and inspired the development of the first German tank in the 1930s, the light Pzkw I, which played an important part in the campaign in France in 1940.[32] Between 1926 and 1927, thirty-nine German officers were attached to the Red Army and thirteen (or fourteen) Red Army officers to the German army.[33]

Charles de Gaulle's *The Army of the Future*, published in 1934, sets forth the author's vision of the place of armor in future wars (and in the French army). The book was translated into German in 1935. De Gaulle states that he knew that Hitler and his associates, especially Goering, "devoured" the book.[34]

Guderian claimed that books and articles by Fuller and Gifford Martel (a British military theorist) sparked his interest and gave him much food for thought.[35] In 1937 Guderian wrote in *Achtung Panzer!*: "After mature consideration it was decided that, until we had accumulated sufficient experience on our own account, we should base ourselves principally on the British notions, as expressed in *Provisional Instructions on Tank and Armoured Car Training, Part II, 1927.*"[36]

It will be recalled that the main reason for German interest in foreign armed forces was the Versailles Agreement that forbade Germany to develop its armor potential; therefore, it had no choice but to learn from others. The British and French, on the other hand, tested their own equipment and tactics and were naturally less dependent on external sources of information. But the German army continued to learn from others, even after it developed its own Panzer branch. Dupuy has shown that the German general staff did not suffer from the Not Invented Here (NIH) syndrome and that it conducted a systematic assessment of other countries' technical and tactical capabilities in peacetime, as well as in war.[37] In this respect the German army seems to illustrate a classic case of a military establishment with no qualms about trying to obtain relevant information from foreign armed forces and learning from it. The influence of this culture will be presented in the context of learning under fire in the chapter dealing with Germany's response to the surprise of the T-34 tank.

On the other hand, according to Brian Bond:

> British officers displayed little interest in military developments in European countries or signs of having read articles in foreign military journals. Where was the British equivalent of Guderian eagerly translating and circulating *avant-garde* articles from Britain, France and other countries?[38]

Achtung Panzer! was translated into English in its entirety only in 1992. The editor notes: "Yet *Achtung Panzer!* [. . .] seems never to have been fully translated into English and has not been widely read outside Germany."[39] Interestingly, the German 1933–1934 Army Regulations has only recently been translated. Although the United States Army used this manual during the war (the first part was inaccurately translated),[40] a complete English translation was made available only in 2001.[41] This is not to say that the British armed forces were unfamiliar with the German regulations. Although they followed developments in Germany (and circulated their findings in classified internal publications), interest in their most dangerous enemy was rather limited. The consequences of this aversion to learning from bitter experience is discussed (see Ch. 10) when we examine Britain's slow recuperation from the surprise of German anti-tank warfare in the Western Desert.

ONE-DIMENSIONAL AND MULTIDIMENSIONAL DOCTRINES
A Comparison of the Degree of Flexibility in the IDF and Wehrmacht Prewar Combat Doctrines

If combatants on one side are unfamiliar with a particular form of combat because of their negligence or underestimation of it, it is the same as though the enemy initiated a doctrinal surprise, whether a defensive or offensive one. The following examples illustrate how a doctrine based on a single form of combat can debilitate the forces using it from dealing properly with doctrinal surprise, and why a doctrine in which a particular form of combat receives priority—but not exclusivity—allows the forces to cope more effectively with an unexpected situation.

This section broaches the problem of dogmatic doctrine, presenting it in a different light from the previous discussion. The first type of dogmatism discussed was the result of imposed uniformity of opinion and conceptual inhibition, both of which obstruct critical review, whereas the type of dogmatism now under discussion is characterized less by thought repression—since the dominant doctrine has virtually been consecrated—than by "group think" based on the belief that certain elements in the doctrine are absolutely correct. Another feature of this type of dogmatism is its basic assumptions that preclude the development of flexibility, as in the case of French doctrine before WW II (see Ch.12).

A comparison is made between the degree of flexibility in the offensive doctrines of the German army and the IDF. One methodological caveat should be mentioned at the outset: the memoirs of German commanders relate to a long period of fighting (six years) and recall incidents from various engagements. As a result, their descriptions are less focused (and may suffer from biased recollection, such as blaming Hitler's decisions for military setbacks), whereas Israel's events took place in short time spans that are relatively easy to isolate and analyze.

Since its inception, the IDF has operated according to an offensive doctrine that ensures lightning-fast victories. This strategy still holds. The reasons are obvious: Israel's geographical reality (a depth too shallow for defensive maneuvering); the intolerable social and economic pressures that would be caused by an extended war of attrition; the availability of high-quality manpower that gives the IDF an advantage over Arab armies in battlefield improvisation; and the use of offensive tactics as a power multiplier. However, these factors and their proven justification are not treated in the present work. Instead, I deal with specific issues, namely, the consequences of this approach when it was taken to extremes prior to the Yom Kippur War and the single-form offensive doctrine that hampered the IDF's ability to switch from the offensive to the defensive that was required in the first days of the fighting.

The classic case of the "cult of the offensive" was the doctrine of the major European armies before WW I. According to Stephen van Evera: "The gulf between the perception and the reality of warfare has never been greater than in Europe during the years before World War I." Even though defensive warfare was at its most effective, with such original inventions as the machine gun and barbed-wire, the Europeans still believed that the offensive would carry the day (on the assumption that mind over matter would prevail).[42] Jack Snyder describes how the offensive concept, which developed in Germany, France, and the Soviet Union, had catastrophic results for each of those countries;[43] the most extreme case being France, where a unique doctrine of the "unconditional offensive" evolved, which "came very close to submitting France to German invasion and was largely responsible for the terrifying human cost incurred by the French in their futile attacks on German positions during the early phase of the war."[44] Some Israeli scholars have also used the term "cult of the offensive" to describe the IDF's pre-Yom Kippur War doctrine.

The unrivaled status of offensive warfare in the IDF began in the 1948 War of Independence and peaked after the 1967 Six-Day War. Ariel Levita, an expert on Israeli military doctrine, states that the IDF's victories, which were based on the offensive approach, "created a sense of disparagement, indifference, and even disdain toward the defensive elements in the military doctrine."[45] He also claims that "the conceptual enslavement to the offense resulted in a fatal blow to the IDF's ability to fight defensively . . . [This was] a warped, arrogant perception of the defensive battle, to the point of completely abandoning it, which eventually led to inexcusable unfamiliarity with its salient features."[46]

The section in the Agranat Yom Kippur War Investigation Commission Report entitled "Emphasizing the Offense and Neglecting the Defense" states that although the IDF doctrinal literature related evenhandedly to different types of warfare, in reality offensive combat was taken much more seriously than defensive warfare. General (Res.) Avraham Adan described the doctrinal development that led to the construction of the fortified positions along the Suez Canal before the Yom Kippur War: "Between the War of Independence and the Six-Day War the IDF lost its ability to deploy defensively since it had prepared solely for a war of movement in which it seized the initiative as much as possible."[47]

Dov Tamari has termed the approach that dominated IDF thinking after the Six-Day War the "cult of armored warfare." The IDF envisioned the likelihood of an Egyptian preemptive strike, but the proper defensive action against it "was not perceived as a primary campaign with internal unity. Nor was it linked in any way to the ensuing offensive except perhaps as its 'catering service.'"[48]

The Israeli scholar Shimon Naveh refers to this phenomenon as the "cult of the offensive," explaining that since offensive warfare is the optimal and natural way of realizing the tank's tactical capabilities, the view that posited the tank in the eye of offensive warfare turned into a misguided obsession between 1968 and 1973. This conceptual asymmetry diminished the IDF's defensive capabilities at the tactical level and as a basic form of operational maneuver to the point where they were eventually all but forgotten.[49]

How did this approach affect training? As shown above, doctrine "is . . . the path chosen for attaining victory, and determines the development of the army's organization and weapons systems." The warfare doctrine, the structure and organization of military units, and the weapons systems are designed and tested, inter alia, through military exercises—a relatively accurate measurement of a doctrine's chief features. Sometimes the results of the exercises provide a more reliable measure than the military manuals.

A look at IDF exercises held only one year before the Yom Kippur War reveals that of the twenty-eight major training objectives (in the sense of expected tasks the participating forces must fulfill) in skeletal exercises carried out by Ariel Sharon's and Avraham Adan's divisions, only one objective was related to defense and even that one was not clear. None of the armored brigades' training objectives pertained to defensive combat. The infantry brigades eked out a single defensive objective: "hasty defense deployment" but this was not the major training objective of the exercises.[50]

The structure of the units was also influenced by the single-form doctrine. The standing division (originally intended to be made up of various branches and capable of fighting in every form of combat) became a one-branch tank division without any of the infantry, mortar, or artillery capabilities essential for offensive combat, let alone for defensive fighting. The Six-Day War relied overwhelmingly on armor rather than the other branches. The Agranat Report's conclusion of the impact of unit structure on operational flexibility stated: "This is a case where a tank division trained for armored combat in mobile battle was incapable of holding ground or deploying for self-protection after dark."[51]

Similar problems with weapons and the "cult of the offensive" can be seen in France's pre-WW I underestimation of machine guns and heavy artillery as offensive weapons. The French hoped to make rapid advances on the battlefield,[52] but when this failed to happen, these weapons became absolutely necessary for waging a defensive campaign.

The upshot of the single-form doctrine was that the ground forces became unskilled in defensive warfare. The emphasis on offensive combat, which the

IDF had always advocated, grew even more extreme after the lightning victory in 1967. But by 1973 the waning of conceptual and doctrinal flexibility seriously impaired the ability of the regional command and the forces in the field to deal with the Egyptian units crossing the Suez Canal. Instead of deploying defensively and waiting until enough forces arrived before going on the offensive, IDF units attacked relentlessly. This best describes the ill-conceived attacks against the Egyptian bridgehead on October 8 and the attempts to cross the canal before conditions were ripe. The lack of flexibility resulted in the excessive waste of regular forces in the initial days of the war and of reserves at a later stage. In *A Storm in October*, battalion commander Eliyashiv Shimshi has a chapter entitled "Learning How to Hold with Mobile Forces." In it he discusses the difficulties in waging defensive combat: "This was a critical stage—the Egyptians were on the offensive, and not everyone on our side knew what to do."[53]

Doron Almog analyzed the fighting on the Golan Heights, noting that the errors included: insufficiently dense minefields and obstacles; defining the rear defense line as the border; and throwing reinforcements into the fray at the very outset of the war.[54] The accumulation of blunders was the result of having neglected the lessons of defensive warfare while the force was being designed and developed before the war.

A key element in defensive combat is preparing commanders for mental difficulties they might have to face, such as ordering a withdrawal. The IDF's unreadiness for this eventuality was blatantly apparent in the shock and confusion felt by officers of all ranks,[55] including the defense minister who allowed himself to refer to the situation on the Golan and Sinai fronts on the second day of the fighting as being on the brink of the "destruction of the Third Temple."[56]

This is an extreme case of a single-form doctrine, the outgrowth of a conceptual fixation on one particular type of warfare (the offense). The effect of the linear mindset on training and unit structure led to reduced flexibility because of the enervated capacity to cope with unexpected situations—in this case, the inability to engage in defensive combat.

The IDF recovered quite quickly considering the manifold problems stemming from the single-form doctrine. The recuperation may be attributed to basic characteristics in the Israeli concept of command (see below).

An example of a multidimensional doctrine that enabled battlefield flexibility is the German offensive doctrine before WW II, which stressed offensive combat and lightning-fast victories for many of the same reasons that induced the IDF to adopt a similar doctrine. The need to avoid fighting on two fronts against states with greater human and economic resources forced the Germans to build a military force that could quickly overcome the enemy and avert a war of attrition.[57] This explains Hitler's obsession with swift military operations.[58] "As the army considered itself first and foremost an offensive force, its doctrine and training were geared accordingly."[59]

"Unit Command Regulations" *Truppenführung—HD-300* (published in 1933–1934) reflects the effect of strategic concept on doctrine. A difference of opinion exists among scholars regarding the impartiality of the regulations toward all the forms of warfare.

The Israeli scholar Martin van Creveld compared the fighting capabilities of the American and German armies in WW II and reached the conclusion that Germany's offensive strategy led to a doctrinal emphasis on the attack. He found evidence for this in the index to the regulations where the entry "attack" appears fifty-three times and "defense" only thirty-four.[60] Despite van Creveld's claim that the offensive was the doctrinal preference, the German army manual is in fact quite balanced. For example, it notes: "The changing situation of combat often requires transition from one type of engagement to another. The transition from attack to defense may occur when holding a position that has been taken, or if necessary, under enemy pressure."[61] This is a classic example of the dialectical relationship between the defense and the offense that Clausewitz's theory describes.

Naveh argues: "Unlike its successor the blitzkrieg, which adhered exclusively to the offensive, the manual reflected a balanced approach to offensive and defensive, seeing both as essential and complementary forms of operational manoeuvre."[62] The reason for such divergence of opinion among scholars may lie in their contrasting views of the different levels of war. "Therefore, despite encouraging the offensive spirit at the tactical level, the manual indoctrinated its users to exercise a multi-dimensional, combined and balanced approach at the operational and strategic levels."[63] Even if we reject Naveh's claim that the blitzkrieg concept did not extend beyond the tactical level, and that for all practical purposes it undermined the impartiality of the manual, this does not refute his argument that the manual was balanced at the operational and strategic levels. This balanced doctrine would play a key role in the coming events.

Military regulations, it will be recalled, do not always express the dominant views of the army. In the case of the "Unit Command Regulations," we can analyze views that are independent of official theory.

A good example of awareness of the dangers in inculcating a dogmatic, one-tracked choice for the offensive appears in a postwar criticism of American field regulations. The critics in this case were the German generals themselves, first and foremost, the former chief of the general staff General Franz Halder. The German officers call attention to the overemphasis of the regulations on the attack. "Nevertheless we are opposed to elevating the offensive to the level of a dogma and to any exaggeration which teaches 'attack at any price'. We do not undervalue daring and boldness in action, but we do not want an officer who is uncertain about his next move to be misled into taking refuge in attack just because attack is 'more soldierly.'"[64]

Steiger, who has studied German armored tactics, claims that prior to the war, German military officers were aware of the importance of defense and role of the tank in defense.[65]

In 1937 General Wilhelm von Leeb (later, army group commander in Poland, France, and Russia) wrote that "defense" was the strategy best suited for Germany. He understood the tendency to build an army based primarily on maneuvering and quick victories, but he warned against a military doctrine geared solely to the offensive:

> However, in this renewed movement, there is another danger. It is possible that it will not lead to maneuvering tactics and strategy, particularly in operative defense [...] We Germans have to look to defensive as an important essential method of conduct of war and conduct of combat, since we are in a central position, surrounded by highly equipped nations. Defensive should not be kept in the background, as before the last war. It must be given its deserved place in education, the formation of the cadres and troops [...].[66]

What triggered the publication of the book was the debate over a series of von Leeb's articles that appeared in the German military journal *Militärwissenschaftliche Rundschau*.[67] This is an example of defensive military thinking that not only integrated into the dominant doctrine but even challenged it.

How was the doctrine applied in training exercises and the research and development (R&D) of weapons? Unlike the uniformity, "single-trackedness" of the IDF's offensive doctrine, the German army was fully cognizant of the importance of defensive exercises. Defense was a key feature in Germany's large-scale, pre-WW II training programs. In the autumn maneuvers of 1937, the last major exercises before the war, the army drilled in both offensive and defensive operations. The defending force's missions included early identification of enemy intentions, blocking enemy intelligence operations, and, if necessary, building a concentrated defense array.[68] The offense commenced operations with defensive procedures to amass sufficient strength for the attack. In short, the exercise dealt equally with defensive and offensive capabilities.

Although the German arsenal was intended primarily for the offense and had developed it according to an integrated, combined arms concept, it also included defensive weapons such as anti-tank and anti-aircraft guns. In *Achtung Panzer!* Guderian argued that the key to victory lies in attacking with concentrated armor. He discussed the development of defense capabilities (anti-tank weapons, minefields, obstacles, and flat trajectory artillery fire) to be used against forward armored units. He also envisioned an anti-tank integrated force made up of engineers, machine guns, and artillery, capable of arresting surprise armor attacks, penetrations, encirclement, and flanking maneuvers.[69]

Given its primarily offensive doctrine, how did the German army cope with the need to shift to the defense? According to Naveh, in the period between the writing of the 1933–1934 manual and the war, German combat theory went from a balanced, systemic approach to a narrow offensive concept geared to the tactical level. This change meant that when the Wehrmacht had to fight defensively, it was at a loss since no one in the military hierarchy had been drilled in defensive tactics or withdrawal and delay operations. German military planners believed that defense and withdrawal reflected defeatism, and defeatism must be avoided at all costs.[70]

Naveh may be wrong in his limited view of the blitzkrieg as an exclusively tactical concept[71] that dominated German military thinking and imposed an unbalanced offensive approach. Be that as it may, even if Naveh is right, the testimony of German commanders proves that he misjudged the Wehrmacht's ability to handle sudden changes on the battlefield.

Evidence of the Germans' difficulty in adapting to defensive combat appears in their debriefings on the British attempt to break through the German-Italian "Siege of Tobruk."[72] On the other hand, numerous reports testify that the Germans adapted relatively easily to the new situation.

Many German officers who fought on the Russian front recall their difficulties with defensive warfare when the German offensive ground to a halt in the winter of 1941. They usually lay the blame on Hitler's orders that expressly forbade German withdrawal and ordered deployment for "static defense." This directive prevented field officers from engaging in "mobile defense" since it would have meant ceding territory, even though "mobile defense" was more in line with the German combat doctrine. Other descriptions written by German commanders express no surprise at having to fight defensive battles or any special problems associated with them. Their memoirs contrast with those of Israeli officers in the 1973 war, when the latter realized that the campaign would take longer than expected and be mostly defensive.

General Frido von Senger und Etterlin states that "the 'halt order' and the resulting discrepancy between wishing and doing cannot be attributed to Hitler's ideas about the First World War. For these ideas could also be found in the interwar German regulations, which required that on the conclusion of a battle in the deep zone, the forward edge of the battlefield must again be in the defender's hands—which means that the line must be held."[73] This limitation notwithstanding, von Etterlin describes how German units succeeded in shifting to the defense. "Thus the armored divisions, originally organized as purely offensive formations, had become the most effective in defensive operations."[74]

Guderian mentions the transition from attack to defense while maintaining control over the forces and a meticulously drafted plan for withdrawal to a prepared defense area. He describes the mental effort involved in arriving at such a decision, but the impression is that it was made professionally, without the tension and strain characteristic of surprise situations.[75] When he countermanded Hitler's order forbidding retreat, instructing some of his units to withdraw to areas most advantageous for defensive deployment, he was dismissed from command of the Second Armored Army.[76]

Colonel Hans von Luck, commander of a reconnaissance battalion on the Eastern Front, also depicted the orderly withdrawal to a defense layout, where his battalion deployed as rear guard. He too expressed no surprise at being ordered to carry out a defensive operation.[77]

The military historians Bruce Condell and David Zabecki argue that contrary to the accepted view that the German army was skilled only at offensive maneuvering, it also fought adeptly and tenaciously in defensive situations. As an example, they note Field Marshal Walter Model, who repeatedly proved himself extremely capable in defensive battles.[78] Liddell Hart, too, claims that Rommel's ability to improvise anti-tank defenses in the Western Desert shows that the Germans were not only masters at putting offensive theory into practice in armored warfare, but also at defending themselves against it.[79]

All told, the Germans' offensive approach did not detract from their ability to conduct effective defensive operations. The reason for this seems to be their multidimensional approach to training, weapons R&D, and theoretical military literature outside the official doctrine. When they lost defensive battles in Europe, it was not because of their failure to adapt to battlefield conditions but because of the combination of general erosion of strength and Hitler's strategic directives that often hampered the implementation of an effective defense.

A comparison of the IDF with the German army shows that while both held to similar offensive concepts with major implications on force planning, the German concept offered much more leeway than the IDF's for a multidimensional approach. Maintaining flexibility calls for a doctrine that provides solutions to given, basic situations but at the same time, enables the army to devise appropriate responses to events that far exceed those that the doctrine chiefly deals with. Put differently, a doctrine should be multidimensional: it should include education, training, and the organization of units that furnish a basic answer for every type of combat, even if on a limited scale. Another example of a dogmatic, single-form approach is France's preoccupation with defensive warfare before WW II (see Ch. 12).

In conclusion, flexibility on the battlefield should be the outgrowth of a doctrine and approach that infuse:

Toleration of ideas that diverge from the official doctrine; the opportunity to publish these ideas; promotion of their supporters; and unreserved willingness to learn from other armies. This is probably one of the hallmarks of the "culture of flexibility."

A multidimensional approach to doctrine, training, and weapons development. Such an approach is vitally important even if the strategic inclination leans toward a single-form approach.

3 ORGANIZATIONAL AND TECHNOLOGICAL FLEXIBILITY

THE SECOND STRATUM IN THE FLEXIBILITY COMPLEX deals with the structure and organization of an army. Structure and organization can be treated on a number of levels, three of which are discussed in this chapter. The first level is the qualitative and quantitative relationship between basic elements, such as firepower and maneuverability, defensive and offensive fighting capabilities, decision and attrition. The second level is the organization and structure of military units: weapons diversity in units and the level where combined arms are employed. The third level refers to the weapons themselves, including their operational capability and technical properties. This chapter discusses flexible military organization on these three levels.

The characteristics of flexible military organization and structure include: a balance between basic force elements, units and weapons, weapons redundancy designed to answer basic operational problems, and the technological versatility and changeability of the weapons.

I contend that force planning must consist of balance, diversity, versatility, and changeability, which are based inter alia on the concept of "alternative objectives," which lies at the heart of Liddell Hart's strategy of indirect approach. Planning alternative objectives enables a commander to conceal the direction of his advance for a rather lengthy time period, thereby diminishing the effectiveness of the enemy's defense. Liddell Hart cites the eighteenth-century French military commander Pierre-Joseph de Bourcet: "Every plan of campaign ought to have several branches and to have been so well thought out that one or other of the said branches cannot fail of success."[1] On the principle of alternative objectives, Liddell Hart had this to say:

> For, the principle of alternatives is simply the power of variability, which, in turn, means a state of adaptability—and if anything is adaptable it cannot be rigid. One might therefore say that the idea of having alternatives is perhaps the

only one in the realm of warfare that is safe to turn into a principle, because it is the only one which is fool-proof against being made rigid in application.[2]

He also believed that "[a] plan, like a tree, must have branches—if it is to bear fruit. A plan with a single aim is apt to prove a barren pole."[3] The principles of balance, diversity, redundancy, versatility, and technological changeability are based on Liddell Hart's concept of multiple objectives, which brings it into the field of force structure. According to this principle, if a military body is organized around a single branch or weapon, it runs the risk of finding itself up against a concentrated response that can neutralize it like a "barren pole" (Liddell Hart's expression). To avoid such a situation, force organization must include a number of branches, that is, it must diversify its weapons and the capabilities of its units. In combat, when the enemy discovers a way to neutralize one branch, the others will flourish.

Before providing a detailed explanation, I elucidate some of the principles of organizational theory. One approach to organizational analysis—the systems theory—states that organizations are inherently "wobbly" entities; in other words, the interdependency of their elements makes them unstable. Thus, an alteration in one element produces a change in the rest. Also, different parts of the organizational system are capable of operating independently even though they are integers of the same system. One theory that exemplifies the systems theory is called the contingency theory or the theory of structural stipulation. This theory is based on the principles of open systems, one of which is equifinality. According to equifinality, systems are not necessarily predicated on a single-value proportion between goals and means since the same results can be attained in numerous ways. In other words, no one pattern is suited to every organization.[4] This approach and theory have obvious implications regarding the ability of the military organization to deal with technological and doctrinal surprise.

I will not delve into the issue of whether the force must be organized in a balanced, diversified way and possess redundancy in garrison, or be task-organized only prior to battle. This issue takes into consideration many factors that should not influence the need for the abovementioned qualities.

BALANCE

According to the balance principle, an army that tries to meet the challenge of technological and doctrinal surprise must maintain a balance among its basic capabilities. This is not a matter of specific weapons capabilities, but of elements in force planning that support basic capabilities: defense and offense; shock troops and staying forces (in the narrow, military sense of a fighting echelon as opposed to logistic support echelon); firepower and maneuverability.

Simpkin viewed the importance of balance in force planning in this light:

> This balance is in fact the safeguard against the commander or the planner being wrong. And the degree of imbalance acceptable diminishes up through the levels as the margin of error grows. The efficient allocation of troops to tasks probably demands high tactical imbalance. The same is partly true of operational imbalance; this is acceptable so long as the basic assumptions over the scope and duration of the operation or campaign hold. If they fall, as they did for Operation Barbarossa when the campaign extended into the Russian winter, disaster may result unless the imbalance can be corrected. Strategic imbalance is surely most dangerous.[5]

Imbalance results in a disproportionate susceptibility to surprise and cripples an army's ability to recuperate.

A balance between the elements of firepower and maneuverability is indispensable if flexibility is to be attained. Any serious divergence from this balance detracts from the army's ability to cope with surprises that neutralize any of its key elements.

The second category of balance is that between defensive and offensive capabilities. Many states have been (and still are) under constraints that require a choice between offensive and defensive force planning. Sometimes the constraint is amplified by a single-form doctrine (see Ch. 2). Military experience shows that a force, whose organization lacked the capability to implement a form of war considered of secondary importance, is invariably hard pressed to recover from technological and doctrinal surprise.

The third balance required for flexible force structure is between the fighting echelon (maneuvering and combat support units) and the logistical support echelon. Their relationship is generally based on the army's concept of the ability demanded for a battlefield decision, as opposed to dealing with a lengthy war of attrition.

The Balance Between the Maneuvering and Firepower Elements

Handel states that the cycle of modern military technological development "has shifted the advantage from the defense to the offense, and back, a number of times since the start of the technological revolution."[6] This phenomenon seems to come from the changing relationship between firepower and maneuverability, both of which are directly influenced by technology.[7] An understanding of this point has critical meaning for force balance. The difficulty in predicting the characteristics of the future battlefield can result in an imbalance that overexaggerates offensive maneuverability or defensive firepower. If the ORBAT radically upsets the balance between maneuverability and firepower, the army is liable to be caught by surprise and at a loss to adjust to the new situation. The following is Handel's view on this plight:

> Such changes are not always perceived before the outbreak of war. For example, despite numerous indications from the Boer War and the Russo-Japanese War, in which the growing advantage of defensive over offensive weapons was clearly demonstrated, most European armies before the First World War emphasized the development of exclusively offensive doctrines; similarly, the Israeli army before the 1973 war misread the technological trends favoring the defensive in anti-tank and anti-aircraft weapons and consequently relied on an exclusively offensive doctrine. Conversely the French, before the Second World War, learned the lessons of the First World War so well that they overestimated the power of defense.[8]

An example of imbalance between the maneuvering and firepower elements can be seen in the German army's preparations for Operation Barbarossa. The army attributed great importance to artillery and close air support at the tactical level. Its campaigns in Poland and France were carried out with superlative coordination and support. Naveh claims that the systematic failure in comprehending the importance of the firepower element, above the tactical level, was not strongly felt in these campaigns because of their short duration and the relatively limited geographical dimensions of the fighting theater. In addition to this conceptual gap, German priority of maneuverability resulted in a concentrated effort on tank production rather than artillery mechanization. As German units advanced deeper into the vastness of the Soviet Union, the conceptual and weapons shortcomings brought heavy damage to their operational capability. According to Naveh, some of the Wehrmacht formations found themselves without any artillery support in the bitter winter of 1941–1942.[9]

Soviet artillery and air superiority gradually wore the Wehrmacht down until it lost its operational initiative. The forced, unplanned move to the defensive in the winter of 1941—because of the halt in offensive maneuvers—increased the importance of firepower, whose absence was sorely felt, and reduced the advantage of armor maneuverability, on which German force planning was based. This fundamental deficiency impeded the ability of the Germans to wage a defensive fight in a war of attrition. In other cases the opposite happened; maneuverability, instead of firepower capability, became crucially important. A balance between the maneuverability and firepower elements generally enables armies to recuperate from the surprise of having a key element in their warfighting doctrine neutralized.

An example of such an imbalance—and its correction—occurred in the British Eighth Army's structure in the African Desert in WW II. The military author J. Nazareth noted the need for dynamic thinking as the key to effective military command. Nazareth held that "Montgomery believed that an important cause of the defeats of the Eighth Army was the lack of the balanced force because there was no strong mobile arm."[10] The British lacked a mobile branch comparable to the German mechanized, armored Afrika Korps. One of Montgomery's first decisions upon taking command of the Eighth Army was to develop this branch in order to restore balance to the British forces.

Nazareth states that the Battle of Alam el-Halfa in the summer of 1942, in which the British succeeded in blocking the German attempt to break through to the east, testifies to the application of the concept of balanced force. Rommel's effort to outflank the British layout and attack the Alam el-Halfa layout from the south met with delaying operations by the British Seventh and Tenth Armored Divisions. The Afrika Korps was wasted and forced to withdraw.[11] The restoration of balance to the Eighth Army by creating a maneuvering element in the form of a powerful mobile branch enabled the effective British response. Even if Nazareth exaggerates the importance of the mobile branch in this case, and the main British layout in the Battle of Alam el-Halfa was static, and the main element in the British success was firepower based on air superiority, Montgomery nevertheless introduced a balance between firepower and maneuverability in which the Eighth Army had been deficient until then.

A recent example that illustrates the problem expected from an imbalance between maneuverability and firepower (stemming from overdependence on precision-guided munitions) can be seen in the 2003 war in Iraq. The success rate in destroying the armored vehicles of the Republican Guard divisions was minimal when the targets were static and hidden. Only after the Americans launched the ground offensive and the Iraqi divisions deployed for battle, did the vehicular movement enable the Americans to identify targets, which led to their high destruction rate from the air.

The American helicopter attack of March 24, 2003 against the armored vehicles of the Iraqi Medina Division—which had survived the previous air strike and moved into a built-up area—failed, and most of the Apache choppers suffered damage.[12] Another case of the difficulty in destroying static or concealed armored combat vehicles from the air was the American operation in Kosovo in 1999. The bombing of a small Serbian force (less than 25,000 soldiers) for several weeks led to 2 percent losses rather than 25 percent as expected.[13] One of the American lessons from the 2003 war in Iraq was to "try to force the defenders to deploy outside of the cities."[14] To do this, an army's ability to pose a threat must be preserved in the form of maneuverability. The overreliance on firepower can create a situation in which an army is caught by surprise, and its firepower fails to provide the answer. In such a case, an army will have a hard time recovering.

The Balance Between Defensive and Offensive Fighting Capabilities

The balance between the maneuvering and firepower elements has a direct influence on another crucial equilibrium in the military structure: the balance between defensive and offensive fighting capabilities. The force must be organized so that it can cope with these two basic forms of war.

> Military organizations at its [*sic*] several peaks in history has [*sic*] been based on the combination of a defensive pivot and mobile offensive wings. The first afforded the stability from which the decisive mobility of the second could be developed most effectively and securely.[15]

There are many examples of armies, caught by surprise, that found it difficult to fight defensively because of their organization, which was disproportionately based on offensive elements.

The book *Icebreaker* by Vladimir Rezun (pen name, Viktor Suvorov) claims that since Stalin intended to attack and "liberate" Germany in the summer of 1941, the Red Army was built for an offensive war that emphasized paratroopers and armored units and had phased out its defense-oriented units, such as sabotage and guerilla units.[16] For example, the immense obstacle layout known as the "Stalin Line," which could have provided an effective defense against the German invasion, was dismantled.[17]

According to Rezun, once the Soviet Union was attacked, its army failed to engage in effective combat because its defensive elements had been neutralized. On the other hand, Richard Overy, an expert on the Third Reich and

WW II, is convinced that Stalin did not plan to attack Germany unless he had a definite pretext. Rather, he intended to implement preliminary offensive moves as part of his defense concept, but they never materialized because of the sudden onslaught of Operation Barbarossa. According to Overy, even if Stalin did not intend to launch a war, the Red Army still would have been forced into a defensive war that it had not systematically prepared for.[18]

Beth Gerard writes that prior to Stalin's purges of the military establishment (beginning in the spring of 1937 when Marshal Mikhail Tukhachevski was accused of treason), the Soviet doctrine tended to the offensive but after the purges, it inclined toward the defensive. The author bases this argument on the disappearance of proponents of an offensive doctrine, whose places were filled by supporters of a defensive one. This led to a number of changes in force planning, such as the subordination of tanks, which until then had been organized in mechanized corps as independent brigades and battalions—to infantry divisions—and the cutback in the pace of heavy tank and aircraft production.[19] In 1940 the priority for defense yielded to offense, but:

> The efforts made in 1940 and afterwards to regain an offensive force posture were insufficient to prepare the Soviet armed forces to wage the kind of battle that they needed to wage in the first year of war. The Red Army was equally and tragically incapable of pursuing a strategic defense, largely because the thrust of its strategic theory had always been offensive.[20]

Gerard cites General Andrei Eremenko, an army commander and front commander, according to whom "[s]ome of the general ideas of the Soviet military doctrine, though essentially correct, could not be carried into effect in the initial stage of the war due to . . . flow in our defense preparations."[21] The failure in organizing the Soviet forces for combat, much of which revolved around the question of defensive and offensive capabilities, limited the Russians' ability to recover swiftly from a battlefield surprise when it had to fight on the defensive.

The section on doctrinal flexibility described how the IDF was organized in an imbalanced fashion at the outbreak of the Yom Kippur War. It regarded its main weapon, the tank, as a "wonder weapon." Although the tank was basically a multipurpose weapon, the IDF perceived it as the chief agent of the offensive doctrine. Since the IDF was built as a ground force with an overwhelming emphasis on the offensive, insufficient resources were allocated to the construction of infantry, mechanized infantry, and engineering layouts, even though the warfighting doctrine prescribed them as defensive mainstays.

Mortar and artillery support layouts, too, were insufficiently developed.[22] These layouts could have compensated for armor deficiency in the defensive battle when armor was unable to make the most of its mobility. All of these shortcomings increased the vulnerability of the IDF (see below).

Balance Between the Fighting and Support Echelons

Another basic element is the balance between fighting and logistic elements. Over half a century ago, General Archibald Wavell stressed logistical difficulties:

> ... movement and supply. These are the real foundation of military knowledge, not strategy and tactics as most people think. It is the lack of this knowledge of the principles and practice of military movement and administration—the "logistics" of war, some people call it—which puts what we call amateur strategists wrong, not the principles of strategy themselves, which can be apprehended in a very short time by any reasonable intelligence.[23]

Van Creveld states that research has long neglected military logistics and that "[h]undreds of books on strategy and tactics have been written for every one on logistics."[24] The danger of an imbalance between the fighting and support echelons in the event of technological and doctrinal surprise results in the inability to respond properly when a mission is not attained and a logistical move must be made for its assistance. When an army is unevenly built, usually with an overemphasis on the fighting echelon, it will be hard pressed to meet changes that require a major logistical solution.

The organization of German forces in Operation Barbarossa is a classic example of this type of imbalance. The operation continued much longer than planned and was seriously impeded by weaknesses in the Germans' logistical capability. In this case, the unexpected length of the fighting may be considered a doctrinal surprise since the Germans' warfighting doctrine was based on a swift decision. The Germans built their army according to their doctrine with a far greater emphasis on the fighting element than the logistical one that was required for extended combat.

In his book *Supplying War: Logistics from Wallenstein to Patton*, van Creveld analyzed the Germans' logistical problem. He does not blame Barbarossa's failure on logistical deficiencies, but charges that the paucity of resources and bungled management of logistical resources played a major role in the failure of the operation.[25]

The German army launched Operation Barbarossa in June 1941 with fuel reserves for three months, a basic supply of ammunition carried by the troops, and a shortage of spare parts for no less than two thousand different types of vehicles. Van Creveld writes that "[i]t might be expected that these shortages would have induced the German leadership to reconsider the rationale of the whole campaign. Instead of so doing, however, they managed to convince themselves that what they had originally estimated would take five months to achieve could, in fact, be achieved in four, or even in one."[26] Due to the miscalculation of the length of the war, the German army was unprepared for winter fighting and the results were catastrophic.

Posen notes that the Luftwaffe was geared for short-term wars—not for campaigns of attrition—having no reserve aircraft and only a small stock of spare parts. The Luftwaffe was intended as a "striking force"—not a staying force. Already in 1940 the Luftwaffe proved that it was capable of intensive operations for only short periods. Even the relatively brief Battle of France wasted some Luftwaffe units by half. The German air force was unsuited for the type of campaign demanded in the Battle of Britain,[27] let alone a campaign of several months in Russia.

Operation Barbarossa began with a series of tactical and operational victories. But the influence of the imbalance between the strengths of the fighting and logistical support echelons, combined with weather conditions (early rains that impeded movement, and later, snowstorms and freezing cold) severely undermined activity at the fighting echelon and resulted in aborted missions. Steiger studied German armored tactics on the Russian front and concluded that the main reason for Operation Barbarossa's failure was the logistic plight of German supply lines. Toward the end of the operation, shortages of fuel, oil, and spare tank parts had a cumulatively deleterious effect on the tactical situation and the strategic gains of Wehrmacht armored units no less than that of the battle conditions, weather, and the enemy's situation.[28]

According to Naveh, the German logistic failure resulted in the heavy loss of life and fighting equipment and in many cases the inability of tactical units to carry out their missions.[29]

What was the origin of the problem (excluding the enormous objective difficulties)? The source of the imbalance seems to lie in the basic concept that depreciated the logistical element, misjudged the nature of combat in Russia, and miscalculated the length of the coming war. In the introduction to the English translation of the 1933–1934 German field manual, James

Corum asserts that logistics was the major "blind spot" in German military thinking. For example, many civilians filled the place of logistics officers at the divisional level, and the proportion of instructors to cadets in German military academies in logistics sciences was 1:120 compared to 1:20 in tactical studies.[30]

The German logistical capability proved itself admirably in the difficult conditions it faced; nevertheless, it lagged far behind the army's assault capability. The asymmetry between the Germans' fighting capability and logistical support capability severely hampered the army's capacity to deal with the surprise of prolonged fighting and weather conditions.

A positive example of balance between the fighting and logistic support echelons was the IDF's "logistic revolution" prior to the Six-Day War that granted the fighting forces a large degree of independence.[31]

What is the right balance? Probably the one that enables an army to adapt even when a basic assumption on future war proves to be unfounded: the need to defend while prepared for offensive operations, the need to conduct a long war when a quick decision is not available. Balance here is determined in the strategic level of force planning, and a failure in this level can result in a catastrophe.

DIVERSITY

According to the principle of diversity, force structure should include a wide range of weapons and units so that if some are incapacitated by countermeasures, the commander can still devise a solution with the remaining resources. For example, if a choice must be made between force planning based on one type of weapon—no matter how promising its effectiveness—and force planning based on a variety of weapons—each of which is less than first-rate—the second option should be taken. The origin of the need for diversity in dealing with technological and doctrinal surprise can be found in the field of biology! Biological diversity enables life on earth to survive unexpected changes, for example, the meteor that struck the planet sixty-five million years ago and caused the mass extinction of many species.[32] Recuperation from the catastrophe allowed the species that survived to flourish—especially the mammals. An army that wants to avoid "mass extinction" by its enemy must develop a diversity of weapons and capabilities so that in the event it encounters a surprise designed to neutralize it, it will still have various means that enable new solutions to be improvised.

Why is it dangerous for an army to rely on a limited number of systems that it regards as "wonder weapons"? A comprehensive answer to this question lies in Fuller's two concepts: "dominant weapons" and the "constant tactical factor."[33] The first concept states that every historical period is characterized by a particular weapon system with an overwhelming advantage that dominates the battlefield. This advantage may be in range, destructive power, precision, and so forth. Examples of this are the battleship, aircraft carrier, submarine, and airplane. Given the dominant weapon's definite advantage, a tactical and operational doctrine develops around it. Different types of land, air, and sea "dominant weapons" can exist simultaneously.[34]

Be this as it may, according to the constant tactical factor, the dominant weapon is destined to lose its status. For example, the machine gun, trench, and barbed wire—dominant weapons in WW I—gave way to the plane and tank. The battleship yielded to the plane that took off from the aircraft carrier, and so forth. History has shown that armies are "slow learners" regarding the declining status of the dominant weapon (e.g., the French fixation on artillery fire power in the face of the tank and the Israeli belief in the tank's superiority despite the rise of anti-tank missiles). The conclusion is that a wide variety of weapons must be introduced into the arsenal. This is the diversity that is currently termed the "combined arms warfare concept."

Kober suggests expanding the concept from the tactical context to the dominant weapon at the operational level and "dominant branch" at the strategic level. After reviewing the historical development of the dominant branch, he concludes that as integration matured on the modern battlefield, the importance of discerning between various military branches and weapons systems—in the context of their contribution to firepower and maneuvering—is receding.[35]

Leonhard observed friction in force planning when the question arises of whether to arm with a "super weapon" (defined as a weapon against which defense is impossible) or develop combined arms warfare (based on a wide range of weapons). He illustrates this with the present-day case of the precision strike concept that some Americans see as a super weapon. Leonhard contends that although precision weapons exist, the precision strike concept is absolutely ridiculous since the enemy will adapt to it by adopting defensive measures.[36]

In his book *Strategy: The Logic of War and Peace*, Edward Luttwak states that weapons whose effectiveness are based on a very specific capability cannot withstand diverse countermeasures. He illustrates this with the example of the torpedo—the major threat to battleships prior to WW I, but of minor importance during the war because of the diverse countermeasures that were developed against the torpedo boat and the torpedo itself. These included projectors, rapid-firing cannons, and underwater defenses, such as steel nets to protect anchored ships. An example of failure when a weapon is based on a narrow area of proficiency is the anti-tank missile as an answer to the tank.[37] The effectiveness of a limited-purpose weapon is only temporary; the greater its effectiveness, the faster it will disappear.

Rosen sees super weapons in this light:

> A strategy for military technological innovation that seeks as much flexibility as it can buy might be better than one of trying to buy the one weapon that would perform the best if it could be built to specifications at the expected cost and if it eventually turned out to be the weapon which was actually needed.[38]

The discussion on a dominant weapon, super weapon, or special-purpose weapon leads us to the solution of diverse-weapons integration. Before presenting examples to illustrate the importance of diversity for coping with technological and doctrinal surprise, I briefly explain the current concept for force employment on the battlefield that integrates a wide range of weapons.

In the introduction to *Combined Arms Warfare in the Twentieth Century*, Jonathan House defines the term "combined arms" in the following three ways:

> The idea that different combat arms and weapons systems must be used in concert to maximize the survival and combat effectiveness of the others. The strengths of one system must compensate for the weaknesses of others.
>
> The organization of combined arms and a C2 (command and control) structure brings different weapons together for combat. This may include permanent, peacetime organizations and ad hoc (or "task organized") combinations of the different elements in wartime.
>
> Combined arms tactics and operations are the actual roles performed and techniques applied by these different arms and weapons in supporting one another in battle.[39]

Today's armies accept the combined arms warfare concept as a basic paradigm in force planning and its employment. Some armies, such as the American army, even try to reach beyond this to a higher level of integration. This is a relatively recent phenomenon. The concept crystallized over many years at different rates in different armies, and its present status developed through trial and error. History teaches that armies that maintain a quantitative superiority, but employ a doctrine other than combined arms warfare, have suffered on the battlefield when facing armies that were quantitatively inferior but that applied this concept.

According to Leonhard, combining forces has a dual role in combat. First, every fighting layout (armor, infantry, and so forth) has advantages and disadvantages. In the combined warfare concept, the strong aspects of one particular element cover the weak aspects of other elements. Second, the proper use of force in combined arms warfare should enable one fighting layout (or weapon) to force the enemy, who is trying to avoid its effect, to enter the impact area of another layout. As a synthesis between combined arms warfare influence on troop survival and its effect on the enemy, Leonhard suggests the "Alkyoneus" principle. In Greek mythology, Alkyoneus was vanquished after Hercules identified and neutralized the source of his power by forcing him into combat under unfavorable conditions. Leonhard found that the basic principle in combined arms warfare is to force the enemy into a situation where he is most vulnerable to the opposing side's capabilities.[40] Leonhard's first element—the ability to cover the weak points of force components (e.g., as air or naval forces), is a basic response in the event of a technological and doctrinal surprise on the battlefield. This should be the main reason for developing diversity.

It is obvious why the concept of combined arms must be applied when facing technological and doctrinal surprise. By its definition, combined arms means integrating various weapons rather than relying only on one type of weapon. I will present examples where the non-application or difficulty in application of this concept meant that the surprised party was at a loss to develop a rapid response, because it did not posses diversity of weapons.

In another book, Leonhard deals with weapons diversity in combined arms in the information warfare era. The basic idea of combined arms warfare is complementation between different weapon systems, which means that if the enemy counteracts one weapon by limiting its effect, it will be forced into the impact zone of another and thus rendered vulnerable. Leonhard believes that this concept, which today applies mainly to fire-systems diversity, should also be applied to sensors. He gives the example of attempts at avoiding thermal detectors, which increase the size of the defended object and its exposure to radar or regular optical spotting equipment.[41]

Regarding technological and doctrinal surprise, the search for a super weapon that focuses on its lethality and tries to prevent at any cost a situation in which the enemy succeeds in neutralizing it, has often proved futile. On the other hand, the combined arms concept assumes that a weapon is important only to the point in which its effect can be avoided by the enemy, that is, the concept is not concerned with the moment the weapon is discovered to be ineffective. Such a view will result in force planning being based on weapons diversity.

Weapons diversity played a key part in the effective response of the Germans to the surprise of the Soviet T-34 tank in 1941, whereas the lack of weapons diversity of Israeli units hamstrung their ability to overcome the Egyptian anti-tank missile surprise in the first days of the Yom Kippur War. Only when diversity was restored was the IDF able to meet the missile challenge effectively. (For details on these and other examples, see below.)

The ideal development of diversity is a difficult task, inter alia, because of the "sectoral perspective" of corps within branches and the branches themselves. For example, Ronald Spector's studies on the development of American military branches in the interwar period note:

> A general appraisal . . . tends to suggest that the army overemphasized the central role of foot infantry and neglected the role of tanks and mechanization; that the navy overemphasized the big-gun battleship at the expense of aviation, anti-submarine, and amphibious warfare; and that the semi-autonomous Army Air Corps tended to overemphasize bombing at the expense of air defense and ground support roles. Only the Marine Corps, with a narrowly defined mission, totally dependent on the larger services for support, appears to have emphasized a balanced all-arms approach to combat.[42]

The Battle of Kasserine Pass that was fought in February 1943 in North Africa illustrates the importance of weapons diversity as the main element in an effective response to technological and doctrinal surprise. This was the first battle between the American and German armies in WW II. The American warfighting doctrine, which had been developed and tested in training fields during WW II, found itself facing the German warfighting doctrine that had been tested and improved in combat in the Western Desert in the two preceding years. This is the case of an army (the American) lacking weapons diversity and finding itself stymied when faced with a technological and doctrinal surprise.

House believes that the Americans' warfighting doctrine and training program taught their armor divisions to emphasize distributed, mobile combat and close armor fighting. Stressing this during training resulted in neglecting study and practice with high-trajectory artillery fire. Furthermore, logistical problems compelled the American army to leave its Second Corps artillery far behind the frontlines in Tunisia. Inexperience in armor fighting, which increased because of lack of weapons diversity (including artillery pieces), produced a crisis in the Battle of Kasserine Pass. American recuperation from the setback came only after weapons diversity was restored.[43]

Martin Blumenson, a military historian who analyzed the battle in detail, claims that the Americans had studied the blitzkrieg in Poland and France and came to the conclusion that the effective solution for blocking German armor was the development of an anti-tank weapon in the form of the tank destroyer. The American First and Thirty-Seventh Divisions, the two main units that took part in the Battle of Kasserine, employed 37-mm, anti-tank cannons mounted on armored vehicles.[44]

The Americans quickly realized that these weapons and their warfighting doctrine had been overestimated. The capability of American armor was also overrated. In early November 1942, the First and Thirty-Seventh engaged French forces in Algeria. The French put up symbolic resistance, which the Americans mistakenly attributed to their own combat skill and took as proof of their ability to triumph in battle against the Germans and Italians. "Confident of their underpowered light tanks with 37-mm guns, trusting the power of the 57-mm and 75-mm guns on their Shermans, they believed themselves to be blooded and tried in action."[45]

The Americans were caught by surprise when it turned out that their anti-tank combat doctrine and weapons didn't stand a chance against German armor. They found themselves in the middle of a technological and doctrinal surprise that resulted from having overestimated their weapons and combat doctrines. According to David Johnson, an expert on the development of tanks and airplanes in the U.S. Army between 1917 and 1945, "One of the first illusions to evaporate in the heat of battle was the effectiveness of the light tank."[46]

In a series of battles fought between February 14 and 17, 1943, the Americans lost over 2500 men, one hundred tanks, 280 vehicles, and thirty artillery pieces. "The shock of defeat was visible among the troops."[47]

John Shy notes that American equipment and training were unsuited for tank-destroying operations. The new doctrine lost credibility when the German armor easily plowed through the American lines.[48] The commander of the Allied forces in North Africa Dwight Eisenhower desperately tried to halt the downslide by transporting artillery pieces and tank destroyers from Algeria to Tunisia. The Americans and their allies withdrew to the north and west with the Afrika Korps in pursuit.

How did the Americans manage to recover from the surprise failure of their weapons and warfighting doctrines? After a series of successes, the Germans had planned to proceed through the Kasserine Pass to the north and west. However, they encountered the retreating Allied forces, arrayed in battle positions, and additional Allied forces that had been dispatched from Algeria. The most important factor for the Allied reinforcements was the Ninth Division's artillery under the command of General Irwin Stafford, which included a battalion of 155-mm guns, two battalions of 105-mm cannons, and two companies of 75-mm cannons. On the morning of February 22, the German commander began advancing north in the direction of Thala and came under American artillery fire. Caught by surprise, he halted and waited for the American attack—that did not develop.[49] Later the same day, Rommel withdrew his forces because of stiff American resistance. According to Charles Dunphie, the British armor commander who fought with the Americans at Kasserine, and witnessed the Americans' deficiency in artillery as a factor in their combat capability and later its importance in filling the gap: "Artillery was the one thing we lacked and the only thing we wanted."[50]

This is an example of the Americans' surprise at the ineffectiveness of their weapons and warfighting doctrine against German armor combat. Recovery took place by employing the artillery element that had been lacking in the first stage of the fighting in North Africa. This example reinforces the arguments favoring weapons diversity for fast recovery from technological and doctrinal surprise.

WEAPONS REDUNDANCY

The "diversity" principle calls for many different types of weapons, each with a special operational effect that covers up the shortcomings of another weapon, whereas the "redundancy" principle relates to a number of weapons with similar operational effects that operate according to different application methods. According to the redundancy principle, when a major operational problem arises (such as a need to confront tanks or planes) and the danger exists that the enemy will make a supreme effort to neutralize any countermeasures, then the side facing the danger must develop a few weapons systems, each based on another working principle. Thus, if one weapon is countervailed, for whatever reason, another weapon can still be employed to answer the operational challenge.

The tank illustrates an operational challenge that has occupied many of the world's armies since WW I. From its first appearance on the battlefield until the present day, various measures have been developed to counteract it—beginning with anti-tank mines, anti-tank rifles, and anti-tank tanks, and culminating in precision-guided munitions. Given the enormity of the threat, almost every army has had to deal with tanks on the battlefield, and a wide range of countermeasures have been devised. For example, some anti-tank mines are activated by direct pressure, others by magnetic sensors, acoustic sensors, or other means.

Heilmeier presents the "holy trinity"—strategic bombers, intercontinental missiles, and submarine-launched missiles—as a possible solution to technological surprise in the age of nuclear warfare.[51]

Luttwak deals with redundancy as it relates to weapons and counter-weapons. He postulates that the link between the success and failure of new weapons is not sufficiently clear to the designers and operators of the weapons systems, but given the "paradoxical logic" inherent in strategy, the more effective the weapons are, the faster the enemy will neutralize them. "Paradoxically, less successful devices may retain their modest utility even when those originally most successful have already been countered and perhaps made entirely useless."[52] He gives the example of British chaff known as "Window." A British experiment showed that where the relatively outdated airborne radar Mark-IV was able to cope with countermeasures, the upgraded version Mark-VII failed.[53]

Luttwak claims that equality between weapons utility and weapons performance exists only in the minds of the scientist and engineer, since it is seemingly impossible, if not patently absurd, that a weapon of superior performance is less useful than a weapon of inferior performance. The truth is that this sometimes happens because of an enemy's efforts to neutralize weapons of extreme effectiveness that are being used against it. Regarding the implications for force planning and the redundancy concept, Luttwak writes:

> The prescriptive implication for nations at war is clear: when scarce development resources must be allocated between competing scientific concepts and engineering configurations, it is unwise to rely on the judgment of scientists and engineers. As such (although they can be wise strategists too), scientists and engineers

are unlikely to see merit in the diversion of resources to develop second-best equipment alongside the best. But that is precisely what prudence demands.[54]

He adds that "as soon as a significant innovation appears on the scene, efforts will be made to circumvent it—hence the virtue of suboptimal but more rapid solutions that give less warning of the intent and of suboptimal but more resilient solutions."[55] Further examples (see below) demonstrate that when force planning is based on numerous alternatives—some less than optimal—recovery from technological and doctrinal surprises on the battlefield is quick.

Luttwak also discusses another aspect of redundancy, from the point of view of the need for homogeneity of weapons, sometimes derived from economic constraints. He contends that the phenomenon of battlefield counterresponse prevents weapons homogeneity, which is, financially speaking, the optimal solution. Homogeneity that lowers the cost of weapons systems during production, and later reduces maintenance costs because of their number,[56] goes against the nature of fighting concerning counterresponse. Nothing prevents the purchase of uniform administrative equipment, such as boots and trucks, but:

> For military equipment that must function in direct interaction with the enemy . . . homogeneity can easily become a potential vulnerability. If, for example, antiaircraft missiles are standardized on a single homogenous type, in order to obtain economies in production, maintenance and training, the resulting savings could be very large as compared to an array of several different missile types. But in war a competent enemy will identify the weapon's equally homogenous performance boundaries and then proceed to evade interception transcending those boundaries . . . But the target presented to the enemy's countermeasure effort will be unitary, enabling him to focus all his efforts . . .[57]

Despite the economic curbs in the force planning process and the economic advantages of weapons uniformity, Luttwak notes that the lack of uniformity (redundancy) is of critical importance for military effectiveness.

An example of recovery that was made possible because of alternatives created in force planning is the IDF's use of fording equipment in the Yom Kippur War.[58] The IDF began planning for the crossing of the Suez Canal water obstacle immediately after the 1967 victory. The initial solution was to procure "unifloat" pontoons designed for civilian purposes. The first rafts arrived in Israel in late 1967, but it was soon discovered that too much time was needed to convert them to military use.[59] Thus, the Gillois amphibious tank-carrier (IDF codename: "alligator") was acquired, but this equipment was relatively antiquated, and the IDF had a hard time repairing and modifying the "alligators." Sixty-two "alligators" were sent to the Merkavim factory for modification and from there to army supply depots at the rate of three a month beginning in December 1971.[60]

In 1972, in the midst of the development of the "unifloat" transportation equipment, the IDF began working on a new invention—the roller bridge. This bridge, made of rollers filled with plastic foam, was regarded as the "magic" solution to the water-crossing assault challenge. Bridging a water obstacle the size of the Suez Canal could be done in minutes and damaged parts replaced quickly. The bridge's main disadvantage lay in its mobility. It took between five and seven tanks for the task, and the roller bridge was hellishly cantankerous to maneuver because of its weight and bulk. The IDF tested all of the crossing apparatus in a simulation canal that was dug in the Roifa Dam in Sinai. In May 1973 the ministry of defense gave priority to the roller bridge, thus freezing the upgrading of the "alligators." How, then, was the IDF caught by surprise?

> The problem was that the ORBAT, weapons, and doctrine were based on the assumption that the IDF would be crossing an area clean of enemy forces, an area under IDF control, and not an area under heavy fire where bitterly contested battles would be waged. In effect, this was the main reason why the roller bridge—the best and fastest means of an assault crossing that the IDF had at its disposal—failed to reach the canal first. Arriving before them were the mobile, rapidly-operable "alligators" and "unifloat" pontoons, but getting them into the water and assembling them as bridges involved a tremendous effort and heavy casualties.[61]

According to Dov Tamari:

> The assumption that "the standing army would halt" at the lines of contact created an additional blunder, namely, that after a brief defense the IDF offensive could proceed according to prearranged parameters: the Suez Canal crossing could begin at the water line. The crossing doctrine and nearly all of the fording equipment were adapted to this assumption (but in reality, the IDF had to engage in heavy combat, launch an attack, and haul the bridging equipment a long distance to the canal while fighting the whole way).[62]

The "alligators" were the first pieces of equipment to ford the canal. Sixteen of them reached the crossing point, and tanks began lumbering across on the morning of October 16.[63] On the following evening, the tanks started rolling on the "unifloat" pontoon bridge. The roller bridge was finally assembled on October 19 in the wee hours of the morning, three days after the "alligators" had begun operating.

In this case force planning applied redundancy in the form of three alternatives to the operational task of crossing the canal, and this is what enabled the Israeli commanders to overcome the operational problem, even though they were surprised by the strenuous conditions they had neither planned nor trained for. Even if the IDF had not planned it this way, because of the late decision to concentrate on developing the "wonder weapon"—the roller bridge—the alternatives for the canal crossing had not been minimized, or worse, eliminated. Had a few more years passed, the development of the roller bridge might have advanced and resulted in a reduction of alternatives. The IDF's canal crossing in the Yom Kippur War is an excellent example of the principle of weapons redundancy that is vital for coping successfully with technological and doctrinal surprise on the battlefield.

Britain's struggle against Germany's bombing navigational devices—known as the Battle of the Beams—is another example of the need of weapons redundancy. In this case, the Germans used three different navigational methods—the "Knickbein," the "X-Gerät" system, and the "Y-Gerät" system in response to British countermeasures designed to interfere with the devices.[64]

TECHNOLOGICAL FLEXIBILITY

This section deals with the nature of a weapon itself (putting aside its integration with other weapons). Technological flexibility has a number of basic characteristics:

Versatility—the multifunctionality that enables the weapon to be used against various threats.

Changeability—the ability to quickly change the technological components of a basically non-multipurpose weapon.

Flexibility—in terms of weapons development—enabling adjustments in the choice and purchase of weapons.

The third type of flexibility deals with uncertainty by significantly reducing the time between the procurement and the use of the weapon.

Versatility

According to Jones, a multi-role weapon is inferior to a weapon developed to answer a specific challenge, and its development will take longer than usual because usually it is tested on different types of targets before winning approval. This also means it will have a shorter life span after it becomes operational. Whatever the case, Jones presents a number of examples of multi-role weapons that proved their effectiveness:

> The American Mustang airplane performed superbly in high- and low-altitude bombing missions, night fighting, and patrols.
>
> The German 88-mm cannon (originally designed as an anti-aircraft weapon) proved more effective against tanks than planes.
>
> The anti-aircraft proximity fuse was adapted for ground targets.

Jones argues that these weapons were not planned to be multifunctional. Each was designed for a specific purpose and later adapted to other uses. He proposes a similar modus operandi in the future.[65]

The logistics scholar Moshe Kress employs the term *technological flexibility* to describe the multi-role of logistic means. He also gives examples in which versatile use was made in the wake of an unexpected operational necessity: the modification of Soviet ammunition trucks to water tank carriers in the Manchurian War of 1945; and the Persian Gulf War when American buses, with wooden seats for evacuating the wounded, were remodeled for prisoner transportation.[66]

Rosen deals with the attempt to reduce the inherent uncertainty in force planning. One of the two tracks (for the second, see below) that he proposes for attaining flexibility is to acquire multipurpose weapons. The American army specifically planned to acquire a multipurpose weapon because it was impossible to predict fighting conditions, and it was prepared to pay the high price for such a weapon rather than acquire a specific-purpose one. He illustrates this by citing the acquisition of aircraft carriers, which are more expensive than cruise missile ships but also far more versatile because they can carry different types of planes. Another example is the procurement of bombers. Bombers are more expensive than intercontinental missiles, but they have proven their effectiveness in conflicts where missiles could not be used. Another possibility is to acquire specific weapons for every scenario and build different units around the weapons. This kind of flexibility, however, comes with a very high price tag.[67]

The B-52 strategic bombers were planned as backups for the intercontinental ballistic missile (ICBM) fleet and submarine-launched missiles. Although they were not employed for this purpose, they were found to be effective for more conventional needs. They had a wide range of use in the Vietnam War where they carried out the strategic bombing of the North Vietnamese cities of Hanoi and Haiphong, operational bombing of the Ho Chi Minh Trail, and close tactical support for the marines in the defense of Khesanh.[68]

Gordon claims that the concept of weapon versatility comes from the prophet of air warfare, the Italian strategist Giulio Douhet (1869–1930), who envisioned the bomber as a platform for destroying ground targets that defend themselves against enemy fighters. Not all air forces accept this approach. According to Gordon, the Israeli Air Force (IAF) adopted this principle for the airplane and application of air power. He brings the example of the Phantom (F-4E) that was employed in the Vietnam War, Israeli-Egyptian War of Attrition, and Yom Kippur War as an all-weather interceptor and multi-role attack plane.[69] The meaning of this for force planning is the advantage of procuring a wide range of weapons, munitions, and missiles capable of destroying diverse targets.[70]

More examples of versatile weapons: the proximity fuse—developed jointly by the British and Americans—was designed to shoot down aircraft and German V-1 rockets but was also used against ground targets;[71] the German 88-mm anti-aircraft cannon was considered the best anti-tank weapon in WW II. The latter served both purposes throughout the war. Rommel made wide use of the 88 against tanks in the Western Desert campaign. ". . . the Germans had the one weapon—the 88-mm dual-purpose (anti-tank and anti-aircraft) gun which was always envied for its versatility; and remained unequalled in this respect."[72]

Changeability

Ben-Israel's view of the technological lessons learned from the Yom Kippur War:

> The ability [of modern technology] to adapt weapons systems during the fighting has had a considerable influence on the principles of war. Flexibility, improvisability, and changeability play a major role in modern combat. The Yom Kippur War, for example, proved that building electronic warfare systems against known enemy threats was insufficient. Instead, they had to be built for dealing with changes that the enemy introduced into the systems' electronic parameters during the fighting.[73]

The Israeli developer of military technologies Eli Shvili suggests dealing with uncertainty in countermeasures by developing graded or modular technological solutions. A graded solution is prepared ahead of time in a number of stages that can be altered by creating external contact surfaces in the system. The modular solution is a partially preplanned solution, with few stages, allowing part of the system's elements to be modified. For this type of solution "to work," the system's volume, output, contact surfaces, and so forth must be preplanned. Transition from stage to stage must be done in such a way that it causes a technological gap in the enemy's response.[74]

An example of a graded solution is presented by James Constant, an expert in strategic weapons planning. Constant describes the importance of flexibility in missile warheads, given the missile's numerous possible trajectory paths, types of target, incoming data, and likelihood that the target will be changed while the missile is in flight. Under these circumstances, the warhead's initiative system must be capable of being updated while the missile is airborne.[75]

An example of a modular solution as protection against battlefield contingencies is the development of German tanks in WW II. These tanks were built so they could be adapted to future requirements, even though these requirements were still unknown. The Panzer Mark-3 tank was designed in three main sections: the lower hull that contained the engine and steering transmission and was connected to two other sections—the forward and rear upper structure. The upper structure could be modified without designing an entirely new tank. In the beginning of Operation Barbarossa, when the Germans realized that the 37-mm gun was ineffective against Soviet T-34s, they replaced it with a shorter 50-mm diameter barrel within a few months and later with a longer barrel of the same diameter.[76]

Another example of modular development is the British Centurion tank. Changes in the hull, armor thickness, turret structure, motor, weapons, fire-control system, and night-vision systems were introduced. From 1945 to 1975 over thirteen main models appeared.[77]

Luttwak describes the modular development in the Israeli Merkava tank over the years:

> They progressively introduced new engines to increase its agility while leaving armor and gun unchanged; then substituted a much more powerful 120 mm gun for their original 105 mm, while leaving engine and armor unchanged; then added a low-light TV tracing device (for anti-helicopter use); then increased

armor protection against antitank missiles; and so on. Each time, these changes were complicated by the need to fit the preexisting design, but they also presented the opportunity to respond quite quickly to new threats and developments, incorporating the lessons not only of field exercises and technical tests but also of combat experience.[78]

Even if Luttwak is not precise on details and does not discuss wartime modifications, this example of adaptability demonstrates its potential when responding to technological and doctrinal surprise. Shvili claims that the modular approach was applied to large platforms like planes, tanks, or missile ships in post factum because of their size. He also notes that the most recent technological breakthroughs—based on miniaturization—enable the introduction of graded changes in a platform the size of a missile. This has already been done in the shoulder-fired Stinger missile.[79]

FLEXIBILITY IN WEAPONS DEVELOPMENT

This section discusses flexibility in weapons development, where the development process can be quickly adapted and modified, thereby providing an effective solution to an updated threat. Although our study concentrates mainly on responses during actual combat, this type of flexibility is looked at briefly.

Rosen terms this "type II flexibility," which is based on delaying the decision on the procurement of weapons to the last minute. This goes far in diminishing the omnipresent uncertainty factor regarding the relevancy of a particular weapon on the future battlefield. A decision can be delayed by acquiring knowledge of the wide range of possibilities during the parallel development of a number of weapons until the final prototype is ready and then waiting until its production cost and effectiveness are relatively clear. In this way the decision on the mass purchase of the weapon (or weapons) is made only at the last possible minute.[80]

The success of the approach is illustrated in the American development of long-range ballistic missiles in the 1940s and 1950s, a period characterized by great uncertainty over rapidly developing technology, relations with the Soviet Union (the wartime ally that went from friend to foe), and basic questions on the use of nuclear weapons. At the technological and operational levels, it was not clear which technology was preferable for the strategic bombing, how and when to employ a nuclear device, and the best way to adapt it to the missions of the three military branches. The committee dealing with these issues acknowledged that given the diversity of strategic bombing uses—bombing cities, striking limited military targets, air defense, submarine warfare, and so forth—the correct approach was to study as large as possible a number of basic technologies and not to commence procurement without a solid factual basis. This decision was made in 1946. By the time of the first hydrogen bomb tests in 1953, delivery called for a missile much smaller than what was previously required for a nuclear warhead. Hydrogen bomb technology led to the decision to develop a fleet of intercontinental missiles as the preferred solution.[81]

In dealing with the development of military layouts, Bonen tried to find solutions to the problem of developing weapons systems at a rapid rate—now possible due to modern technology—and overcome the lack of feedback in peacetime. His solutions included guidelines for the desired ratio between investment in equipment and R&D in the same field and also ways to shorten the development of military layouts.

Part of the solution lies in an attempt to optimize weapons development so that even though "decisions on changes and additions to the military layout generally tend toward short-term optimization," a development track will be taken that provides flexibility that diverges from this trend each year. The upshot being that after a number of years, parallel to the acquisition of weapons, other directions of research may emerge that provide answers to the latest operational problems. This type of development process has a relatively good chance for dealing with changes on the battlefield (assuming they are identified in time). However, since the process does not guarantee the prevention of going astray in different directions and perennial delay in the development of layouts, Bonen suggests shortening the development processes by examining the possibility of employing new weapons in small layouts (experimental military units) to provide rapid feedback in the development of weapons and doctrines. Bonen concludes by saying that the telescopic process that enables an overlap between the stages of development, integration into a small layout, and afterward a large layout will shorten development time.[82] This process has the potential for greater flexibility because decisions regarding development in this telescopic process can be made in a relatively shorter time than in the vertical process and can be adapted more quickly to change.

4 COGNITIVE AND COMMAND AND CONTROL (C2) FLEXIBILITY

THE THIRD LEVEL IN FORCE PLANNING, the cognitive and C2 stratum, is based primarily on the conceptual and doctrinal flexibility that appears as the theory's first stratum. In other words, armies whose basic concept or doctrine is flexible tend to demand cognitive flexibility of their commanders, and tend to develop a C2 system that grants a large degree of leeway to commanders on the battlefield. Conversely, armies characterized by conceptual and doctrinal dogmatism tend to eschew cognitive and command flexibility.

Unlike the second stratum—the organizational and technological level—which comes under the rubric of force planning and military science, cognitive and command flexibility belong to that area in the art of war that deals with force implementation on the battlefield. Many military theoreticians and most military doctrines deal with this issue. Since the stratum is based on relatively detailed knowledge, a brief survey of the theoretical and applied literature on the subject is presented in four areas: military theory, military doctrine, military history, and future warfare. (The application of cognitive and command flexibility in response to technological and doctrinal surprises will not be discussed because this field, to a large degree, constitutes a black hole in military research.) After reviewing the extant literature in the four aforementioned areas, I present the characteristics of the flexible C2 concept and doctrine that are found to be effective in countering technological and doctrinal surprise. This response stands in stark contrast to the rigid C2 doctrines that demonstrate limited recovery, at best, from surprise. The chapter concludes with three brief examples illustrating this point.

Cognitive flexibility is a psychological term that refers to the trait of an individual—in this case the military commander. In the context of technological and doctrinal surprise, it refers to the commander's ability to respond quickly to battlefield contingencies by improvising solutions that result in rapid recovery. Mental dogmatism is the antithesis of cognitive flexibility since it denies that a surprise has occurred and insists on solutions that invariably fail.

Command flexibility grants the commander the freedom to make on-the-spot decisions (within the boundaries of the army's command method). The level of command flexibility can be defined as the line between decisions that the commander is allowed and required to make independently and those that demand permission from the superior level. In our context, the commander is given the freedom to implement creative solutions to technological and doctrinal surprises on the battlefield. Command rigidity, the opposite of command flexibility, is characterized by rejection of new ideas and suppression of any deviation from conventional responses.

A close connection—though not necessarily a two-way one—exists between the two types of flexibility. Command flexibility is predicated on cognitive flexibility since a flexible command system cannot exist unless the commander's thinking is flexible. However, neither is the opposite always true. A system based on rigid C2 can respond flexibly if the commander is flexible in his thinking. This type of flexibility generally is less effective, as in the case of the Soviet response to Operation Barbarossa, than when cognitive and command flexibility operate synergistically. The measurement of cognitive and command flexibility lies in the number of changes that commanders introduced in response to surprise in general and technological and doctrinal surprise in particular.

Cognitive and command flexibility is the threshold state for recovery from technological and doctrinal surprise. Sometimes devising a new fighting method is sufficient, even if weapons diversity is limited or the doctrine that was drafted prior to the war proves unfeasible. On the other hand, when cognitive and command flexibility is absent, the other strata in the flexibility concept will be of little value.

The army's C2 doctrine is the formal expression of the levels of cognitive and command flexibility. A flexible doctrine, based on the awareness of the universality of uncertainty and the inability to prepare for it by drawing up fixed courses of action, recognizes the difficulty in providing ready-made solutions in the form of principles and directives for operational problems on the battlefield. A flexible C2 doctrine develops original, creative thinking in the commander and relegates decision making to the lower levels of command. When a commander operating within a flexible C2 system comes face to face with a technological and doctrinal surprise, he finds it easier to devise an immediate solution because he has been trained to improvise and does not have to go through the chain of command to get approval to deviate from the original plan. The Israeli military psychologist Dan Sharon believes that to inculcate in commanders the ability to cope with surprise, they must be trained in cognitive skills other than those ordinarily employed in armies. The basic cognitive skill for dealing with surprise is creative thinking since only this type of mental activity can improvise effective solutions.[1]

A rigid doctrine will be at a loss to respond effectively to sudden change or surprise because it tries to provide a preplanned answer to every contingency and seeks to control events on the battlefield through a centralized command system. A commander trained in the principles of a fixed, prearranged doctrine will be hard pressed to cope with a situation outside the proverbial box. When the commander is faced with a technological and doctrinal surprise and manages to find a solution for it, he will still be obstructed from implementing it because of the centralized command system that requires him to obtain permission for his moves.

Flexibility or rigidity in the C2 concept and doctrine can be gleaned from the doctrinal literature and the way the chain of command operates in the following areas:

> The degree of emphasis on independence and initiative of junior commanders.
>
> The tendency toward a decentralized or centralized C2 doctrine.
>
> The attitude toward commanders who take the initiative and improvise on the battlefield.

COGNITIVE AND COMMAND FLEXIBILITY IN LITERATURE AND RESEARCH

Many military thinkers throughout history have attributed great importance to the commander's cognitive flexibility,[2] but have dealt less with command flexibility. The reason for this seems to lie in the intrinsic nature of past armies that operated within rigid hierarchies.

Helmuth von Moltke (1800–1891), who is considered the father of the flexible method of command (today known as "mission command"), recognized the need for flexibility in the C2 doctrine and thinking patterns of commanders in order to deal effectively with the fluctuating nature of war. Although he did not explicitly use the term *flexibility*, he realized that war required such a tool. Therefore, he provided guidelines for the command system, commander education and training, and military planning. His ideas became the foundation of the German doctrine and have remained so till today.

A survey of the war principles in the armies of the world shows that although surprise, as a war principle, appears on all the lists, its solution—flexibility—is generally absent. The reason for this may be the Clausewitzian concept, according to which the solution to battle uncertainty and surprise lies in the concentration of mass and the determination and tenacity of the commander. Many war principles submit to this logic: the IDF's devotion to mission; objective in the American army; target selection and preservation in the British army, attack (which, according to Harkabi, is found in all of them), concentration of force, mass, and so forth.[3] Other solutions to battlefield surprise are the flexibility principle, which as noted, appears in only a few armies, or the emphasis on flexibility in other ways.[4] In conclusion, military doctrines have a range of views regarding flexibility—whether cognitive, planning, or command flexibility—whose goal is to furnish solutions to sudden changes on the battlefield.

Much research in military history has been carried out on the influence of flexibility on battles and wars.[5] Some military historians have studied the effect of command aptitude and leadership on the results of military confrontations.[6] Other scholars who specialize in command and leadership emphasize the importance of flexibility in the command system.[7] Cognitive flexibility is universally recognized as a required trait in the future commander.[8]

Military literature emphasizes the importance of the commander's cognitive flexibility to cope with battlefield reality. This is a frequently studied issue as can be seen in the writings of military thinkers, military doctrines, and research on the commander of the future. Command flexibility is also considered, albeit to a lesser degree, a factor in "quality military organization." Cognitive and command flexibility is the basis for dealing successfully with technological and doctrinal surprise (see below).

THE DECENTRALIZED (MISSION) COMMAND METHOD AS A KEY TO RECOVERY FROM TECHNOLOGICAL AND DOCTRINAL SURPRISE

This section deals with decentralized C2 as a concept conducive to doctrinal flexibility. Decentralized C2 developed in Prussia and came to fruition in the German army's modus operandi in WW II. Today we call it mission command (the translation of *Auftragstaktik*) or speak of "flexible command."[9] Its proven success has made it the command doctrine of the majority of the armies in the world (on the theoretical level at any rate).[10]

Moltke is regarded as the father of the German mission command doctrine. Searching for a solution to the C2 problem for mass armies, he realized that although the strategic level could direct operations via the telegraph, it did not provide the necessary flexibility at the operational level. On the commanders' need for independence on the battlefield, he wrote:

> Because of the diversity and the rapid changes in the situations in war, it is impossible to lay down binding rules. Only principles and general point of view can furnish a guide. Prearranged designs collapse, and only a proper estimate of the situation can show the commander the correct way ... Only if leaders of all ranks are competent for and accustomed to independent action will the possibility exist of moving large masses with ease.[11]

Moltke believed that the command method must be based on commanders whose training emphasized the correct command principles and who were given operational independence. These officers would carry out the will of the supreme commander, even when the latter was unable to convey his will because of time constraints or unfavorable circumstances. Moltke claimed that an army's plan did not have to go beyond the first encounter. Successive plans should always be based on new situations created in the wake of the previous engagement.[12] This method was predicated on training the German field commanders in a clearly defined doctrine, after which they would be capable of initiating moves and making the "right" decisions on the battlefield. The concept that was taught in the command and staff colleges enabled the commanders to channel their efforts to a clear mission and aim.[13]

A retreat from Moltke's concept began when Alfred von Schlieffen (1833–1913) served as chief of staff. Von Schlieffen believed that Moltke had erred in basing the command system on general guidelines rather than on detailed orders.[14] Schlieffen claimed that the command system had to be more centralized, that deviations from the prescribed, detailed plan could be made only after approval was transmitted by telephone from the supreme commander whom Schlieffen termed the "modern Alexander."[15]

Even if Schlieffen's approach did not go completely against the tradition of decentralized command but only added the additional factor of detailed planning, after this factor proved unsuccessful on the battlefields of WW I, a revived emphasis was placed on the mission command approach. The renewed interest was largely due to the efforts of Hans von Seeckt, the German chief of staff until 1926. The clearest expression of the place of uncertainty in the German C2 system appears in the first four sections of the 1933 German Army regulation, *Truppenführung—HD*-300. Six years later Germany went to war with this regulation:

1. War is an art, a free and creative activity founded on scientific principles. It makes the very highest demands on the human personality.
2. The conduct of war is subject to continual development. New weapons dictate ever-changing forms. Their appearance must be anticipated and their influence evaluated. Then they must be put quickly into service.
3. Combat situations are of unlimited variety. They change frequently and suddenly and can seldom be assessed in advance. Incalculable elements often have a decisive influence. One's own will is pitted against the independent will of the enemy. Friction and errors are a daily occurrence.
4. Lessons in the conduct of war cannot be exhaustively compiled in the form of regulations. The principles enunciated must be applied in accordance with the situation.[16]

According to Condell and Zabecki in their introduction to the manual's translation: "Its purpose was not to give German military leaders a 'cookbook' on how to win battles, but rather it was designed to give them a set of intellectual tools to be applied to complex and ever-unique war fighting situations."[17]

The difficulty in condensing the essence of war down to general principles based on an understanding of the uncertainty factor can be seen by the fact that the German army has never published an abridged list of war principles, even though an unofficial list can be inferred from German doctrinal literature.[18]

The principles of mission command were developed and expanded before and during WW II and often can be tracked in the writings of Guderian, F. W. von Mellenthin, and Manstein. The manifold successes of the German army in coping with technological and doctrinal surprise are not presented at this point. (They appear sporadically in the theoretical section of the book and are described in greater detail in the empirical section below.) To stress the importance of the mission command system as perceived by the German commanders in WW II, what follows are their own views. In 1980 General William DePuy, the first commander of the United States Army Training and Doctrine Command carried out a study on the form of C2 that the German commanders had employed during the war. General DePuy asked: "Did you receive the same responses to situations, opportunities, and initiatives from individual company, platoon commanders, even squad leaders that you received from divisional, brigade, and battalion commanders?" General Hermann Balck's reply:

> Yes, and you must understand how this was achieved. The senior German commander almost never interfered with subordinate commanders unless they made a terrible mistake. In this way they taught them to develop individual initiative. They gave them room for carrying out initiative, and censured them only in dire cases. This approach was used down to the individual soldier who was praised for displaying initiative . . . [19]

In his book *Genius for War*, the military historian Trevor N. Dupuy writes: "If any one aspect of military performance was emphasized more than any other in the General Staff and in all German military training, it was encouragement of individual initiative."[20] An interview with German Colonel Hans G. Pestke on training appeared in the American *Marine Corps Gazette*. Pestke answered General DePuy's query on how junior commanders were trained in decision making:

> You must let young officers make decisions. They must be given the chance to make on-the-spot decisions, through war games, that is, terrain models and map exercises. When a young soldier makes a decision, you must praise him. He might not have found the perfect or *right* solution, but he made a decision. Praise at this point must be forthcoming. You must do this again and again. That is how we give our young officers and NCOs confidence.[21]

The three elements of decentralized command (mission command) that Moltke developed and that the German army incorporated in WW II became the standards of the C2 concept and doctrine that granted commanders maximum flexibility: emphasis on a junior commander's need for independence and initiative, a decentralized C2 doctrine (as opposed to a centralized one), and a favorable attitude toward commanders who took the initiative and improvised on the battlefield. Rommel had this to say on C2 flexibility:

> From the outset it was our endeavour to turn our army into an instrument for the most rapid improvisation and to accustom it to high speed of manoeuvre. Officers who had too little initiative to get their troops forward or too much reverence for preconceived ideas were ruthlessly removed from their posts . . . [22]

A classic example of the Germans' ability to improvise can be seen in the large number of creative solutions that appear in their tactical doctrine, organization, weapons, and combat-support means on the Russian front. After the war, the German generals who fought on the eastern front prepared a manual for the American army with scores of examples of improvisation.[23] Bruce Gudmundsson, who studied German innovations of the stormtroop tactics during WW I, relates it to Germany's highly decentralized, mission-oriented organization that permitted frontline commanders to tailor their tactics to the situation.[24]

It will be recalled that recovery from technological and doctrinal surprise cannot occur without cognitive and command flexibility. This is why every example of recovery presented so far includes this type of flexibility.

Rommel's operations provide a good example of cognitive flexibility integrated within command flexibility in recovery from technological and doctrinal surprise. Rommel's superb skill in improvising is illustrated in the Battle of Arras, France in 1940 when he used 88-mm anti-aircraft cannons as anti-tank guns against the surprise appearance of British Matilda tanks whose armor plating was impenetrable to German tank fire.

Other instances where cognitive and command flexibility played a major part in recuperation are the German army's recovery from the T-34 tank and the IDF's recovery from anti-tank and anti-aircraft weapons in the Yom Kippur War. Failures resulting from a poor level of cognitive and command flexibility are the British inability to employ 3.7-inch anti-aircraft cannons as anti-tank weapons and France's paralysis in the face of Germany's blitzkrieg onslaught. (A detailed analysis of these cases appears in the historical part.)

Sometimes recovery from technological and doctrinal surprise has resulted almost solely from cognitive and command flexibility. The following examples illustrate this:

The first case is the German army's initial encounter with Soviet Stalin 3 tanks in the battle near Târgul Frumos in Romania on May 2, 1944. There the Grossdeutschland Mechanized Infantry Division under the command of General Hasso von Manteuffel deployed defensively against Soviet armor. Manteuffel recalled: "We heard tank fire of the heaviest caliber from a great distance go past between our cars."[25] The Germans realized that enemy tanks were firing at a range of three thousand meters from static positions. Since the Germans had never encountered heavy Soviet tanks of this type, Manteuffel's first thought was that they were a German Tiger tank company that had gotten lost. He then ordered a company of Tigers to open fire on the new Soviet tanks but their shells bounced off the armor plating.

The German battalion commander on the scene ordered the tanks to attack. When the tanks came within 1800 meters of the enemy, they again commenced firing. Four Stalins were hit, and three more withdrew from their positions. At this stage Manteuffel ordered a company of the Panzer IVs to outflank the remaining Stalins. In a swift move, the German tank company drew up behind the Stalins and fired at a range of 1000 meters. These Stalins were also hit.[26]

Manteuffel writes that "this encounter was shocking, as previously our 88 mm guns had destroyed Russian tanks with direct hits at maximum range without difficulty. Now the Tiger crews could destroy the enemy tanks only with the greatest difficulty and suffering heavy losses."[27]

Although there is no direct link between the diameter of the gun and the weight of the shells and muzzle velocity, it should be noted that the Stalins had 122-mm cannons, whereas the Tigers had 88s and the Panzer IVs had 75-mm guns.[28]

The British Command and Staff College summary of the lessons learned from the battle, concluded that the Germans displayed superb skill, inter alia, because of the absence of prearranged plans that, had they existed, would have probably stymied their initiative and proved counterproductive to the actual situation. The British editor also mentions the German commanders' flexibility and determination that enabled them to take full advantage of the maneuverability of a relatively small force and wipe out a quantitatively superior force in a surprise move.[29]

This was a case of recovery from technological surprise—in the form of heavy tanks with superior range and defensive armor—by the flexible use of existing weapons. This is an example of dealing with technological surprise through cognitive and command flexibility that came to expression in quick understanding that this was a new threat and in providing a swift answer by improvising a battle technique and taking the initiative at the subordinate levels.

The Red Army's organizational restructuring after the start of Operation Barbarossa is a case of cognitive flexibility without command flexibility. This case is of special interest because it illustrates outstanding flexibility at the senior level of command and rapid adaptation to change in the Soviet army's situation. On the other hand, the high command ordered the changes but not on the basis of a flexible command system. C2 rigidity, as described in the section dealing with doctrine, remained unchanged.

Operation Barbarossa was launched on June 22, 1941. The results of the fighting testify to the manifold shortcomings in the Soviet military organization, but the response to the need to reorganize came extremely fast. On July 10 the front was realigned into three strategic theaters.[30] On July 15, the Soviet High Command (Stavka) issued Order No. 1 for sweeping organizational changes which included:

1. Cancellation of the corps level. This resulted in the creation of smaller army formations consisting of five to six infantry divisions, two to three armor brigades, one to two light cavalry divisions, and a number of artillery regiments. Reorganization provided the relatively experienced army commanders and their staffs with direct control over the rifle divisions.
2. Organizational change of the rifle division into a more simple one. Anti-tank, anti-aircraft, armor, and field artillery were removed. The new situation gave better, centralized control over those scarce elements in accordance with operational needs. The number of troops in the rifle divisions was cut back from 14,500 to 11,000.[31] Compared to April 1941, the numerical strength of the divisions decreased by 30 percent, cannons and mortars by 52 percent, and motor vehicles by 64 percent. Between July and December 1941, 124 rifle divisions were dismantled because of their exhausted combat capability due to the high casualty rates. At the same time, 308 new divisions were established.[32]
3. The cancellation of the mechanized corps was due to the shortage of experienced commanders and modern tanks. Most of the mechanized divisions in these corps became rifle divisions.
4. The reduction in the number of tanks in the armor divisions meant an increase in the number of divisions.
5. A significant expansion in the number of cavalry units and the creation of thirty new cavalry divisions.

House and David Glantz (the latter an expert on the war on the Russian front) note in a positive vein the wisdom of the Red Army's senior command, which was able to draft and implement sweeping organizational changes in the midst of the prolonged crisis caused by the German advance.[33] House calls attention to the Soviet leadership's "coolness under fire."[34]

In the course of the war, further changes[35] were introduced that enabled the Red Army to adapt to the changing operational situation. The previous example illustrates exemplary cognitive flexibility at the highest level of leadership that was a key factor in the Soviet solution to the Germans' doctrinal surprise—the blitzkrieg.

In conclusion, despite the advantages of the decentralized command method, it has taken the armies of the world a long time to adopt it. According to the Israeli C2 researcher Hanan Shai, even when adopted, almost no army has managed to apply this method effectively because of its complexity.[36]

THE INFLEXIBILITY OF THE CENTRALIZED COMMAND METHOD LIMITS RECOVERY FROM TECHNOLOGICAL AND DOCTRINAL SURPRISE

The rigid approach to C2 is the antithesis of mission command and is based on meticulous planning, detailed directives, and a centralized command that has the sole authority to permit any deviation from the approved plan. This approach has dominated most armies for ages.

The impact of this method on recoverability from surprise can be seen in the Red Army's low rate of success at the tactical level against German mobile armored units; its rigid C2 doctrinal has already been discussed (see above).

Nathan Leites, an expert on Soviet fighting methods, notes the double standard in the official Soviet analysis of battles and campaigns: on the one hand, the authorities demanded that the commanders draw up meticulous plans and stick closely to the initial orders; on the other hand, they admitted that a heavy price was paid for proceeding according to original decisions. Commanders who modified plans had to defend themselves before superiors who disapproved of such conduct. Leites states that the emphasis on adapting plans in the Soviet art of war is a counterreaction to the tendency of fighting blindly according to a predesigned plan and doggedly maintaining the plan even when it proved unworkable.[37]

In short, the Soviet command system in WW II consisted of detailed orders with an emphasis on meticulous preparation (inherent in this type of command system) and the fear of taking the initiative (a natural by-product of Stalin's purges). All of these factors combined to implant inflexibility, which had always characterized Russian warfighting.[38] Shai noted that the Soviet army's limited flexibility in WW II rendered it vulnerable to surprise.[39]

The Soviet command culture may have been aware of the uncertainty factor in battle, but it did not encourage flexibility. This predilection was responsible for part of the Soviet army's difficulties when it was forced to recover from the surprise of German mobile armor maneuvers in the summer of 1941.

The American army also exhibited difficulties in responding quickly to surprises in WW II. Before the war, the American approach for dealing with uncertainty differed from that of the Germans. After the war, the American army invited a group of German generals, headed by the former chief of staff Franz Ritter Halder, to carry out an in-depth critique of United States Army's operation regulations (FM 100–5). The Germans declared that the American system tried to prescribe a formulaic answer for "every" contingency, thereby limiting the commander's initiative.[40]

The result of the different approaches to the uncertainty problem can be seen in DePuy's analysis that compares the effectiveness of the German military to that of the British and Americans in battles where the two sides fought each other. The results showed that the Germans were 20–30 percent more effective when both sides were numerically equal.[41] In *Fighting Power*, van Creveld analyzed DePuy's findings in an attempt to explain the operational effectiveness gap between the German and Allied armies. He discovered, inter alia, different approaches to the phenomenon of war and the C2 concept that had implications on the commanders' flexibility and independence. The author summed up the German generals' criticism of the American army:

> As compared to the German conception of war, the American regulations display a repeated tendency to try and foresee situations and lay down modes of behavior in great detail. This procedure limits a commander's freedom of action and, by rendering him incapable of handling in accordance with the actual situation, robs him of a very important prerequisite for victory ... the American regulations display a marked tendency to underestimate the importance of surprise, maneuver and improvisation ... Owing to the attempt to foresee each situation in advance, the regulations tend to be stereotyped.[42]

The American army implemented fundamental changes in its C2 concept after 1973. Shai claims that not only was the decentralized C2 system absorbed by the Americans, but that it was one of the key factors for their success in the 1991 Gulf War.[43] The centralized command system, as exemplified in the Soviet and American doctrines of WW II, impeded recovery from technological and doctrinal surprise. France's capitulation to the blitzkrieg is discussed in detail (see Ch. 12).

In periods of low-intensity conflict and stability operations, when the number of the troops that conduct fighting simultaneously is small, and tactical operations often influence the strategic level, there is a tendency toward centralized command. This modus operandi can erode the decentralized command system and create difficulties in coping with uncertainty in an all-out conflict when close supervision by the senior level of the lower levels is not possible.

The rigidity of American submarine commanders at the outbreak of WW II is an example of a lack of cognitive flexibility. The long string of failures against Japanese merchant ships was the result of the submarine commanders' training (indoctrination) before the war, and it was not solved by lesson learning and improvement. The solution came in Machiavellian style: given a situation where a change in the modus operandi must be made and the skipper is unable to change his nature and thinking patterns, the skipper must be replaced. Thus, in 1942, 30 percent of the U.S. submarine commanders were replaced, 14 percent in 1943, and a similar number in 1944.[44] If the submarine commanders' change of mission, from attacking the Japanese war fleet to sinking its merchant fleet, can be described (from those commanders' point of view) as a doctrinal surprise, then this case illustrates the negative influence of warfighting and training doctrines that inculcated dogmatic thinking and whose end result was the inability to deal with surprise.

In conclusion, the C2 concept and doctrine is a major influence on recoverability from technological and doctrinal surprise. An abundance of literature on various levels has been written about flexibility, but not in the context of recovery from surprise in general or technological and doctrinal surprise in particular. The command method that develops the optimal ability to recoup is the decentralized command (known as mission command). An army that employs a centralized command system hamstrings its commanders' ability to cope with technological and doctrinal surprise. In addition to conceptual flexibility (see Ch. 2), cognitive and command flexibility constitutes a key element in the "culture of flexibility."

5 THE MECHANISM FOR LESSON LEARNING AND RAPID DISSEMINATION

THE FOURTH STRATUM IN THE FLEXIBILITY CONCEPT is learning lessons from war and various military encounters and disseminating them rapidly. This is regularly done in military organizations. Countless books and articles have been written on lesson learning in interwar periods and the results of the process in the next round of fighting. However, many historical cases show that applying lessons from past conflicts to meet future technological and doctrinal surprises and basing force planning on lessons from the past has very limited success. For example, the Soviets learned of the effectiveness of armor in the Spanish Civil War (1936–1939) and subsequently dismantled their armored divisions, subordinating the tanks to infantry divisions. They reversed the decision only after witnessing the speed with which German armor units rolled through Poland and France, but by then it was too late. Another lesson that proved fatal was France's realization, despite tank development in the interwar period, that defensive firepower would halt the German armor thrust.

The fourth stratum deals with real-time learning, not with learning lessons from past wars. This study shows that an important factor in recoverability from technological and doctrinal surprise is the ability to derive lessons while the surprise is taking place, that is, devising immediate solutions and circulating them throughout the army.

Karl Weick and Kathleen Sutcliffe, in their book *Managing the Unexpected: Assuring High Performance in the Age of Complexity*, set down a number of rules in organizational activity designed to induce optimal performance in unexpected, surprise situations. The book analyzes organizations that succeed in coping with daily challenges that have a high potential for failure. One such example is the crew of an aircraft carrier.

Weick and Sutcliffe discern five patterns that characterize these organizations:

1. Dealing constantly with major and minor foul-ups, making no attempt to gloss over them.
2. Coping with complex situations, making no attempt to simplify them.
3. Sensitivity to an activity's details, constantly trying to learn from potential malfunctions.
4. Commitment to flexibility. Developing the ability to identify problems, deal with them, and recover from them.
5. Respect for expertise rather than for hierarchical status. Encouraging a wide range of expertise that enables complex situations to be dealt with effectively.[1]

These yardsticks are generally characteristic of military organizations that succeeded in recovering from technological and doctrinal surprise. The following discusses the first and fourth yardsticks—the heart of recoverability.

Contingency containment has been accomplished mainly by a "commitment to flexibility." This means developing the ability to cope with unexpected situations that are, a priori, unavoidable. This ability uses the organization's collective knowledge and rapid learning to enable people to deal with all types of surprise. Organizations that succeed in coping with surprise situations broaden their knowledge, develop their ability to learn quickly, improve the speed and accuracy of communications, increase areas of expertise, sharpen their proficiency in recombining existing tasks, and work comfortably with improvisation.[2]

According to research findings, armies that have recovered from surprises have done so with the help of three elements:

1. A culture and a mechanism that encourage learning from mistakes.
2. A mechanism for quickly relaying information from one unit that encountered and overcame a surprise situation to other units in a similar situation.
3. Coordination and communication between the army and defense industry so as to provide rapid technological solutions based on lessons learned on the battlefield.

A CULTURE THAT ENCOURAGES LEARNING AND DRAWING LESSONS

Rosen believes that deriving lessons and devising innovations in the course of war may seem relatively easy due to the lack of obstacles such as conceptual fixations, lack of intelligence, and so forth. However, regarding the ability to develop a response to new battlefield events, we can identify an organizational and intellectual difficulty in understanding the significance of the new situation and in encouraging the response to it. Rosen further claims that the army must learn how and what to learn and that this is an extremely difficult task.[3]

What are the elements in the military culture that enable lessons to be learned?

Weick and Sutcliffe who, as stated, studied organizations that successfully cope with a constant state of danger, identify four elements in organizational culture that assist them in dealing with unexpected situations. They define the first element as "the culture of report"—where people are willing to report their mistakes and feel secure in doing so. The second is "the culture of justice"—treating guilt and punishment in a way that does not deter members in the organization from admitting their mistakes. The third is "the culture of flexibility"—where information flows swiftly to professionals who have the authority to make decisions. In this culture the status of hierarchy is lower than that of expertise. The fourth component is termed "the culture of learning" (or "the learning organization")—where new knowledge is absorbed into the "old body" of knowledge. Learning takes place in a free and open discussion of the new knowledge and its importance.[4]

An organization infused with a culture that values learning from its mistakes, and that is not blinded by success, imparts in its members the understanding that their prediction ability is limited, their knowledge is limited, and there is a strong likelihood that they will be caught in surprise situations. This type of organization encourages error reporting, endeavors to learn the details of every event, and tries to view them as windows of opportunity for the entire system.[5] The Agranat Commission Report had this to say on the practical results of such a culture in the army:

> One expression of effective leadership throughout the chain of command and staff work in the field of training is the transformation of "awareness and curiosity" into practical activity at every level for every soldier. This means being open and receptive to changes in the enemy's weapons systems, unit structure

and organization, and level of cooperation gained through exercises. The test of this awareness is the speed and quality of tactical response to enemy action.[6]

Gal Hirsh[7] analyzed the way the IDF met the challenge of a rapidly changing combat situation in the West Bank and Gaza Strip in the beginning of the twenty-first century. He claims that among the challenges that the IDF must deal with is "an enemy and environment that is changing at a dizzying rate." The IDF must develop the ability to identify changes in the current situation, conceptual coherence, and a dynamic unit-community atmosphere that allows and encourages learning, and so forth.

An example of an organization that failed to draw lessons during the fighting due to "cultural" inadequacy of the type just described is the British army's encounter with Rommel's armor and anti-tank tactics in the Western Desert. Examples of a culture of openness and a genuine propensity to learn from mistakes can be seen in the way the Wehrmacht dealt with surprises (see below).

A MECHANISM FOR RAPIDLY CONVEYING LESSONS

Even if the army has a culture of learning from its mistakes, it will be of little use in recovery from technological and doctrinal surprise if the lessons learned by one unit are not disseminated throughout the army. A mechanism must be established for circulating the lessons. Richard Downie, who studied U.S. military *learning from conflicts*, recommended that military organizations "establish a systematic process to rapidly transmit and disseminate doctrine to units in the field."[8]

In the book *Men Against Fire*, S. L. A. Marshall discusses the transfer of newly learned information:

> During war, it oftentimes happens that one company, by trial and error, finds the true solution for some acute problem which concerns everyone. But when that happens to a company, I can assure you that it is the exceptional company officer who takes the initiative and passes his unique solution along to his superiors even after he has proved in battle that the idea works. A good company idea in tactics is likely to remain confined to one company indefinitely, even though it would be of benefit to the whole military establishment. Such omissions are not due usually to excess modesty or indifference on the part of the officer, but to his unawareness that others are having the same trouble as himself.[9]

He further notes that:

> Constant change is the natural order of war. New situations frequently call for hitherto undreamed solutions. The maximum efficiency therefore comes of a system of thought which assures that each new and sound idea will be circulated promptly.[10]

Looking at organizations that successfully cope with surprises, Weick and Sutcliffe observed the speed involved in learning from mistakes and transmitting this knowledge to other elements in the organization. Such organizations increase their knowledge reservoir by encouraging and rewarding self-confessed mistakes and error reporting.[11] Hirsh identified the need for developing a communications system that enables the rapid transfer of new information received from one element in the military organization to the rest of the organization.[12]

One example of ineffectiveness and tardiness in recovery from technological and doctrinal surprise is the IDF armor's confrontation with Egyptian Sagger missiles in the Yom Kippur War. In this case a problem arose in conveying the lessons learned by units that had been vying with the missiles to fresh units entering the fray.

James Carafano describes in his *GI Ingenuity* U.S. army difficulties in conveying lessons learned between two divisions that fought in the same area during the same month.[13]

The British and Americans' ignorance of the Japanese Zero fighter planes is an example of surprise originating, inter alia, from communication breakdowns that caused the faulty use of information already in the system. During WW II the Zero was the backbone of the Japanese air force and navy. All escort missions for long-range attacks—Pearl Harbor, Singapore, and the Philippines—were assigned to these planes. Despite the key role that the Zeros played, American and British commanders and pilots were almost totally in the dark about them, confident of their own planes' alleged superiority.

In the Battle of Singapore (between December 7, 1941 and mid-February 1942), Singapore and the Malay Peninsula fell to the Japanese invasion force. An important element in Allied weakness was the British air force stationed in Singapore, which was significantly inferior in size and capability to the Japanese and was swiftly destroyed by the Japanese air force.

A Japanese Zero was brought down in China in May 1940. On September 29 the Allies' joint intelligence branch in Singapore relayed details of the plane's armament and fuel supply to Air Command. But a glitch at Air Command resulted in the report being filed away and nothing further being done. Allied pilots received no information of the approaching disaster. The general impression was that the Japanese plane consisted of "rice paper and bamboo stalks." In addition, the Australian and New Zealand pilots lacked combat experience, whereas the Japanese pilots were already experienced airmen from the fighting in China. The Allies had 141 planes in Singapore, about half of which were very much inferior to the seven hundred high-quality Japanese aircraft.

The British air force in Singapore relied on antiquated American Brewster-Buffalos for air defense. The British air commander in East Asia, General Robert Brooke-Popham,stated: "We can get on all right with Buffaloes out here, but they haven't got the speed for England. Let England have the Super-Spitfires and Hyper-hurricanes. Buffaloes are quite good enough for Malaya."[14] In effect, the Brewster-Buffalo's maximum velocity was 295 mph compared to the Zero's 325–335 mph. Only the Hurricanes were able to compete with the Zeros but suffered from poor maneuverability under ten thousand feet.[15]

The British overrated the ability of the Royal Air Force (RAF) to tackle the Japanese in the belief that the enemy had not yet been up against a high-quality air force.[16] Another fallacy that the British learned too late was the range of Japanese bombers. In his memoirs, Churchill recalls the fleet commander's error in ordering the battleships *Repulse* and *Prince of Wales* into open waters without air defense. Both vessels were sunk by Japanese planes that flew from Kuantan, four hundred miles north of Singapore. The British had no inkling that Japanese torpedo bombers were capable of such a long range. According to Churchill: ". . . at this date no attacks by torpedo bombers had been attempted at anything approaching this range. The efficiency of the Japanese in air warfare was at this time greatly underestimated both by ourselves and by the Americans."[17]

In this case, disparagement of the enemy, combined with the organizational problem of conveying information to the "consumers," seem to have aggravated the surprise encountered by the pilots defending Singapore when they engaged the Zeros and the sailors of the *Repulse* and *Prince of Wales* when attacked by Japanese bombers.

On the same day that Pearl Harbor was attacked, the Americans based in the Philippines suffered another setback. Japanese planes struck the Clark and Iba air bases on the Philippine island of Luzon and destroyed almost all of the Allied aircraft still on the ground. General MacArthur's air commander requested Flying Fortresses for attacking the island of Formosa (Taiwan), the

Japanese base approximately 450 nautical miles (850 kilometers) north of Luzon, but permission was refused. Due to a false alarm regarding an impending Japanese attack, American bombers took off and remained airborne, protected by fighter aircraft.[18] But soon after they landed, the Japanese bombers struck, with 108 Zeros escorting them.[19] The American fighter aircraft were P-40s. "Lieutenant Moore and his two companions were trying to give chase. They found to their amazement that the Zeros were faster and more maneuverable, climbing at an astounding rate. They had been assured there was no such thing as a good Japanese fighter plane."[20] The Japanese air assault brought down all of the P-40s, in addition to thirty intermediate-range bombers, and the whole fleet of Flying Fortresses (excluding three). All of the Japanese bombers returned to base safely; seven fighters were lost.

The United States War Department received details of the Zero from Colonel Claire Chennault, the commander of the Flying Tigers—an American volunteer air unit that helped the Chinese in their struggle against the Japanese,[21] and from other sources. Some pilots even developed tactics to counter the Zero prior to Pearl Harbor, "but many refused to take it seriously."[22] Had the American pilots in the Philippines internalized the lessons, they probably would not have been caught by surprise.

The Americans, like the British, underestimated the Zero's range, a vital factor in a bomber attack's escort, and failed to see that it was capable of making a return trip flight from Formosa—almost 900 nautical miles. This case illustrates the need for a mechanism that conveys information relevant to recovery from technological and doctrinal surprise.

The American and Israeli armies have developed mechanisms for quickly disseminating information. The American army's Center for Army Lessons Learned (CALL) was established in 1985 by General Gordon Sullivan, the U.S. Army's chief of staff.[23] The rationale for setting up CALL was that the basic changes being introduced into the army in that period required formal processes for learning from mistakes and learning lessons, both from training and later from combat. Established with the assistance of academic scholars, CALL remains one of the most advanced agencies in the world for leveraging knowledge and information and sharing, permeating, and integrating it into the fighting forces. It took part in operations in Haiti in 1994 and since then has been involved in almost all American ground forces actions. "Learning officers" and "learning teams" are dispatched to collect data, which is then analyzed to gain practical insights for changes in the army. The lessons derived from the operations are either transmitted directly to the forces or integrated into the training programs, doctrine, and weapon systems development. This sophisticated mechanism gives the Americans the flexibility to respond relatively quickly to changes on the battlefield. Similar processes, though smaller in scale and differently structured, exist in the IDF[24] and were implemented in the Second Lebanon War and Operation "Cast Lead" (Gaza 2009).[25]

A MECHANISM FOR RAPID LINKAGE BETWEEN THE ARMY AND THE DEFENSE INDUSTRY

The results of the army's lesson learning in wartime are usually in the adaptation of existing capabilities to new situations. This is done generally by modifying the combat doctrine and, in exceptional cases, the new use of a weapons system or existing weapons. To develop a technological response to a new problem, it is necessary to integrate elements that are often outside the military but are involved in weapons development and production.

As a solution to technological surprise, Heilmeier suggests creating an atmosphere of partnership and knowledge-exchange between technological experts and military commanders and establishing intimate working relationships between military scientists, the academic research community, and industrial engineers.[26] He proposes this in connection with the prediction and assessment of new technologies prior to a confrontation. I claim that recovery from technological surprise also calls for the ability to provide a quick technological solution in the course of the fighting and that this ability is dependent on the relationships between the abovementioned components.

Examples of fast and effective cooperation in dealing with technological surprise are the German military industry (vis-à-vis the surprise of the Russian T-34 tank) and the solution of the Israeli Authority for the Development of Weapons and Military Technology (RAFAEL) to the SA-6 missiles in the Yom Kippur War. These examples are described in detail in the empirical section below.

Another example of productive cooperation between army and industry is the 1982 Falkland War. The official British report on lessons from the war, in the section on acquisition and improvisation, states:

> In the exceptional circumstances of the campaign our procurement processes proved adaptable to meet the wide variety of military needs against very tight timescales. New operational demands were satisfied in record times through the

ready availability of a broad spectrum of scientific and engineering expertise in the Ministry of Defense research establishments and the cooperative resources of the United Kingdom's defense industry . . . rapid development of new equipment capabilities; and accelerated introduction of equipment into service. Examples included the development of important, often vital, equipment by combining existing items in new ways, such as the creation of AEW (Airborne Early Warning) equipment using Nimrod Searchwater radars in Sea King helicopters (in only 11 weeks); the invention, production, proving and delivery in record time of many new equipments, including man-portable radar jammers (10 days from order to delivery) . . .[27]

For the British War Office, one of the most important lessons from the 2003 War in Iraq was the importance of flexibility in the response of British industry to the army's needs in real time. A report on the lessons learned from the war states that the British War Office approved more than 190 urgent operational requests. "Industry responded magnificently to the surge of requirements in the build up to the operation, proving the value of the partnering approach that the MOD [Ministry of Defence] has developed over recent years."[28]

What is the driving force behind the flexibility concept? Is it enough for a commander to demand flexibility from his subordinates, or must the entire military system be based on flexibility? Like other qualities in complex systems, more seems to be required than just issuing an order from above and expecting the entire system to adapt to a cognitive characteristic as complex as flexibility. The secret seems to lie in the organizational culture of the military system as it comes to expression in the way commanders relate to unprecedented circumstances, unfamiliar and unconventional ideas, and the need to learn and improve. The following historical test cases illustrate how flexibility was first applied by fighting units that improvised methods of dealing with surprise, and only later, if at all, how the military leadership introduced local, flexibility-based solutions into the official, updated doctrine. The two cases dealing with German military flexibility describe its development by field commanders and application and incorporation into new doctrinal instructions. In the Israeli case, the results of flexibility may have remained in lower echelons because of the brevity of the war.

Flexibility culture is the silver thread running from the concept and doctrine stratum—which perceives uncertainty as a fact of life and is receptive to a broad range of possibilities that can be developed in wartime, to the fourth stratum—the requirement and capability for an ongoing learning process as the key for coping with change.

As stated, signs of such a culture appeared in the German and Israeli armies in certain cases when they succeeded in dealing with surprise. The British and French armies, on the other hand, displayed signs of the opposite culture: a conservative mindset that stifled discussion that deviated from the official line, suppressed initiative, and tried to solve uncertainty by preplanning and the use of firepower.

The theory that has been presented is made up of four strata of flexibility that, when taken together, create a unified concept. The next part of the book analyzes seven historical test cases in which flexibility was employed when militaries were challenged with technological and doctrinal surprise.

Part Three

RECOVERY FROM SURPRISE— A HISTORICAL VIEW

This part of the book presents seven historical test cases that lend empirical support to the main theory of the research. The test cases were chosen according to four criteria.

The first is the effectiveness of the solution. In some cases one party was caught by surprise, but after it recovered, it created a new operational problem for the other party; in other cases, recovery failed altogether. The second criterion is the number of elements of flexibility or inflexibility discernable in the recovery. Each example illustrates several elements. The third criterion is surprise at various levels of war: from the technical-tactical level (the German reaction to the Soviet T-34 tank) to the strategic level (the German response to the challenge of British chaff). The fourth criterion is the various reasons for the surprise.

Two criteria were used to determine the extent of recovery: the time required (which is relatively easy to measure) and the assessment of battlefield effectiveness (which is more difficult to gauge) because of the complexity of various factors related to the outcome. Still, the effectiveness of the response is based on certain elementary criteria. The best solution results in complete recovery and generates a new problem for the enemy. The second level of solution neutralizes the problem without creating an operational counter-challenge for the enemy. The third solution minimizes the amount of damage caused by the surprise. The lowest level of effectiveness is the failure to recover from the surprise.

As stated, the conceptual-doctrinal element is of paramount importance since all the other elements depend on it. The least important element is learning the lesson of the surprise and quickly passing it on to the rest of the armed forces. This is because an army can remain flexible even though its means for coping with the surprise do not filter down to other units. However, in such a case, flexibility will have a smaller impact on the army's overall effectiveness.

6 THE GERMAN RECOVERY FROM THE SURPRISE OF BRITISH CHAFF

THE GERMAN RECOVERY after Britain's surprise use of chaff in July 1943 is a classic example of this book's thesis. This example illustrates the case of a strategic surprise that was met swiftly, employing all the elements of flexibility so that the enemy's intentions were counteracted, causing it unexpected operational problems.

SURPRISE: SOURCES AND IMPACT

Before the war and immediately after hostilities erupted, Luftwaffe commanders contended that the only way to protect Germany from British bombing raids was through the massive use of anti-aircraft guns. When the British began attacking German industrial targets in May 1940, it soon became clear that this was not the case: the German anti-aircraft layout proved to be ineffective.[1] Germany's few fighter planes failed to locate British night-bombers even when the latter were detected and tracked by long-range radar.

Until the summer of 1941, the heads of the British Bomber Command believed that if Britain produced enough bombers and trained enough aircrews, Germany would be dealt a fatal blow. But by late 1942, they realized that Germany's delayed reaction to Bomber Command's early successes had actually caused Britain intolerable losses.[2]

The Germans reacted by setting up a night-interception fleet of 250 twin-engine planes and a massive ground support system.[3] In early July 1940, German planes employing the Freya radar system proved themselves capable of intercepting British aircraft. In late July, a Freya radar station spotted a British destroyer 100 kilometers out at sea and directed the Luftwaffe to it.[4] German planes promptly sank the vessel.

The British named the German anti-aircraft layout the "Kammhuber Line" after Josef Kammhuber, the commander credited with setting up the German night-fighter defense system. The German modus operandi worked as follows. A Freya radar system would scan a relatively broad sector. When it spotted an enemy bomber, it would relay the data to a Würzburg radar system, which commenced tracking. A second Würzburg radar system homed in on the German fighter that had already scrambled and was circling a radio beam and light beacon. When the bomber came in range, the ground controller guided the fighter to the interception vector while the tracking radar illuminated the target with three giant projectors. Other projectors continued tracking the bomber until the fighter located and intercepted it.[5]

In the winter of 1941–1942, the first British countermeasure—a simple technique—was employed against the Kammhuber Line. This was a tactical solution that involved large, dense, fast-flying bomber formations that enabled some of the aircraft to slip through the line unscathed while the German controllers concentrated on intercepting a limited number of enemy planes. The method achieved a modest degree of success. However, keeping the planes in tight formation proved tricky, especially on the flight back when discrepancies in speed and navigation caused the planes to scatter. Indeed, the lion's share of British losses was sustained on the return trip. The Germans responded by beefing up their radar layout. In April 1941 the British discovered that the Germans had installed airborne, ultra high frequency (UHF) radar known as the Lichtenstein[6] that significantly enhanced German interception capabilities. The denser concentration of radar on the Kammhuber Line and the addition of airborne radar led to heavy British losses.

The British had various methods for coping with German radar on their night bombing raids. Two of the secondary methods were: (1) a warning device that became operational in the spring of 1943 and informed the pilot when his aircraft had been identified by enemy radar, thus giving him time to initiate evasive action; and (2) a transmitter (operational in August 1942) that simulated the echo of a larger airborne formation. But the two main methods for thwarting enemy radar were jamming and emitting false signals. Jamming was risky business because the source could be picked up by enemy fighters equipped with the proper receiver. Using chaff to produce false signals was simpler and contained the additional element of deception. The British codename for chaff was "Window." After a series of experiments, the optimal length of the metal strips, bundle size, and distribution were finally determined. By April 1942, when sufficient chaff had been produced for operational use, the project was abruptly halted. Fearing that the Germans would copy the method in order to attack the British Isles, Night Fighter Command halted its use before its effect on the interception capability of British fighter planes could be tested. R. V. Jones, head of scientific intelligence on Britain's air staff, disagreed, claiming that the chances of the Germans mounting the same kind of massive attacks were very slim.[7]

When Jones obtained a report in October 1942 indicating that the Germans were already familiar with chaff, his opponents' arguments were proven wrong. (The Germans experimented with chaff but preferred not to use it for the same reasons that the British kept the idea strictly secret.)[8] In May 1943, a German airplane carrying a Lichtenstein radar device fell into British hands. Chaff proved effective against this type of radar[9] but began to be used operationally only in July 1943. Hamburg was heavily bombed on the night of July 23–24. "Out of 791 aircraft, only 12 had not come back, a loss rate of 1.5% compared with 6.1% for the six previous attacks on Hamburg. It could thus be argued that Window on that single night had saved between 70 and 80 aircraft."[10] Hamburg was mercilessly bombed for an entire week: four RAF night attacks and two smaller-scale day attacks by the Americans. The RAF flew over three thousand sorties against the city.[11] The bombs created a fire storm that devastated the city, killing forty thousand people and demolishing nearly three hundred thousand buildings.

Up until late 1942, the German Air Force had been satisfied with the success rate of its radar-based anti-aircraft layout; the Hamburg raid and British chaff-jamming techniques took it completely by surprise.[12]

The chaff severely interfered with the operation of the ground-based Würzburg system and the airborne Lichtenstein radar device. "A Study on the Present 'Window' Situation" was the name of a Luftwaffe report written immediately after the attack on Hamburg:

> Since July 25, the enemy, first at night—in isolated cases in daylight too—combined with the raids into Reich territory, the dropping of "Hamburg bodies" (German codename of the chaff). The technical success of this action must be designated as complete [...] by this means the enemy has delivered the long awaited blow against our decimeter radar sets both on land and in the air.[13]

By early August 1943, the British jamming technique brought the effectiveness of Germany's night fighters to near impotency.[14]

Surprise was intentional since the British made prodigious efforts at dissembling.

Regarding this point, it is important to understand the relationship between the surprise and the level of war on which it was supposed to have an impact. The surprise was intended to play a major role in Britain's strategic plans that had been drawn up by the RAF before the war: Germany's defeat by destroying its industrial capacity through strategic bombing.[15] But this prewar concept suffered a serious setback once war broke out. Daytime bombing turned out to be unfeasible, and the technology for precision nighttime bombing was still non-existent. The idea, however, was revived by Air Marshal Arthur Harris, who became head of Bomber Command in mid-February 1942 and immediately began full-scale preparations. Cologne was the target for a massive air attack. The amount of destruction made such a strong impression on the British War Cabinet that it decided to provide Bomber Command with more aircraft.[16] At the Casablanca Conference (February 1943), the British and American bomber commands were given the following assignment: "Your primary object will be the progressive destruction and dislocation of the German military, industrial, and economic system, and the undermining of the morale of the German people to a point where their capacity for armed resistance is fatally weakened . . ."[17]

The Allied leadership regarded the deep-penetration bombing as one of the chief strategic means for winning the war. Its implementation proceeded in three main stages. The first consisted of attacks on the heavily industrialized Ruhr region (March to June 1943). The next stage began with the bombing of Hamburg (see above). Regarding the use of chaff at this stage, Churchill recalled: ". . . and for the assault on Hamburg an additional device called 'Window,' which we had long held in reserve, was used for the first time."[18] In June Churchill authorized the use of chaff at a staff meeting that he convened specially for this purpose.[19] The effect that the attack on Hamburg had on the German High Command can be appreciated from the comment made by one of the heads of the Luftwaffe, Erhard Milch: "If we get just five or six more attacks like these on Hamburg, the German people will just lay down their tools, however great their willpower."[20]

The third and last stage in the plan to defeat Germany from the air was the nine bombing raids on Berlin between November 1943 and March 1944. On November 3, 1943, Harris wrote to Churchill: "We can wreck Berlin from end to end if the U.S.A.A.F. will come in on it. It will cost us between 400 and 500 aircraft. It will cost Germany the war." In a letter to the Air Ministry dated December 3, he wrote: "It appears that the Lancaster force alone should be sufficient but only just sufficient to produce in Germany by April 1, 1944, a state of devastation in which surrender is inevitable."[21] Churchill's view of the attack on Berlin was: "If this great industrial center could have been paralyzed like Hamburg, German war production as well as morale might have been given a mortal blow."[22]

What prevented Harris from bringing the war to an end with strategic bombing? And why did this strategy fail? The answer probably lies with the losses suffered by Bomber Command. The reason that British losses soared and even surpassed pre-chaff levels can be ascribed to German recovery from the surprise.

RECOVERY FROM THE SURPRISE

German recovery is attributed to two elements of flexibility that characterized the German military in general: conceptual and doctrinal flexibility and command and cognitive flexibility.

Recovery can be divided into three constituent elements. The first is tactical: combat methods known as "Wild Boar" and "Tame Boar" that were developed immediately after the chaff surprise. The second element is a relatively simple technological device: guns mounted on interceptor aircraft enabling them to attack Allied bombers from underneath their bellies, which were a blind spot for bomber crews. The third element was the result of the development and rapid deployment of radar (especially Lichtenstein SN-2s) capable of detecting enemy aircraft even in the presence of chaff. All three elements demonstrated the Germans' willingness and ability to learn from enemy surprises, develop countermeasures, and distribute them among the units in as short a time as possible. (The German army's capability for this is discussed in Ch. 7.)

Still reeling from the Hamburg catastrophe, the Luftwaffe cooperated fully. According to Alfred Price, a scholar of aerial combat in WW II: "There was no conflict between the General Staff and war industry, no rivalry between bombers and fighters; only the one common will to do everything in this critical hour for the defense of the Reich..."[23]

Command and Cognitive Flexibility

The Luftwaffe's immediate response was to introduce new fighting methods. The Germans were fortunate in having started the testing of a new method for downing enemy nighttime bombers a few months earlier: "The wisdom of relying solely on Kammhuber's system of rigidly-controlled interceptions had been questioned even before the Hamburg attacks."[24] The main problem with this method was that it offered almost no protection to the target sites, focusing instead on hitting the bombers before they entered German airspace.

Since April, Captain Hajo Herrmann, an outstanding bomber pilot working at the Luftwaffe's Staff College, had supported the use of single-engine fighters originally intended as daylight interceptors (single-seat Messerschmidt 109s and FW 190s) for nighttime operations above the targets themselves. He argued that the light from the anti-aircraft projectors, and the illumination produced by ground fires and the flares discharged by the lead bombers to mark the targets, would assist interceptor pilots in obtaining eye contact with the enemy aircraft.

Kammhuber, the commander of the nighttime interceptors, considered the idea preposterous. Herrmann infuriated him by going over his head and gaining permission from Hubert Weise, the commander of the Reich's anti-aircraft forces, to perform a small experiment with the new tactic.[25] In a late June report, Milch supported Herrmann's method of employing interceptors. Herrmann had been organizing and training even before the Hamburg raid[26] but after the first bombing raid on Hamburg, his method naturally took on strategic significance. Since the procedure did not require precision surface or airborne radars, it was immune to chaff. On July 25, 1943, Air Marshal Hermann Goering ordered Herrmann to prepare at once a wing of single-engine interceptors for target defense. The new unit was called Wing 300, and the tactic received the appellation "Wild Boar." Herrmann assembled posthaste a wing of sixty planes. Although he had never regarded the tactic he had been touting as anything more than a supplement to conventional nighttime defense procedures, he was now ordered to put it into practice as a major element in night defense.[27]

The first opportunity to test the new combat tactic against British chaff took place on the night of July 27—the second bombing raid on Hamburg. Herrmann sent up a dozen planes, and British losses began to rise.[28]

At the same time, the twin-engine night interceptors (mostly Messerschmidt 110s and Junkers 88s) that could no longer be directed by ground control had to change their modus operandi. On July 29, Victor von Lossberg, a former bomber pilot and then a member of the Night Interceptor Staff, proposed a new method codenamed "Tame Boar."[29] The twin-engine interceptors would no longer defend a sector as before; now they would take off and zero in on the bombers, guided by radio beacons. The ground station would direct the aircraft to the area where precision radar had picked up the highest concentration of chaff. At this point the pilot would make eye contact with the enemy aircraft. Von Lossberg hoped to intercept as many enemy bombers as possible on their flights to and from Germany. When a twin-engine interceptor identified a bomber in the target area, it was given the green light for "Wild Boar" (freelance night-fighting).[30] The next day the method was examined and approved by a committee that included Milch, Weise, Kammhuber, and Adolf Galland. On August 1, Goering signed orders for engaging the enemy with the "Wild Boar" and "Tame Boar" tactics; he also decided to give top priority to the development of an interceptor force. Weise instructed the Reich's entire nighttime defense to operate according to Herrmann's method instead of the Kammhuber method.[31]

The "Wild Boar" and "Tame Boar" methods had an impact on almost every aspect of the Reich's air defenses. Various methods were employed to help the interceptors identify their prey in the target area. Whenever the cloud layer was not too dense, projectors were lit to create uniform illumination from above, thus enabling German interceptors, flying over the bombers to spot the latter's silhouette. The fuses on the anti-aircraft shells were calculated to go off at an altitude band of approximately fifteen thousand feet so as to reduce the risk of anti-aircraft fire hitting friendly planes. Above that height, the interceptors had free rein to engage the enemy.

Night interceptor crews needed time to adapt to the new methods. Previously the airplanes stayed within a fifty-mile radius from their base and always operated under the guidance of a group of radar stations, which made navigation relatively easy. The new method revolutionized the entire procedure. Flight crews had to fly all over Germany hunting for prey, landing at different airbases when their fuel ran low. The new tactics also called for a retrofitted type of ground control. Under the former *Himmelbett* ("sky bed") system deployed on the Kammhuber Line, the battle had been directed from the radar stations. Air division headquarters continued to receive reports but were assigned a minor operational role. The new method reversed the roles. The ground stations used radio beacons to direct the forces the length and width of Germany until they reached the battle zone, at which point air division HQ was given command of all the aircraft that entered its sector.[32]

The first time the Germans fully employed "Tame Boar" was on the evening of August 17 (three weeks after the Allied devastation of Hamburg). The effort to protect the V-1 rocket research and development facility at Peenemünde ended rather poorly (twenty-four bombers were shot down over the target and another sixteen on the return flight). This was due to a number of reasons.[33] But six nights later, on August 23, the new method made an impressive showing. German interceptors bagged fifty-six of the

727 bombers sent to Berlin, nearly 8 percent of the attackers, the highest loss rate until then. The British pilots were caught by surprise; they had not expected enemy fighters to appear over areas where anti-aircraft batteries were operating. The returning crews reported that German aircraft had attacked them from above and that projectors had been of greater support to the interceptors than to the anti-aircraft guns. Despite the heavy losses, Air Marshal Harris, the head of the British Bomber Command, was not discouraged. He had no illusions that it would take several bombing raids to destroy a city the size of Berlin. Thus, he ordered additional attacks on the nights of August 31 and September 3. These too encountered fierce resistance. Bomber Command lost a total of 123 airplanes in three nights, or approximately 7.5 percent of the attacking force,[34] in addition to 114 damaged aircraft. As stated, these were the heaviest losses sustained on a mission so far, and ironically, these occurred just when the British believed that German defenses had been broken. The Germans recovered astonishingly fast from the chaff surprise.

The Germans adopted another method to lessen the damage to their cities. The British had their lead bombers discharge flares above the target to mark it for other planes that were flying without navigational equipment. The Germans counteracted by firing off their own flares above the lead bombers. This forced the lead pilots to avoid flying in too close a formation, which meant a greater dispersal of bombs in the entire attack. But no matter how hard the Germans tried to improve their tactics, "Wild Boar" suffered from a glitch that exacerbated as winter set in. The single-engine planes that had been transferred from their original daytime missions to night operations had difficulty landing in bad weather. Fearing a fatal landing, an increasing number of pilots were ditching their planes. A more basic problem was that ground controllers had to locate targets at night without radar assistance, relying solely on their experience and intuition, which frequently failed, especially in bad weather. As a result, the efficacy of "Wild Boar" gradually decreased. On March 16, the 30th Air Division was dismantled.[35] For the remainder of the war, "Tame Boar" was the main German air defense method.

Developing Redundancy in Weapons Systems

The second element in German recovery was the use of organizational and technological flexibility to accelerate the development and deployment of new radar systems. As the effectiveness of the new combat methods decreased, the Germans speeded up their efforts to introduce new equipment in order to consolidate their initial successes. On July 28, only three days after British planes dispersed chaff for the first time, Field-Marshal Milch instructed the electronics industry to accelerate production of anti-chaff apparatus.

The solutions were mainly in the field of airborne radar, although some of the problems were solved using ground radar for directing anti-aircraft guns.[36] The visual search method that worked well on bright summer nights proved ineffective on overcast winter nights. Direction-finding equipment was needed to overcome the chaff. The first two devices, "Naxos" and "Corfu," had been ordered prior to the chaff crisis. Their purpose was to home in on the British H2S transmitter fitted in the British Pathfinders. Once they latched onto the lead aircraft, German ground stations could track the course of the entire British bomber stream. Moreover, since the Mosquitoes being used for deception were without the H2S, the Germans had an easier time identifying the real bomber force.[37]

The Naxos radar could zero in on the H2S radar emissions at a distance of fifty-six kilometers.[38] The neutralization of the British H2S radar by German radar dealt a serious blow to British bombing capabilities. Walter Thompson, a British pathfinder pilot, described the mounting of German airborne radar onto twin-engine fighters:

> Now the fighters had such a marked superiority over the bombers that bomber crews were instructed not to use H2S, or to use it only for short periods when vitally required for navigation or for blind marking on long range targets. There was no other technique available. But the accuracy achieved against the Ruhr and against Hamburg could not be achieved on long range targets such as Berlin, and the attrition from fighters became too great to sustain.[39]

Another radar system that helped the Germans deal with chaff, one that was developed before the chaff crisis, was the Flensburg, a homing device designed to track the Monica, a tail-positioned radar device that enabled British bombers to identify German fighters trying to detect them with radar. The Monica proved to be counterproductive, if not outright lethal for the British. After a German aircraft carrying a Flensburg radar device made a forced landing in Britain in July 1944, Harris immediately ordered the Monica removed from all bombers. The Germans' trump card in the battle against chaff was the Lichtenstein SN-2 radar, developed in the summer of 1943, which operated at a relatively long wave length and was unaffected by chaff. The Luftwaffe immediately launched a crash program to produce these devices.[40] They were installed in night fighters by early October and within three months entered general use. At six kilometers above ground level, the radar had a range of about six kilometers. This meant that the moment an aircraft was guided to the bomber stream by the radar, it could engage the enemy without ground assistance.[41] The twin-engine night-interceptors were again able to locate bombers flying into and out of Germany.

The high point in the use of Lichtenstein SN-2s came on March 30, 1944 when British bombers dropped their payloads on Nuremberg and suffered the heaviest losses of the war. "But the fact remains that this same force, six months after its technical 'knock-out,' in the summer of 1943, now represented a crucial threat to the whole continuation of the enemy bomber offensive. That it could do so was thanks, in large measure, to its 'Lichtenstein SN-2' airborne radar—and to its 'slanting music.'"[42]

Jones refers to the British effort to counter the Lichtenstein SN-2 radar "an intelligence fiasco."[43] It took ten months to develop and produce a successful jamming device (January 1944). The same amount of time was required until a navigation and target location alternative to the H2S, the "G-H," became operational.[44]

Louis Brown, a historian who has studied the use of radar in WW II, claims that Germany's two systems (Naxos type-Z, which detected H2S radar and was able to guide interceptors to the bomber stream without ground support, and the Flensberg radar that homed in on the Monica) complemented the Lichtenstein SN-2 and created a lethal combination.[45]

The Germans turned the cards on the British. Their technological development was so successful that the initial surprise use of chaff ended up posing a more formidable problem for the British that took months to identify correctly. The German counter-surprise not only neutralized chaff but also caught the British off-guard by creating a hornet's nest of problems in navigation and target location that heretofore had been non-existent.

Technological Versatility

A third facet of German recovery was technological versatility. This refers to the development of a positioning of the planes' guns so that British bombers could be attacked at an angle outside their crew's field of vision. Less than a month after the bombing of Hamburg, a number of German planes had a pair of 20-mm cannons mounted on the roof of the aircraft. The guns could fire at a ten- to twenty-degree angle relative to the vertical. Identification of enemy aircraft was made possible with a projector on the roof of the plane above the pilot's head. This device enabled the interceptors to fly below the bombers and attack their unprotected belly. The device was called Schräge Musik (Jazz) because of its noise when fired. After the war, one German fighter pilot testified that the standard technique was to fire upward when the interceptor was approximately fifty to seventy meters beneath the bomber, even though this endangered the German aircraft if the bomber immediately exploded.[46]

The British did not catch on to this technique until the end of 1943 for two reasons: most British aircraft lacked a lower turret, so the approaching interceptors stayed inside the bomber's blind spot;[47] and the 20-mm cannon did not fire tracer shells, which meant that the British crews never saw what hit them.[48] If the interceptor shot at the bomber's fuel tanks, the aircraft would immediately go up in flames, so that neither the crew nor any neighboring bombers knew what actually happened. Thompson states that since no tracer shells were observed, some bomber crews thought they were witnessing dummy explosions set off by bombs released from above to "spook" the pilots. Only in early 1944 did the bomber crews see the new installation with their own eyes. According to Thompson, Bomber Command should have recognized that a bomber's vulnerable underbelly was a "juicy" target for a German fighter pilot. It is inconceivable that the bombers were not outfitted with belly turrets at an early stage in operations.[49]

Another author relates that British pilots were at a loss as to why their planes suddenly blew up and as a result, their morale plummeted: "Bomber Command intelligence did not realize what the real danger was for too long."[50] Other accounts by British bomber pilots further substantiate that the German technique of attacking the planes' soft underbelly was a bitter surprise. Murray believes that this technique had a major impact on the high rate of losses in the last stage of the battle over Berlin.[51]

The weapon had been developed in the field based on an invention by Paul Mahle, an aerial gunnery sergeant. During his visit to a weapons testing facility, Mahle noticed a Dornier 217 bomber with guns mounted at a steep angle for protection against enemy interceptors. He became obsessed with the idea. He experimented with a pair of 20-mm guns on a wooden mount that he installed above the cockpit along with a projector. The pilots in his squadron had doubts about its effectiveness. The first tryout of Mahle's brainchild came on the night of August 17–18, 1943 and proved successful beyond all expectations: two pilots shot down six bombers above Peenemünde. On October 2,

1943, the commander wrote in his log that his planes have brought down eighteen enemy aircraft to date using this method, without any loss or damage to themselves. The news quickly spread to other units, and Mahle became a celebrated figure among pilots, who asked him to install the new device on their planes, too. The gunnery sergeant's initiative turned out to be an indispensable weapon that eventually became standard issue in the Reich Air Ministry. Mahle received a letter of commendation and monetary award. By 1944 almost all of the night interceptors had been retrofitted with the new device.[52]

THE CONSEQUENCES OF RECOVERY FROM SURPRISE

The main consequence of the German recovery from the surprise use of chaff was that the British attempt to end the war through strategic bombing fizzled out.

Bekker claims that this turn of events was the result of three air battles. The first was the bombing of Leipzig on the night of February 19–20 when 78 out of 823 bombers were shot down. The second was the raid on Berlin on the night of March 24–25 when seventy-two British bombers were lost. And the third was the attack on Nuremberg on the night of March 30–31 when 95 British airplanes out of 795 failed to return.[53] According to the official British history:

> Thus Bomber Command was compelled, largely by the German night fighter force, to draw away from its operations by apparently less efficient means than hitherto. This situation, in view of the fact that Berlin was by no means destroyed, meant that the Germans had already won the Battle of Berlin.[54] Moreover, in the operational sense the Battle of Berlin was more than a failure. It was a defeat.[55]

The offensive against Berlin lasted five months (November 1943–March 1944), during which Harris lost 1128 bombers. Compared to the first two stages of the raids against the Ruhr region and Hamburg, where the missions succeeded despite the losses, Berlin was a different story. Mission failure, together with the heavy loss of aircraft, had a profound effect on Bomber Command morale.[56] In April 1944 Harris called a halt to the aerial offensive due to "casualty rates which could not be sustained in the long run."[57]

THE SOURCES OF RECOVERY

The Germans' recovery capacity is attributed to their ability to implement all the strata of flexibility. The theoretical part of this book describes German conceptual and doctrinal flexibility, and examples were given of doctrinal diversity and tolerance toward ideas that deviated from the official doctrine. The thinking in the German Air Force prior to WW II was greatly influenced by the Wehrmacht,[58] where initiative and improvisation were looked upon favorably. The Luftwaffe and the Wehrmacht operated according to similar doctrines and shared the same ideas about uncertainty.

Some of the sources of flexibility appear in the Luftwaffe's basic instructions. Procedure No. 16, "Fighting an Air War," dated 1935, contains a number of basic rules that seem to have counseled the Luftwaffe in its recovery from the chaff surprise. Regarding uncertainty, the introduction states:

> Section 3: The leadership and battle of the Luftwaffe are decisively influenced by technology. Aircraft models, weapons, munitions, radios, etcetera, are in constant development. The means of attack are in constant competition with the means of defense. During the course of a war, discoveries and improvements in material can have an enormous effect upon the state of hostilities . . .[59]

The chapter dealing with leadership principles states:

> Section 32: Air strategy and air tactics are at the beginning of their development. A practical test of the principles of air war is difficult to develop in peacetime. Technical advances in peacetime can suddenly call into question recognized principles, and in wartime these principles can be found impractical. Such developments must be foreseen, and their influence correctly evaluated and quickly utilized.[60]

The Luftwaffe's theory of command and control, based on the foregoing concept and doctrine, encouraged cognitive flexibility, initiative, and original thinking, which made it possible for relatively low-ranking individuals (such as Herrmann and Mahle) to bring forth creative ideas and obtain their superiors' backing for their implementation. The Luftwaffe was an organization that was willing to countenance overstepping the chain of command, as in the case of Herrmann and Kammhuber:

> Section 33: . . . Leadership means correctly applying principles in an uncertain situation; this makes leadership an art.

> Section 37: The situation in the air war changes quickly and without warning. The commander must take this into account . . . Recognizing the correct moment for changing a decision is difficult . . . The leader's perseverance should not become rigidity. Perseverance must be combined with flexibility.[61]

The chapter dealing with aerial operations states:

Section 230: [...] Military technological development, however, may also lead to totally new bombing and attack methods. The more the units are trained with an offensive spirit, the more rapidly they will be able to employ these new methods effectively.[62]

Section 250: Fighter forces are to be allowed as much independence as possible [...] In defense districts, fighters and anti-aircraft artillery should be organized under unified command. The fighter force commander is also to be allowed the greatest possible freedom.[63]

Most scholars of Britain's strategic bombing offensive and the German reaction give high marks to the Germans' ability to improvise and introduce immediate changes. Jeremy Howard-Williams, a former British night-fighter pilot who flew Mosquito aircraft that were outfitted for electronic warfare, described the German response to chaff: "The Luftwaffe reacted with typical ingenuity to adapt themselves to the fresh problems."[64] German organizational and technological flexibility came to full expression in the number of radar devices developed after the jamming of the Würzburg and Lichtenstein radar systems. Employing the 20-mm guns creatively is another example of technological versatility.

The next chapter elaborates on the Germans' readiness to learn. The present case, which illustrates the speed with which improvised solutions were integrated into the official doctrine, also reflects the Germans' willingness to learn and their ability to disseminate the information. Despite their considerable success in countering the chaff, their incessant drive for continuous improvement led Goering to offer financial prizes upward of 300,000 marks for new solutions to the chaff problem. It is not known if any proposals were submitted.[65]

Another element of flexibility was the close ties between Luftwaffe commanders and the German arms industry. Brown writes: "In early 1944 the Luftwaffe could mark a victory over the attackers . . . They had stripped the cover of the darkness from Bomber Command through the ingenuity and industriousness of their radar engineers and fliers."[66]

· · ·

In conclusion, the preceding case study illustrates the Germans' impressive capacity to recover from a major surprise. All four elements of flexibility were present. The German doctrine stressed the uncertainty factor on the battlefield and as a result, it encouraged a form of command and control that emphasized initiative and independence. Added to this were: the cognitive flexibility of officers and soldiers; a variety of assets capable of serving as the basis of improvised solutions; the ceaseless drive to improve and learn from mistakes; and the close ties between the Luftwaffe and the arms industry. When all of these elements were present, they contributed to an extremely effective recovery that not only neutralized the initial surprise but created an unexpected operational problem for the enemy. The Germans' successful recovery reversed Britain's strategic plans. The speed of the German response was remarkable given the complexities involved in developing countermeasures.

7 THE GERMAN RECOVERY FROM THE SOVIET T-34 TANK SURPRISE

THE SECOND CASE STUDY is the German recovery from the surprise encounter with Soviet T-34 tanks at the beginning of Operation Barbarossa in the summer of 1941, which was the result of their supreme confidence in their armor capability. The T-34 was far superior to the German tanks in nearly every technical parameter. Since the Germans were just as surprised by the heavier Soviet tank—the KV-1—it may be inferred that the written accounts of the T-34 also refer to the KV-1.

The German recovery illustrates the near-perfect application of the flexibility concept, which was based on conceptual and doctrinal flexibility, cognitive and command flexibility—trademarks of the Wehrmacht—as well as considerable organizational flexibility and lesson learning, including the mechanism for relaying lessons learned quickly to the military industry.

THE SURPRISE, ITS SOURCE, AND DEGREE OF INFLUENCE

The Russians developed the T-34 as part of their T tank series. It was technically superior to the German Panzer III and IV in every aspect.[1] Even before the war, the Germans knew that the Soviet Union was producing tanks in large quantities. In June 1941, the number of German tanks stood at 3500; the Germans estimated the number of Russian tanks at 10,000.[2] The German generals believed that their tanks had a technical advantage over the Soviet ones that would compensate for their quantitative inferiority. However, the Germans were in for a big surprise in the opening rounds of Operation Barbarossa.

A German company commander who took part in the fighting on the Russian border describes an encounter with T-34s on the second day of the operation (June 23). The Soviet tanks knocked out the German tanks relatively easily because the German shells proved ineffective against the T-34 even in flank fire at 300 meters. Eventually the Soviets broke contact and withdrew.

The German company commander took a close look at the damaged tanks that had been abandoned on the battlefield and realized that he had not been given any information on the T-34, whose armor plating, munitions, and trafficability were superior to those of any German tank. "This was a shocking recognition to the German tank and tank destroyer units and our knees were weak for a time." Eventually the Germans discovered the Soviet tank's weak points: a poor communications system and slow crew performance because only two crew members were positioned in the turret.[3] Nevertheless, the sudden appearance of an impenetrable enemy tank was a sharp blow to the German tank crews' morale:

> Half a dozen anti-tank gun fire shells at him (T-34) which sound like a drumroll. But he drives staunchly through our line like an impregnable prehistoric monster . . . It is remarkable that lieutenant Steup's tank made hits on a T-34 once at about 20 meters and four times at 50 meters, with tank Grenade type 40 (caliber 5 cm), without any noticeable effect.[4]

The Germans' awareness of the inferiority of their tanks precipitated a serious problem at the tactical command levels. Guderian wrote of the impact of the problem on the commander of the Fifth Armored Brigade: "For the first time during this exacting campaign Colonel Eberbach gave the impression of being exhausted, and the exhaustion that was now noticeable was less physical than spiritual. It was indeed startling to see how deeply our best officers had been affected by the latest battles."[5]

The Soviet tank surpassed the German tanks in firepower, armor plating, and trafficability. Its drawback lay in narrow angles of vision for the commander; a turret manned by only two men, one of whom was the tank commander who also served as shell loader; and shoddy C2 radio equipment. German tanks could fire two to four shells for every one shell of the T-34. The Germans were fortunate that a relatively small number of these Soviet tanks (approximately 1000) were deployed in June–July 1941. The Soviets spread them out along the entire front in groups of two or three, thus enabling the Germans to deal with them in piecemeal fashion. Only in October did the Soviets realize the full extent of their threat to the Germans' capability and the substantial impact of the T-34s on the German defeat in the Battle of Moscow. By then the Soviet tanks were deployed in groups of twelve to twenty and could influence the dwindling leads of the Panzer divisions between October and December.[6]

On October 7 at the approach to Moscow, the Fourth Soviet Armored

Brigade (numbering twenty-two T-34s, seven KV-1s, and thirty-one lighter tanks) clashed with the Fourth Panzer Division (attached to Guderian's Second Armor Group that was part of Fedor von Bock's Central Army Group). On that day alone, the Germans lost forty-three tanks and in the entire battle, 133 tanks, forty-nine artillery pieces, and an infantry force the size of a regiment.[7] In effect, the Panzer division was wiped out. Guderian recalled this battle in his memoirs:

> Numerous Russian T-34s went into action and inflicted heavy losses on the German tanks. Up to this time we had enjoyed tank superiority, but from now on the situation was reversed. The prospect of rapid, decisive victories was fading in consequence. I made a report on this situation, which for us was a new one, and sent it to Army Group; in this report I described in plain terms the marked superiority of the T-34 to our Panzer IV and drew the relevant conclusions as they must affect our future tank production.[8]

In addition to the technological superiority of the T-34, the success of the Soviet armored brigade can also be attributed to improved training and combat leadership.[9]

At the tactical level, the tank caused a serious problem for German armor crews; however, despite its proven technical superiority, its performance in the field failed to achieve a decision in the battles and campaigns of Operation Barbarossa. The tank made a strong impression on the German military commanders. Some generals, like Mellenthin[10] and Ewald von Kleist,[11] claimed in 1941 that the Germans lacked a fighting vehicle comparable to the T-34. This tank, which would later be produced in unprecedented numbers (close to 40,000 tanks),[12] was one of the biggest surprises that the Germans encountered in the summer of 1941, and it remained a technical-tactical problem until the end of the war. The German response was the Tiger tank, which could effectively challenge the Soviet T-34 technically, but whose production was relatively limited—approximately 1350 vehicles.[13] Despite the lengthy preparations that went into Operation Barbarossa and the German officers' considerable knowledge of Soviet armor, they were caught by surprise when they encountered the T-34.

. . .

How can this be explained? The Germans apparently thought that their new Panzer IIIs and IVs were technologically superior to the Soviet tanks and capable of compensating for their quantitative inferiority.[14] German officers and researchers claim that the Russian tank must have been known to the German military. Some argue that the tank had even been offered to the Germans for sale before the war; others contend that the Germans knew about the tank by the name of a different model, but since they lacked its technical data or production volume, this information was meaningless. The reports of German intelligence, based on interrogations of captured Polish intelligence officers, contains technical details of the T-34 under the name of another model.[15] David Kahn asserts that although hundreds of T-34s participated in border battles between the Soviet Union and Japan in 1938, the Germans did not have an inkling of the tank.[16]

The Germans' failure to see the threat facing them may be attributed to their sense of superiority and invincibility. In the summer of 1941, the Wehrmacht was at the height of its power after demonstrating its armor strength in the conquest of Poland and France. The combination of military and racial superiority toward the Slavs was undoubtedly a contributing factor in their underestimation of the enemy abilities.

Betts holds that the Germans deluded themselves. German intelligence had been under the influence of Nazi ideology and told Hitler that the Red Army lacked leadership. This also may be attributed to the fact that the Soviet military leadership had been decimated in the purges. However, the Germans miscalculated Russian strength in other areas as well: the Russian soldier's fighting ability, the number of divisions in the Red Army, the number of tanks (two and a half times less than the real number), and for that matter, the number of weapons the Russians had produced, including T-34 and KV tanks.[17]

A nexus can be made between the Germans' conception of their superiority over the Soviet Union and the relative advantage concept (see Ch. 1). The spectacular victories in Poland and France convinced the Germans that they had the answer to the Soviet quantitative superiority in the form of relative superiority—in the combined use and coordination of forces, command, and others areas. Because of this and the application of the Clausewitzian concept regarding the dispensability of military intelligence,[18] the Germans were caught off guard by the capabilities of the T-34 although it was manufactured in mass quantities, and its parameters had not been masked by Soviet deception. Surprise of this type may be classified as "self-surprise."

This surprise, however, was felt only at the tactical level, partially because of the Germans' recoverability, and partly because of the exploitation of weaknesses in the Soviet armor doctrine and the low tactical proficiency of Russian commanders.

RECOVERY FROM SURPRISE

German recovery was based on two key strata in the flexibility concept. The first has already been presented in the theoretical section: conceptual and doctrinal flexibility that enables an army to fight on the defensive despite the predominance of an offensive doctrine and that accepts thinking that diverges from official doctrine. The second stratum is decentralized C2 that demands the highest level of cognitive flexibility from the commander. Since these subjects have already been dealt with, the remaining two strata are now emphasized in the context of the Wehrmacht: organizational flexibility and lesson learning and dissemination.

Cognitive and Command Flexibility

The Germans immediately improvised a number of tactical solutions for dealing with the surprise, such as armor-versus-armor techniques, anti-aircraft guns and artillery employed against armor, infantry and engineering equipment used by infantry units, and combined arms.

The German tank crews devised fighting techniques that seem patently absurd but also testify to absolute freedom of thinking. A report from the Twenty-Fifth Armored Brigade (Second Armored Army) dated July 8, 1941, states:

> The German panzer crews had a small chance of success if they could outmanoeuvre the Russian tanks and attack them diagonally from the rear: "Hits against the rear drive-wheel are often successful, along with chance hits on the turret rim." It had come to this. German panzer crews were recommended to climb out of their vehicles, creep up to the enemy tanks on foot and blow them up![19]

Colonel (Res.) Helmut Ritgen of the Sixth Panzer Division recalls that the Germans' first encounter with heavy Soviet tanks at dawn on June 24, 1941 came as a total surprise. Nevertheless, by 12:30 p.m. that day, in another battle, the Germans managed to hit the Russian tanks by concentrating their tank fire on a single tank and aiming at the firing loopholes on its body. Still on the same day, the Germans discovered that if they maneuvered behind the Soviet tank, they could knock it out of action.[20]

This is an example of a response in the quickest possible time (between minutes and hours). Written expression of this response appeared later in updated combat instructions circulated by the German armor commander in the spring of 1941:

> Since the 50-mm gun can penetrate the T-34s sides only at short ranges, the following techniques have proven themselves most correct for fighting: draw the enemy's attention away from the front and pin him down with Panzer III fire. Choose a position that conceals [your] tank's hull or zigzag quickly so it will be difficult for the enemy to hit you; while taking maximum advantage of concealment in the field, two other Panzer III tanks will try to outflank the T-34 on the right or left in order to reach its flank and rear where it can be hit at close range.[21]

The Germans used artillery and anti-aircraft guns in flat-trajectory fire against the T-34. Until long-barreled, 75-mm guns were mounted on the Panzer IVs in March 1942 (eight to nine months after the first encounter with the heavy Russian tanks and six months after it became absolutely clear that this was the problem), artillery and anti-aircraft canon were the only ground weapons capable of penetrating the enemy's armor. On July 26, 1941, one of the German infantry divisions employed its entire anti-aircraft regiment as defense against Soviet armor north of Lvov.[22] A number of descriptions appear in battle diaries:

> Russian heavy tanks are difficult to fight because they mostly open fire at a distance of over 800 meters, so that only field guns of 8.8-cm Flak (anti-aircraft) and upward are able to engage them.
>
> During this period of technical inferiority, the German panzers sometimes took over the role of back-up weapons by trying to "draw the enemy (Russian tank) along behind them by rolling back to one side, so that he comes into the line of fire of the 8.8-cm AA guns and the 10-cm cannon." (4. Pz. Div. Report: 12 March 1942)[23]

R. H. S. Stolfy notes that the Germans employed field artillery (105-mm howitzers) against the T-34.[24] They also used captured 76-mm guns.[25]

The infantry also was employed non-conventionally against the surprise threat. An infantry officer of the Eleventh Armored Division who participated in the border battles (Kiev Theater) states that in clashes with the T-34 and the KV-I, the German infantry used engineering equipment:

> We had close combat and used material from the engineers, mostly mines, and all soldiers were trained in the anti-tank use of these mines. Also, we had in all

rifle companies special troops of two or three soldiers who worked together. One man handled security and a second operated on the blind side of the tank with a mine and tried either to place the mine on the tank's rear hatch or use a hand grenade bundle (held with wire) thrown over the tank gun barrel. A third method involved using a shape charge, which was magnetic, emplaced against the tank. But the Russians then countered this by placing concrete on the armor plate so the mines would not stick.[26]

Forces on the front improvised hand grenades and anti-tank mines. During the static fighting, engineers planted a great number of wooden boxes with anti-tank mines inside that were pressure detonated.[27]

A wide variety of combat techniques that were used against heavy Russian tanks appear in the answers to a German High Command questionnaire that was circulated in March 1942 dealing with lessons learned on the eastern front regarding command, instruction, and organization:

> It is difficult to fight against Russian heavy tanks because they usually open fire at a distance of 800 meters or more ... Heavy artillery and dive-bombers must try to destroy the Russian tanks; alternatively, our tanks must attack and then, by moving backwards and to one side, try to draw the enemy along behind them so that they get within the range of the 8.8-cm AA and 10-cm cannon. If this method is not successful, then there is only one course of action left in order to get at the enemy—assault units carrying 3-kilogram charges.[28]

The Germans' ability to recover quickly from surprise stemmed, as stated, from the capacity to improvise and seize the initiative—and is an expression of cognitive flexibility and the freedom of action given to commanders to change plans and techniques when necessary in accordance with a flexible C2 doctrine.

Weapons Diversity

Another factor in rapid recoverability, based on the stratum of organizational flexibility, was the diversity of means at the German commander's disposal for devising solutions to battlefield exigencies. The German doctrine emphasized the combined arms. Robert Citino studied the German army's combat doctrine and training during the interwar period and came to the conclusion that "[t]he tank was no panacea, no wonder weapon ... It was simply another piece—an important piece, to be sure—of the German army's combined arms puzzle."[29] The importance of combined arms can be seen at the doctrinal level, where it appears in the *Truppenführung*—the German army's manual of 1933–1934[30]—and at the company level in field manuals dealing with force deployment.[31] Weapons diversity was based on the concept of combined arms. This can be seen in the Panzer division's structure in 1935 where armor, infantry (mechanized and motorcycled), motorized artillery and anti-tank weapons, engineers, and reconnaissance were all integrated into a unified fighting force.[32] "Flexibility was its hallmark. It could assault and penetrate an enemy position, break through into the clear . . . withstand the strongest enemy counterattacks, and then regroup and do it all over again."[33] Each element of the division possessed a large variety of weapons.

Infantry battalions and regiments were equipped with machine guns and various types and sizes of mortars and anti-tank cannons. The mechanized artillery regiment included a heavy battalion of 150-mm guns and two light battalions of 105s pulled by halftracks. Engineers had flamethrowers, mines, explosives of all sizes and shapes, smoke equipment, mine detectors, and barbed wire coils in addition to inflatable rafts, pontoons, and two armored vehicle-launched bridges (AVLB) in each division. The infantry also trained in basic engineering skills. Every company had inflatable boats, and the battalion had pontoons and beams for constructing bridges capable of bearing five-ton loads.[34]

A reconnaissance battalion was made up of armored cars and trucks, light artillery (including an anti-tank unit and light-infantry mortars), and a machine gun company; all went to make up what "at first sight, appeared to be somewhat of a menagerie of weapons."[35] All of the elements in the Panzer division were motorized so as not to hold up armor's advance.

The tanks were built for special purposes: armor-versus-armor and armor-versus-infantry. The Panzer III had a high-speed gun capable of firing armor piercing shells. The Panzer IV had a short, thick, low-velocity gun that fired both explosive shells against infantry and armor-piercing shells of lower penetration than the Panzer IIIs.

In addition to its diverse composition, the division was given close air support—"flying artillery"—in the form of Stuka dive bombers. The *Truppenführung* stresses the integration of the air force and ground forces.[36] Another key element was the anti-aircraft units that used their dual-purpose, 88-mm guns. Weapons diversity was found not only in the combat units but also in the headquarters. The headquarters structure of the Panzer regiment in 1939 included a communications platoon on motorcycles, light vehicles and armored-combat vehicles, a tank platoon, and a messenger platoon on motorcycles.[37] The German combat doctrine emphasized cooperation and weapons diversity so that the division could organize its combat teams according to specific contingencies.

Weapons Changeability

Another technological solution was changeability based on the prewar, modular planning of the tanks and lesson learning, combined with close cooperation with the military industry.

A short-term answer was to upgrade the main tank models—the Panzer III and IV. "The Germans were designing their tanks for adaptation to future demands, even though they could not be sure what those demands might be."[38]

The Panzer III was planned with three main sections: a hull that contained the motor, transmission, and steering systems and had the front and rear upper parts connected to it. The upper parts could be changed without having to design an entirely new tank. The 37-mm gun was replaced during the war by a short 50-mm gun and later, by a longer gun of the same caliber. Afterward, the tank was redesigned with a short 75-mm gun and front-armor plating of 30 to 50 mm without the loss of speed.[39] The Germans' rapid response was noted as early as 1942 with the debut of the Panzer III with a long 50-mm cannon and the Panzer IV with a long 75-mm cannon.[40]

The Panzer III and Panzer IV were gradually replaced by the Panther (Panzer V) and Tiger (Panzer VI). The Panzer III hull was later used for command tanks, self-propelled howitzers, infantry-support tanks, and flamethrowers.[41]

The Panzer IV's cannon was replaced, as stated, by a longer gun. The experimental change of the short-barreled 75-mm cannon with a long-barreled 50-mm one took place on August 1, 1941. The reports from the front resulted in a new operational demand for a long-barreled 75-mm cannon. Such a report was received by the Krupp Works and Rheinmetal Company on November 18. On April 15, 1942, Hitler approved of the new gun, and it entered operational service.[42]

Military-Industry Coordination

The Germans' long-range technological solution was to develop new tank models very quickly. Guderian acknowledges that he sent a special report to the army group headquarters in October describing the T-34 and adding his conclusions regarding it. He ordered a committee to be dispatched that included tank designers and representatives from Wehrmacht ordnance, the Reich Ministry of Ordinance, and the companies that built the tanks. The committee reached the front on November 20.[43] The frontline officers believed that the fastest and most effective response to the T-34 would be to duplicate it; however, it soon became apparent that German military industry was unable to produce a diesel-run, aluminum motor like the T-34's on such short notice. The officers wanted the Panzer IV to be immediately fitted with a more effective gun; a long-term request was for a new tank with improved parameters.[44]

The military industry's response to the T-34 was to commence development of new tank models. Four companies were commissioned. In April 1942 the Henschel Company presented a prototype of the Tiger I weighing fifty-six tons. This model was selected, and the first Tigers arrived at the eastern front in August 1942. In March 1942 the M.A.N. Company produced a model of the Panther tank weighing forty-five tons, mounted with a special L/70, long-barreled 75-mm cannon. The first Panthers rolled off the assembly line in November 1942.[45]

A Tiger company in the Grossdeutschland Regiment that was engaged in combat between March 7 and 19, 1943, reported that the Tiger was practically impenetrable to Soviet tank shells and that its 88-mm gun proved highly effective against the T-34.[46] A description, dated July 7, 1943, of the Panther's performance against Soviet KV-1s and T-34s, makes a similar claim: "The main gun is exceptional... The Panther tank is clearly superior in tank-versus-tank combat."[47]

The Culture of Learning From Mistakes and Self-Improvement

Gudmundsson discusses the innovation of stormtroop tactics in WW I where "the habit of questioning, however, gave the German Army a definite advantage when it came to adapting to unforeseen conditions."[48] Millett and colleagues studied the Germans' ability to learn from their mistakes in the war against Poland. They found that during the six months—between the war in Poland and the invasion of France—the ground forces high command developed strict coordination between the training program and doctrinal concepts.[49] Murray asserts that the Germans' ability to learn from the war in Poland is an example of professionalism. He claims that its "willingness to be self critical was one of the major factors that enabled the German Army to perform at such a high level throughout World War II."[50] According to Murray, this approach characterized the German army and the way the Germans studied WW I in the interwar period and critically examined their training exercises. A key to understanding the German approach is the way it dealt with flaws and problems with new equipment and procedures. The Germans regarded error making as a natural part of the learning process, not as events calling for punishment or the suspension of an officer's career advancement.[51]

The German army invested a great amount of effort in training, not only in the interwar period, but also in the course of the fighting. The training was programmed to inculcate lessons from the battlefields of Poland, France, and the Soviet Union. Beginning in November 1939, divisional commanders were required to submit monthly assessment reports of the operational effectiveness of their unit.[52]

According to a company commander from the Seventh Panzer Division who fought in Operation Barbarossa, the flexible C2 and lessons that came to expression in the division's training in France in February 1941 (prior to the invasion of Russia) emphasized:

1. Training and advanced training of junior military leaders, especially their ability to act independently and responsibly.
2. Instruction in the cooperation of all weapons in combat situations as taught by the experience of the Western Campaign.
3. Close cooperation among tanks, armored infantry, motorcycle infantry, engineers, and artillery at company and platoon level.[53]

In addition to the culture of debriefing and learning from mistakes, the German army contained institutional mechanisms for force planning even in the heat of battle. From the beginning of the war, the High Command had infantry, armor, artillery, and engineer inspectors ("corps chiefs") whose task was to assess combat experience and its influence on the warfighting doctrine, training, organization, weapons, and equipment. In September 1944, an inspector of anti-tank weapons was added. The Luftwaffe appointed similar inspectors of fighter aircraft, bombers, and ground-air defenses. The various inspector staffs issued tactical orders on a regular basis.[54]

Thus, German recovery from the surprise of the T-34 may be attributed to a key factor in the flexibility concept: the culture of learning from errors and the mechanism for disseminating the lessons.

In conclusion, German recoverability from the T-34 surprise was due to an application of all the strata in the flexibility concept, and its roots go back to the fundamentals of German military culture: decentralization, self-criticism, and auto didactics. Flexible thinking resulted in the development of a flexible C2 doctrine and the expectation of cognitive flexibility from commanders, both of which enabled the junior-commander level to improvise new methods of operations. Unit and weapons diversity was the platform that the German commanders used for improvisation. The culture of learning from mistakes and the strong desire to improve performance and maintain close relations between the combat units and the weapons industry provided technological answers in the form of improved tanks. Recoverability minimized the debilitating effect of surprise, halting it at the tactical level. The response to surprise was rapid at the field command and military industry levels.

8 THE ISRAELI RECOVERY FROM THE EGYPTIAN SAGGER MISSILE SURPRISE

THE THIRD HISTORICAL TEST CASE is the IDF ground forces' recovery from Egyptian anti-tank warfare in the Yom Kippur War. This type of combat also took place on the Golan Heights[1] but had less of an influence there than on the Suez Canal front.

The following analysis explains how the long-range, anti-tank weapons—Sagger missiles—were dealt with. I argue that they posed a major tactical surprise on the Suez Canal that effected the operational level. The main surprise was not the missile's technical specifications but its method of employment—its vast numbers in Egypt's defense layout. It should be noted that although the physical results of the surprise are still not clear, the missiles seem to have caused only limited damage. Since surprise has a cognitive impact, its effect on Israeli tank crews should not be underestimated, especially when they realized that their prewar armor training was no longer relevant.

Be that as it may, the IDF ground forces managed to recover from the surprise with a fair amount of success. An analysis of the recovery is somewhat complex. It is my contention that although the IDF doctrine was one-dimensional, with an overemphasis on the offensive, and for this reason had become more dogmatic than in the past, this trend had still not fully infiltrated the minds of commanders so that the IDF's basic fighting concept—which relied on innovation and improvisation—remained intact.

Weapons diversity declined as a result of the prewar organization of forces in the force planning process and worsened because of priorities in the mobilization of the reserves. This trend seriously impaired the IDF's ability to recover from the Saggers in the initial stages of the fighting. An important factor in recoverability was the restoration of unit diversity in the midst of hostilities.

The intensity and brevity of the fighting makes discernment of the regular army's awareness of the importance of conveying lessons on the missiles—and its ability to convey them in real time—a difficult task. For all practical purposes, the lessons from the initial encounters with the Saggers were not transferred to the reservist units entering the battle zone.

THE SOURCES AND IMPACT OF THE SURPRISE

The IDF commenced combat operations the afternoon of October 6, 1973 at various levels of surprise. At the strategic level, the entire country had been caught off guard. At the tactical level, Israeli forces were more surprised by the enemy's fighting method and the need to fight defensively than by the unexpected outbreak of war per se.

The Agranat Commission Report reflects the Israeli officers' tactical surprise on encountering Sagger missiles. The following are eyewitness accounts of the IDF's combat readiness based on its warfare doctrine and field exercises.

Testimony No. 8 (Colonel Gabi Amir—Commander, 460th Brigade):

> I think that most of the damaged tanks were hit by missiles and RPGs [rocket-propelled grenades]. I saw hurling balls of fire and didn't understand how missiles could be fired at nighttime. Later I found out they were missiles. During our training we never took into account the possibility of the Egyptian infantry making massive use of missiles. [The Sagger was not equipped with night vision, so the colonel apparently had the wrong impression. Still, this does not detract from the intensity of the surprise].

Testimony No. 1 (Lieut. Col. Haim Adini—Commander, Nineteenth Battalion, 460th Brigade) on the first encounter with Egypt infantry on October 8, approximately three kilometers from the Suez Canal:

> About two kilometers from the "Lexicon-Haviva" Junction we got hit with something I'd never experienced. We started catching hell from missile fire from all directions . . . that later turned out to be Saggers. I gave the order "Shmels!" [anti-tank missiles that the Egyptians used in the Six-Day War] and then I saw my men wounded . . . but still didn't see the Egyptians, only hits by missiles . . . I told them to watch out for the Shmels, knowing they could be evaded as I remembered from the last war . . . I heard nothing and knew nothing about the Saggers.[2]

Based on the report "Israeli Soldiers' Responses in the Yom Kippur War" published in April 1974, the Agranat Commission presented the way individual soldiers dealt with surprise. The chapter "Lack of Readiness for Dealing With the Enemy's Weapons Tactics," states:

> This area of contention refers mainly to armor because of armor's encounters with infantry's new tactics and anti-tank missiles. Thus, the main charge is against the "IDF's unpreparedness for fighting infantry [with anti-tank weapons], a contingency—close-range infantry combat—that we hadn't been trained for and that wasn't taught in field exercises or at armor school."[3]

The following appears in the section "Know the Enemy and Military Intelligence": "The armored corps soldiers stated that they didn't know the enemy, knew nothing about its anti-tank missiles and other weapons. Its anti-tank missiles came as a surprise to them."[4] It will be recalled that the damage to Israeli tanks by the Saggers had not been critical, but the cognitive impact was enormous and seriously eroded the tankers' abilities to fight the way they had been trained to.[5] In this case, surprise occurred at all levels in the chain of command—from brigade commander to buck private.

Egyptian success is also testified to in the Egyptians' own accounts. Mohamed Abdel-Ghani Gamasi, the head of the general staff, recalled the effect of the anti-tank missiles on the first day of fighting. According to the Arab military commentator Hitam Alayubi, the lesson of the war was that the various types of anti-tank units surprised the enemy, causing it unexpected losses.[6] The Egyptians, too, seem to have been surprised by the effectiveness of the missiles on the battlefield.

The Saggers' debut can be classified with a fair degree of accuracy as "self-surprise." The IDF knew about the anti-tank missile's potential for two decades before the war. Already in 1956 it procured French-made, SS-10 anti-tank missiles. The SS-10 was similar to the Sagger but with a range of only 2000 meters (compared to the Sagger's 3000). The IDF trained with the SS-10 and employed it in the 1956 Sinai Campaign, but since its hit rate was less than ten percent, it was withdrawn from service.[7] Shimon Iftach claims that between the Sinai Campaign and the Six-Day War "there was not a single western missile that the IDF did not examine or take an interest in."[8]

Not many IDF tanks were hit by Soviet anti-tank Shmel missiles in the Six-Day War. The Israeli tanks' performance overshadowed the limited damage caused by the Egyptian missiles and eliminated the IDF's need for SS-11 anti-tank missiles (codenamed "Tagar") that had been operational at the time.[9] Again, discussion on measures against anti-tank missiles was postponed even though two Pattons had been knocked out by Shmel missiles fired from the Egyptian side of the canal in July 1967.[10] The lessons from the Six-Day War touch lightly on anti-tank weapons, though they testify to the IDF's perception of their importance as an infantry weapon against an armored attack.

The development of the IDF's understanding of the missile threat shows that there had been many indications and that both the intelligence branch and armor corps were aware of the technical implications. Military intelligence (MI) circulated the first memo on the Sagger's technical specs in July 1970, which noted that the missile was believed to have become operational in the Egyptian army since early that year. This memo was written after a Sagger had been fired from across the canal but failed to explode. The weapon itself reached MI and its technical specs were analyzed.[11] Basing its findings on intelligence reports (but apparently unaware of the abovementioned memo), the Agranat Commission determined that already in early 1972, information was being circulated regarding Egypt's procurement of Saggers and other anti-tank weapons.

An MI report that made the rounds in May 1973 stated that the Egyptian plan for crossing the canal foresaw the securing of captured bridgeheads in the first stage of the operation by means of hundreds of Sagger missiles mounted on amphibian vehicles. Israeli MI reports noted the large number of various anti-tank weapons in infantry companies and battalions. A batch of intelligence reports dated August 1972 mentions the enemy's emphasis on enhancing its anti-tank defenses with anti-tank missiles and airborne anti-tank reserves: "As for these anti-tank reserves, the main mission in the crossing stage of an offensive will probably be to secure the fording area against enemy [IDF] tank reinforcements."

An Israeli intelligence officer, Dani Asher, claims that according to the MI director in the Yom Kippur War Eli Zeira, nineteen briefs on the Sagger missile had been circulated before the war.[12] Lorber states that since the Sagger reports were highly classified material, it was unlikely that they reached all of the relevant quarters.[13] Senior IDF officers who received the reports attributed little importance to them. Interestingly, a description of the missile appeared in the Israeli military journal *Ma'arachot* in July 1970.[14]

On at least one occasion—in late 1969 (or early 1970)—a Sagger missile was fired at IDF forces at the Suez Canal. Little importance seems to have been attributed to this incident, given the intensity of events in the War of Attrition. Since some of the intelligence reports were highly classified, field commanders may not have been privy to them, but material from the armor branch on dealing with missiles did reach officers on the front.

In the winter of 1972–1973, a number of incidents occurred on the Golan Heights in which Sagger missiles were employed against IDF forces. On November 21, 1972, on a battle day codenamed "Capital B," an Israeli tank was hit by an anti-tank missile. The tank units knew about the Syrian anti-tank guns; however, they didn't know about the anti-tank missiles. Lesson learning was carried out, and steps were devised for dealing with the missiles at the tank crew level.

On battle day "Capital D" (January 8, 1973), the Syrians fired between forty and fifty Saggers at Israeli tanks and managed to destroy one of them.[15] Haim Herzog insists that the lessons were quickly learned: "The ramps over which the tanks operated were expanded in order to defend the tanks' flanks, and mortars were allocated to armor units for dealing with infantry equipped with anti-tank weapons." In another missile incident, no Israeli tanks were hit.[16] The final report of the armor corps's doctrine department included information on the missile and instructions for dealing with it at the level of the tank crew.[17]

The tank units of the Northern Command amassed invaluable information on how to cope with the missiles, and this data was relayed to the armor corps's doctrine department. Although the material was circulated throughout the armor corps, it does not seem to have been learned in the Southern Command, let alone internalized in the regular army and reserve tank units. "Although the Saggers were known, the knowledge was insufficiently assimilated. There was even an instruction manual on the Sagger. It was unforgivable shortsightedness not to have internalized the material."[18]

In addition to the Northern Command's operational experience, articles were published warning the armor corps of overreliance on the tank. Concern over the adoption of "armor shock" and "tank superiority" doctrines led Colonel (Res.) Yehuda Wallach to admonish armor officers of the developing threat to their concept. In an article entitled "Eulogy to the Tank" that appeared in *Ma'arachot Shiryon (Armor Campaigns)* in July 1971, Wallach advised armor commanders to carefully read Douglas Orgill's *The Tank—Studies in the Development and Use of a Weapon*, in which the author made a cogent case against the way the IDF had employed its armor in 1967. Wallach called on armor officers to reevaluate their thinking in light of Orgill's conclusions.[19]

In another article, "Is the Tank Dead? Another Nail in Armor's Coffin,"[20] Wallach reviewed a study by the German military thinker F. O. Mikshe entitled "The End of Mobility in Border Defense. Is the Armor Corps Dead?" that was published in Germany in 1971. The article traced the development of anti-tank mining as an inhibiting factor in armor maneuverability. Wallach's conclusion:

> The Israeli tank officer may consider the mining problem as irrelevant and far from his conceptual world . . . If indeed the mining prognosis is correct then the time has come for armor commanders to rethink very seriously the struggle between armor and anti-tank weapons. Ignoring the problem will testify to the "ostrich policy" that Mikshe accuses armor commanders of having.[21]

After the Yom Kippur War, Wallach published an article in *Ma'arachot* in which he admitted that he failed to stimulate a meaningful discourse on the relevancy of Israel's armor doctrine. The responses to his articles did not refer to the impact of the threat; instead, they regurgitated slogans extolling the tank's superiority ("I favor the tank"; "the tank—king of the battlefield"; "the pounding dead").[22] As is known, prior to the 1973 war, there was no criticism on the theory of armor maneuvering based on the lessons from the Six-Day War.

SOURCES OF THE SURPRISE

When analyzing the reasons for the Sagger surprise, it is hard to isolate it from the overall surprise that resulted from the belief that war would not break out. Israel's sense of superiority seems to have been a key factor in the entire war-planning layout, including its attitude toward anti-tank missiles. The reasons for the warning omission are too numerous to be listed here (see Bar-Josef),[23] but a number of factors in the intelligence-doctrinal failure regarding Sagger missiles are noted.

> The shortcoming was not due to the lack of information: It is almost certain that the Egyptians did not intend to trick Israel by using anti-tank weapons. Lorber classifies the Sagger as a surprise caused by misinterpreting the changes on the battlefield.[24] In my opinion the IDF had the right informational background and enough indications to pick up the implications of the developing threat and especially the rising strength of infantry.
>
> The mindset was fixed on a doctrinal concept built on incorrect lesson learning: the IDF believed that armor alone would decide the next war. It ignored the effect of the enemy's use of anti-tank weapons.[25] Even

if the IDF believed, as it justifiably did, that anti-tank missiles were not particularly lethal, then the lessons from the Six-Day War regarding the composition of ground forces and the army's ability to be based exclusively on armor precipitated a decline in the status of artillery, mortars, and mechanized infantry—combat branches that could have supplied a response to the Sagger in the initial encounters and at later stages in the fighting (see below).

"Losing" knowledge due to negligence? The absence of training in defensive fighting may have been the reason why the need for combined arms did not play a more important role in planning. Attack exercises based on the Six-Day War model did not include combined arms. An exercise in defensive combat and the derived forms of combat (delay and ambush) and exercises against infantry equipped with long-range anti-tank weapons might have accentuated the problem armor would face if it found itself without artillery and mechanized infantry support. On the other hand, even if there had been such training, the IDF might have reached the conclusion that the tanks could defend themselves without "outside" assistance. The problem does not seem to have been negligence in anti-tank thinking, whose importance in the Six-Day War had been minimal, as much as it was negligence in combined-arms thinking as a response to surprise.

The gap in small-tactics proficiency stemmed from the nature of the exercises. Armor training seems to have inculcated the ability to fight in two main techniques: firing on enemy tanks from a long range and assaulting infantry troops at close range. This meant that tank crews were primarily trained to "hunt" enemy tanks rather than attack and crush infantry targets and, much less, if at all, to identify and neutralize long-range anti-tank missile squads. Armor's ineffectiveness against the missiles may have largely been a result of this training lacuna.[26]

The Saggers were essentially a doctrinal surprise. Although warning signs had been flashing from many directions, few commanders had enough foresight to plan for the approaching threat. This is a case of force planning performed with copious intelligence data, on the one hand, and interpretation of it on the basis of faulty concepts that included a one-dimensional doctrine, flawed lesson learning, and sense of superiority, on the other hand. Despite the wealth of MI, the presence of these factors caused the training program of the ground forces to almost completely ignore Egyptian force planning and its carefully prepared response to Israel's previous armor success.

Which level of war did the surprise influence? Ben-Israel claims that only a small percentage of Saggers actually struck Israeli armored vehicles.[27] Lanir, too, notes that a postwar check on the number of Sagger hits shows "that the accepted picture of the missiles' effectiveness in damaging Israeli tanks during the war was greatly exaggerated. The Sagger, it turned out, was the least of the factors that caused IDF tank losses.[28] This does not lessen the surprise effect of the missile since effect is measured not only by the physical erosion of the forces but mostly by the degree that the warfighting doctrine is thrown into disarray. Convulsion occurred mainly at the tactical level but was also felt at the operational level because of its cumulative effect on the ability to impede the Egyptian forces that had crossed the canal. It is difficult, as stated, to single out the impact of the Sagger from other battlefield factors, such as shoulder-held missiles, the IDF's unfamiliarity with a defensive doctrine, and other contingencies that armored commanders had not prepared for; however, given the evidence just presented, the ability of the Sagger missile to disorient the warfighting doctrine of armor was considerable.

RECOVERY FROM SURPRISE

The ground forces' recovery from the surprise was demanding and complex. An analysis of the first two strata of the flexibility concept shows how recoverability emerged out of the difficult initial circumstances of the war, which were partially the results of prewar force planning that had constricted flexibility.

Cognitive Flexibility in Spite of the One-Dimensional Doctrine

The "cult of the offensive," as described in detail in the theoretical chapter on the conceptual-doctrinal stratum, reflects the extreme one-dimensionality of IDF doctrine. This one-dimensionality also seems to have had an impact on the IDF's ability to meet the challenge of the enemy's defense layout where the Saggers were deployed.

Before the war, IDF exercises, especially in the armor corps, disregarded the basic-defensive form of battle, where anti-tank weapons had proven themselves a key factor in WW II. The section in the Agranat Report called "Lessons From Foreign Armies, the Six-Day War, War of Attrition, and Other Incidents," states this clearly.[29] The source of the unawareness of the effect of anti-tank missiles when employed defensively may have been due to overlooking the defensive form of battle per se and the lack of defensive training.[30]

A fixation on the offensive doctrine, which led to negligence in defense measures (such as anti-tank guns and missiles), together with armor's prime position seem to have caused the effective employment of anti-tank defenses to be forgotten. Anti-tank defenses had served the IDF well on previous occasions, and scores of them had been built. These were the tank ramps whose anti-tank weapon on the ramps was the tank itself.

Despite the conceptual one-dimensionality that made inroads into the IDF doctrine, Israeli commanders managed to improvise and develop combat techniques for dealing with the Sagger missiles. Eyewitness accounts from officers of all grades point to the wide range of improvisation under various circumstances. General Israel Tal explained to the Agranat Commission the changes in IDF tactics that were introduced in the midst of the fighting: "They just did two things: they learned to gauge the correct range, not to remain in static positions in the killing zone; after [October] eighth they learned not to attack organized layouts with [only] a few tanks when mass and momentum were insufficient."[31]

Eliashiv Shimshi, the commander of the Sa'ara (Storm) reservist battalion stated that one lesson about the Saggers that was learned while engaged in combat was to improve observation of missile squads. It will be recalled, an anti-Sagger technique had been developed during the battle days on the Golan Heights, but Shimshi, the head of the reservist armor branch at the Tze'elim training base and commander of a reserve battalion, described the Sagger as a "non-conventional" weapon,[32] which shows that he was apparently unfamiliar with it.

A tank officer who fought in Sinai reported that while engaged in combat, he devised a drill for a rapid descent from a position when Sagger fire in his direction was identified. Regarding the amount of time it took him to develop the drill, he replied: "From three to six days. When we became fully proficient against the Saggers, the only thing the Egyptians could do was ambush us."[33]

Another officer recalled:

> We learned to beat this problem. More and more we encountered [enemy] missile squads that failed to hit us, while we increasingly got to them ... After such an encounter with the missiles, our men developed the proper kind of drill. Every crew, platoon, and company learned about the missiles, and once you got to know them, they lost much of their danger.[34]

Examples of improvisation unrelated to Saggers include technical responses to close-range infantry fighting by readjusting the tank commander's machine gun mount,[35] or employing captured Egyptian armored personnel carriers (APCs) for transporting Israeli troops.[36]

Despite the doctrine's one-dimensionality and its derivative—dogmatism—the Israeli officer's ability to improvise remained intact. Cordesman and Wagner found that as the fighting progressed, the high level of training and independence of small units was of far greater importance than technology or tank performance.[37] Millet and colleagues claim that the IDF's warfare doctrine tried to take advantage of its superiority in qualitative manpower (relative to the Arab forces) in situations where the Israelis' ability to improvise, make fast decisions, and deploy small, independent units came to expression.[38] Laskov, too, noted the IDF's advantage when it came to improvising.[39] The IDF was famed for its improvising capability. Apparently this trait was still strong despite other doctrinal emphases that had been introduced since the Six-Day War. Levita asserts that the doctrinal standstill in the IDF prior to the Yom Kippur War was also rooted "in the inevitable process of institutionalization that had made the IDF more conservative, even in its successful performance in the Six-Day War and, to less of a degree, in the Yom Kippur War."[40]

Luttwak and Horowitz elaborate on the loss of officer initiative and creativity over the years, claiming that the concept that esteemed technical professionalism and strict discipline in the armor corps that General Tal instilled prior to the Six-Day War "did not detract from armor officers' command initiative" in the Six-Day War.[41] But six years later, in the Yom Kippur War, some of the middle-level officers—battalion and brigade commanders—failed "to demonstrate tactical inventiveness and adaptability. In a certain sense this was part of the price inherent in mechanization."[42]

The IDF may have been fortunate that institutionalization, the by-product of mechanization and a significant increase in force size, did not have sufficient time to completely erode the mentality required for improvisation (that derives till today from Israel's civilian culture of improvisation).

Force Organization—Flaws in Weapons Diversity and Rectification During the Fighting

The IDF entered the fighting with a serious defect in force organization regarding diversity of units and weapons. This was primarily the result of having derived the wrong lessons from the Six-Day War.

Luttwak claims that the IDF recovered from the Yom Kippur War surprise at a heavy cost after a few days due to its use of artillery and infantry that significantly neutralized the anti-tank menace.[43]

Between the Six-Day War and the Yom Kippur War, the IDF adopted a concept that was molded and advanced by General Tal: independent operation of armor units based on highly accurate tank fire, continuous movement, and tank concentrations wielded as "armor fists."[44] The downside of this was that the IDF gradually neglected its artillery, whose number remained unchanged, while the number of tanks rose dramatically. Mechanized infantry that was supposed to fight alongside the tanks was also neglected. The 81-mm mortars that had been attached to tank companies in the Six-Day War remained idle and were taken off the table of equipment in July 1972:[45]

> Instead of a multibranched and multitasked formation, the division became a tank formation, and, as such, depended on air support and was left vulnerable to the enemy's anti-tank infantry layouts.
>
> The mechanized infantry battalions and companies in the tank battalions had their mortars and anti-tank guns removed with the two-fold explanation that the halftracks lacked protective covering like the M-113s and their speed was less than that of the Sho't and Patton tanks.[46]

Another part of the diversity problem was in military education that was supposed to provide young officers with the necessary skills for proficiency in combined arms. This was based on the warfighting doctrine but since the doctrine emphasized the tank in place of other elements, naturally the military education did not relegate importance to interbranch cooperation. The commander of the Officers Training School acknowledged this in his testimony before the Agranat Commission:

> The officers received good professional education. However, I think it was wrong to have such one-sided specialization and leave combined arms unanswered for. The infantry officer was a professional in infantry matters but didn't know how to speak to a tank platoon commander. The tank officer was outstanding in his field but didn't have a clue how to communicate with an infantry officer . . . In retrospect . . . there was no education for combined arms.[47]

According to Cordesman and Wagner, another reason for the bias toward armor may have been the concept that practically metamorphosed into dogma: the infantry soldier would have an increasingly hard time surviving and functioning on the battlefield. The bias grew very strong because of the experience with tanks in the 1956 and 1967 wars and the emphasis on fire and movement in which Israel's outdated mobile infantry fell far behind battlefield requirements. As a result, the IDF found itself critically lacking in infantry, and the little there was lacked training in combined fighting.[48] Even if infantry, mechanized infantry, paratroopers, and reconnaissance forces worked together—and the I.D.F had approximately two armored battalions for every infantry battalion—the fear of infantry's low survivability on the battlefield precluded combined arms.

The lack of artillery was justified, Cordesman and Wagner claim, because of the use of the air force as "flying artillery." This concept created a major shortage in field artillery. The IDF had 300 artillery pieces in 1973—only a few of them mobile—which prevented their integration into armored mobile fighting.[49] Another problem, unrelated to faulty force planning, was the preference for tanks to artillery in the transfer of reservist units from the Beer Sheva area to the Sinai front in the beginning of the war.[50] The upshot of this was that too little artillery and mechanized infantry entered the battle zone at the outset of the fighting.

Cordesman and Wagner contend that because of the gap in the combined arms ability, Israel ceded a number of encounters with Sagger-equipped Egyptian infantry and was unprepared for combat in built-up areas after the canal crossing.[51]

In the failed October 8 counterattack, General Adan's division had only three artillery batteries. When the 113th Battalion commanded by Asaf Yaguri attacked in the Firdan Bridge area, not a single artillery shell was fired by Israeli guns.[52] "The excessive credit that armor received in the beginning of the fighting had a deleterious effect on the ability to employ artillery—the full consequence of which was felt in the October 8 counterattack."[53]

Later, when artillery and mechanized infantry arrived at the front, weapons diversity was created that enabled the IDF to recover from the anti-tank surprise. Israeli troops were able to outflank the Egyptian infantry positions by making use of heavy tank and artillery covering fire, and by increasingly employing the limited mechanized infantry and paratrooper units in assaults or infiltrating attacks.[54] Asher and Asher note that "during the defensive phase, when artillery units began arriving, their importance finally penetrated the minds of the commanders—as a solution to Sagger teams."[55]

The need for mechanized infantry was also quickly understood, and the fighting units began improvising in order to make maximum use of the little they had. General Adan wrote about this reorganization on October 10:

> After analyzing the combat techniques and tactics involved in cooperation with artillery and mechanized infantry, I decided to collect all of the division's APCs (excluding those at the forward command posts), man them with recon troops and mechanized infantry, and spread them out among the battalions. The next day we'll fight with smaller sized tank battalions, but in each battalion's flank there will be mechanized infantrymen riding in pairs or platoons of APCs.[56]

Since the mechanized infantry had been neglected in the interwar period, it was, as stated, of relatively low quality in manpower, equipment, and training. Often the light infantry also assisted the armor forces in dealing with anti-tank weapons. When the IDF units tried to reach the canal for a planned crossing at the "Chinese Farm" area on October 16, they ran into Sagger missiles and lost many tanks. Tanks and APC attacks failed to displace the Egyptian infantry. Adan realized that the Egyptians were trying to lure his forces into a missile ambush; therefore, he suggested to Southern Command that the infantry be employed in night fighting. Southern Command agreed and the clean-up operation was assigned to Uzi Yairi's paratroop brigade.[57] The force failed for reasons unrelated to the Saggers and had to be extricated by the tanks, but an attempt had been made to restore cooperation among the combat branches as the limitations of the tank became realized. The need for combined arms was learned at a heavy price.

Lesson Learning Mechanism

We have seen, then, that a number of lessons at different levels were learned in the course of the fighting. Shimshi described cases of learning and lesson circulation at the battalion level;[58] Adan discussed this activity at the divisional level.[59] On the other hand, almost no cross-unit transfer of information occurred during the fighting, with the result that reserve units entering the battlefield on the third day of the war were caught by surprise by the missiles.

Given the calamitous battlefield conditions, it is difficult to say whether it could have been expected that information on the presence of anti-tank missiles and their impact on the fighting could have been relayed to the regular army and reservist units that joined the fighting later. In reality almost no lessons were conveyed. According to the Agranat Report, the battle picture at the battalion commanders' level—if there was a picture—was hazy regarding Israeli action and the cumulative experience gained in the first two days of the fighting. Lieut. Col. (Res.) Yaguri testified that he knew nothing about the RPG and remembered that "something" had been mentioned about the Sagger. The commander of the Nineteenth Reservist Battalion Haim Adini recalled that he received no information on the enemy's missiles even though the night before the regular army had incurred losses from them.[60] The report stated:

> The lessons from combat engagements before the war, to the degree that they were learned, were learned by few people, were not circulated, and were not included in military education programs. This phenomenon repeated itself after the war broke out: the lessons from the first clashes with the enemy were not immediately relayed—whether in MI reports or operational lessons from encounters and contact with the enemy.[61]

The key element in the problem seems to have been the slapdash entry of the reserves into the fighting, a condition that did not allow for the orderly transfer of lessons gained from the fighting.

In conclusion, the IDF's ground forces managed to recover from the Sagger surprise after paying a heavy price in tanks (though apparently relatively few) and especially in the disorientation of armor's fighting tactics. The recovery took place despite the flawed force planning in doctrine and organization that had constricted flexibility. The army's superb ability to improvise and restore weapons and unit diversity was the main factor in recovery. Recovery from the Sagger missile surprise was part of the IDF's overall recovery at other levels. The effectiveness of the recovery may be judged as mediocre. Damage from the threat was reduced but not entirely neutralized. The failure to relay information between units prevented a greater degree of effectiveness. The speed of the IDF's response at the battalion level was high, sometimes occurring within hours or less. The return of diversity to force organization took longer; beyond the physical arrival of artillery, mechanized infantry, and infantry, it demanded a basic change in the warfighting doctrine.

9 THE ISRAELI AIR FORCE RECOVERY FROM THE ARAB ANTI-AIRCRAFT MISSILE SURPRISE

THE FOURTH TEST CASE deals with the recovery of the IAF from the Egyptian and Syrian surface-to-air missile (SAM) layout in the Yom Kippur War from which it suffered serious damage. The IAF was aware of the SAM threat to its operations from the War of Attrition. Enormous material and human resources, including Israel's top scientists, were diverted to solving the problem. On October 7, Operation Dugman 5 failed to neutralize the SAM layout on the Golan Heights. This omission contained the element of unpreparedness and surprise owing to the IAF's overreliance on a preplanned solution to the challenge. "The missile wrenched the plane's wing in this war—a fact that calls for meticulous lesson learning."[1]

The IDF's official Yom Kippur War history claims that the massive deployment of anti-aircraft missiles was a tactical surprise.[2] I would argue that the surprise had an impact at the operational-strategic level because it undermined one of the cornerstones of Israel's war plan. Nevertheless the air force managed to recover with a certain degree of success mainly because of the cognitive flexibility of junior-grade commanders and their determination and ability to learn from mistakes and convey these lessons at a relatively fast rate. In this case, a limited part of the flexibility strata was used.

SURPRISE: ORIGINS AND INFLUENCE

Egypt's Soviet-made SA-2 missiles (that arrived in 1963) were employed ineffectively and in limited fashion during the Six-Day War. The IDF even captured a number of them. American air force and navy pilots gained considerable experience with these missiles in Vietnam in the mid-1960s, and this knowledge reached the IAF through various channels. Between the War of Attrition and the Yom Kippur War, the IAF took major steps in preparing for the next clash with the SAMs. In his article (that appeared in *Ma'arachot* in September 1971), Shimon Iftach traced the development of the Egyptian and Syrian SAM threat to the IAF, stating that "at present" there were approximately one hundred SA-2 and SA-3 batteries (500–600 missiles) on the Suez front. "This is probably the densest, best organized, and best planned layout ever built."[3]

The IAF-SAM struggle can be divided into three periods: July–December 1969: destruction of most of the Egyptian SA-2 batteries and radar stations; successful evasion of missiles; Soviet concern over damage to their munitions assets; Egypt's repeated request for assistance.

January–April 1970: IAF aircraft penetrate Egypt's depth and bomb the missile batteries; increase in Soviet involvement—deployment of improved SA-2s and SA-3s that, unlike the older versions, were designed to knock out low-flying aircraft; Egypt receives SA-7s; many systems now manned by Soviet personnel.

April–August 1970: Israeli planes clash with SAMs and Soviet-manned Egyptian fighters; IAF planes begin to be hit.[4]

The SA-2s had certain limitations that IAF pilots were quick to exploit.[5] Not a single Israeli plane was hit when the missile was used alone; however, when the SA-3s arrived in Egypt (beginning in mid-1970), the combination of two missile types, plus radar-guided, mobile aircraft cannons, created highly lethal, integrated batteries. The Egyptians, with Soviet help, deployed their newly integrated batteries on the eastern bank of the canal. On June 30, 1970 the first Israeli aircraft was shot down. The IAF tried to solve the problem with American-made deceptive devices that had been developed in the Vietnam War at the cost of planes and pilots.[6] When a bilateral ceasefire was announced on August 7, 1970, the IAF found itself in difficult straits. Even after the ceasefire went into effect, the Egyptians continued to bring missiles into the Canal Zone in violation of the agreement. A year later the missile layout was fully set up and strengthened. After the war, IAF experts found that fifty-five SAM batteries, an SA-7 missile layout, and a large quantity of anti-aircraft cannons had been deployed at the canal on October 6—the day the war broke out.[7]

The Syrian layout on the Golan Heights consisted of nearly twenty-five SA-2 and SA-3 batteries and fifteen first-class SA-6 batteries.[8] In addition, the layout included SA-7s in use with infantry and armor units and hundreds of anti-aircraft guns. This vast array of advanced anti-aircraft weaponry was a stiff challenge to the IAF's offensive capability and its ability to assist ground forces in the Suez Canal and on the Golan Heights. The solution to the SAM threat came in the form of brainstorming and planning that began at the beginning of the War of Attrition and intensified when F-4 Phantoms began being shot down in 1970.

The solution called for a combination of technical jamming and deception measures plus an operational doctrine. The IAF developed and acquired the means to interfere with and counter the SA-2s and SA-3s, but it lacked a technical solution for dealing with the latest-model SA-6s whose electronic system remained undecipherable. Aviam Sela, the pilot who took the first reconnaissance photo of the missile in the summer of 1972, stated that "a monstrous myth was created over this."[9] The missiles were mobile, having been installed on the backs of APCs. Once they were deployed, they were quickly operable and were very tricky to identify after launching since no smoke was emitted in the missile's second (rocket propulsion) flight stage.[10]

A debate ensued regarding the correct operating procedure. The opponents were Aviahu Ben Nun (head of the IAF's planning branch) and Iftach Spector (commander of the 107th Phantom Squadron). Ben Nun relied on massive, complex attack formations based on heavy electronic-assistance systems to "flood" the missile defenses. The plan demanded precision navigation and timing—down to seconds—that left the pilots no freedom of action or maneuvering.[11] The plan was also based on aerial photographs taken on the morning of the attack that confirmed the position of the mobile missile batteries and was also based on IDF artillery fire on the missile batteries near the front, the use of chaff (electronic warfare), and the right weather conditions. Spector balked at the idea of waging war according to such precise planning and demanded greater freedom for his pilots. The plan for eliminating the Egyptian missile layout was codenamed "Tagar 4"; "Dugman 5" was the codename of the air attack against Syria's anti-aircraft missile layout on the Golan Heights. The detailed planning and practice imbued the air force with the belief that its contribution to the approaching victory would be no less than that in the 1967 war, though the price would be upped.[12]

Shmuel Gordon, a Phantom pilot in the war and later a military theorist, aptly sums up the IAF's meticulous preparations:

> The War of Attrition left the IAF pilots and commanders fretful over the results of a protracted struggle against surface-to-air missiles. In no way could they be considered indifferent to the matter. The following years were devoted to developing combat doctrines, weapons, and holding continuous exercises, discussions, and study days. All these labors had one goal in mind: to restore air superiority in the IDF's theater of operations. The addition of a new anti-aircraft missile—the SA-6—intensified the air force's efforts.[13]

The IAF was kept on a high state of alert. Already on Rosh Hashana (the Jewish New Year) in late September, ten days before the war, the air force was put on standby in the event of an Arab attack. A state of high alert was declared at 09:30 on October 5, and the air force was given the green light to call up reservist air crews.[14] On the morning of Yom Kippur (the Day of Atonement), all aircraft and pilots were ready to strike airfields in Syria.[15]

Unlike the ground forces, the IAF's main strength was the standing (non-reservist) layout, and unlike the armor corps, whose tanks had to be brought up from the rear, the IAF's planes were ready for takeoff no matter where the air bases were. On October 5 the air force commander held a meeting and went over the preparations for implementing "Dugman 5" in the likelihood of war erupting on the Syrian front.[16] In a discussion at GHQ on the morning of October 6, Benny Peled, the IAF commander, stated his readiness to carry out a preemptive attack ("Dugman 5") at 11:00.[17] Thus, the air force was prepared for hostilities when war broke out and was not caught by surprise.

Operation "Dugman 5" was planned as a preemptive strike against a Syrian attack. The plan was based on the visual identification of SAM batteries by the pilots. Therefore, when a heavy cloud formation moved across the Golan Heights, the plan was changed: attack the enemy in its air bases deep in Syrian territory where better weather conditions prevailed.[18] Unlike the "Tagar" plan on the Egyptian front, which was based mostly on air-reconnaissance photos of fixed SAM batteries, Syria's widely dispersed, mobile SA-6 batteries precluded carrying out a version of "Tagar." The intended plan was to make a surprise attack that would catch the missiles in their last observed location.[19] As stated, given the weather conditions on the Golan Heights on the morning of October 6, even if Israel had approved of the surprise attack, it would have been impossible to carry out.[20] In effect, the plan went awry because of its overreliance on rigid conditions: political approval for a preemptive strike and ideal weather conditions.

"Tagar" began favorably on the Egyptian front on the morning of October 7, but the realization that the Golan Heights were in jeopardy convinced the IDF chief of staff to order the IAF to halt the operation in its initial stages and send his planes north to assist the forces by attacking Syrian ground forces and implementing "Dugman 5B" ("Dugman" without the element of surprise).[21] According to Peled, "[t]he fact is . . . that we had something like a working agreement that defined how the air force would act in time of war and under what conditions. But the problem was that the agreement was with the IDF and Israeli government while in reality the real partner should have been the Arabs." Peled also describes the IAF's surprise in the change of plans: "Since the enemy was not a signatory to the agreement, some of the senior officers needed time to recover—that's all."[22]

Operation "Dugman 5B" miscarried. It was carried out under the worst possible conditions. Ten Syrian missile batteries had been moved west. Because of insufficient time for air reconnaissance, the new deployment sites were not located. Electronic jamming equipment and deceptive devices that had been installed in the planes that attacked that morning in the south (Operation "Tagar") failed to arrive in time. Some of the Phantom aircrews returning from the Egyptian front did not have enough time for the briefing.[23] The results of the operation were abysmal: the attacking aircraft managed to hit empty dugouts instead of the SA-6 batteries. Sixty Phantoms, whose targets were the SAM missile batteries,[24] knocked out a single battery and damaged another. Six planes were lost.[25]

The air force, whose plans had been based on surprise, now had the tables turned on it when its plans failed to materialize. I would claim that the pilots' surprise at failing to carry out the plan they had trained for was a major factor in the crisis they experienced. In the book *A Dream in Blue and Black*, Iftach Spector describes the squadron's reaction to the surprise: "Yesterday we scrambled for a strike against the Syrian missile layout—and were caught by surprise. We weren't prepared to deal with it. Afterward we were sent to search and destroy bridges on the canal, but were totally unprepared for this mission, too, and again caught by surprise . . ."[26] The author's account graphically portrays the surprise he felt.

The IAF's jet fighters' losses came to 103 planes, forty-six pilots, and forty-one POWs. All told, it lost 25 percent of its aircraft.[27] Most of the losses were from missile hits and anti-aircraft gunfire, especially on close air-support missions.[28] At least 50 percent of the hits were from missiles and 30 percent from anti-aircraft fire.[29] According to Ben-Israel, only one-hundredth of the SA-2s and SA-3s and only 3 percent of the SA-6s that were fired actually hit the planes, but this was enough to have a powerful impact on IDF mode of operations.[30]

Recalling Egyptian successes, the Egyptian minister of defense Muhammad Fawzi wrote that

> [t]he air defense units entered the fighting in October 1973 knowing that they were capable of paralyzing the enemy's air movement and preventing it from maneuvering thanks to the enemy's mobility and supremacy in the air and

theater of operations. They also had technical data, the most important of which was the ability to conceal the gaps in the radar scanning field at low altitudes and integrate air defense elements with fighter aircraft . . . The enemy failed to surprise our forces with new weapons or devices that might have enhanced the capabilities of the enemy's planes without our forces knowing about them.[31]

It is my contention that the IAF was surprised by two basic factors. First, at its inability to implement its warfighting doctrine; second, at the paucity of information on the SA-6 threat.

Gordon claims that the IAF's failure to destroy the anti-aircraft layout on October 7 stems from its inability to implement its doctrine[32] that had been designed to defend Israel's air space from air attacks, neutralize the enemy's anti-aircraft layouts, and only later assist the ground forces.[33] Naturally the doctrine was not realized, but the reason was not only what Gordon claims—the decision by the political level and chief of staff—but also the doctrine's "over-sensitivity" to changes in its basic assumptions. An example of this is the cancellation of Operation "Dugman 5" because of weather conditions. "Dugman 5" and "Tagar" were predicated on the prevalence of good weather conditions that would allow the pilots to identify the batteries. On the morning of October 6, the political level called off "Dugman 5," as stated, though it could not have been carried out even if it had received the go-ahead signal because of heavy cloud formations. The weather factor intervened as a surprise although it should not have. Clausewitz recognized the fickleness of weather as a friction factor: "Fog can prevent the enemy from being seen in time . . . Rain can prevent a battalion from arriving . . ."[34]

A statement like "the air force's failure was due to weather conditions that hampered the realization of its doctrine" would probably not have been accepted as a legitimate excuse. The flaw in the doctrine stemmed from the basic warfighting concept that it was based on, which was too complex and overly stipulated. Operations "Tagar" and "Dugman 5" were based on the assumption that Moltke had been wary of in the nineteenth century: ". . . no plan of operations extends with any certainty beyond the first contact with the main hostile force."[35]

Giora Rom, commander of the 115th Skyhawk Squadron in the war, observed that the surprise element refuted the warfighting doctrine:

> . . . [The outbreak of the war] witnessed one of the watersheds in the history of the air force: technological inferiority. Technological superiority had been one of the cornerstones of the IAF, and in 1973 the air force had to make a great

effort to close the technological gap created by a new type of target, a highly lethal and mobile target. In the meantime the air force tried to compensate for its technological inferiority with combat doctrines that had been developed between 1967 and 1973 and that were based on a series of assumptions, most of which were invalidated during the fighting. Therefore we entered the war at a technological disadvantage and with a combat doctrine that couldn't hold water. For example, we built a combat concept based on attacking select targets (such as air fields or SAM layouts) and on very complex missions that were dubbed "operations" in IAF jargon. These operations included dozens of planes, with navigational timing and attack similar to the air parades on Independence Day . . . In a real war, when all kinds of unexpected things are happening, where gremlins bedevil you at every turn, with countless delays and glitches, the chances of realizing such a doctrine on a combat mission, was minimal at best, and time after time it just didn't work.[36]

The IAF's warfighting doctrine was suited to an exercise under sterile conditions, not to a war fought in the friction and uncertainty of weather conditions, an information gap on the enemy's missile capability, and difficulty in locating the enemy because of the mobility of his anti-aircraft batteries. These were basic inhibiting factors in addition to the enemy's active resistance.

Another aspect of the surprise stemmed from the information gap on the SA-6s.[37] The IAF lacked technical data on the missile; therefore, it had no answer for effective jamming and deception. Price asserts that the missile had not been operational before the war.[38] Lack of information is a familiar problem with new weapons. Ben-Israel states that the Yom Kippur War emphasized the decisive impact of technical intelligence that was illustrated in the technical differences between SA-3 batteries at the end of the War of Attrition and SA-6 batteries in the Yom Kippur War.[39]

Eado Hecht claims that Egypt integrated technological deception into the SAM layout shortly before the war, leading Israel to believe that after the expulsion of the Soviet advisors in 1972, the Egyptians lacked spare parts for their missile systems. Proof that Israeli intelligence received this message has yet to be found.[40]

The third source of the surprise, though this, too, is hard to prove, is the bias following the successful "Operation Moked (Focus)" in the opening hours of the 1967 War. The air force may have been "captivated" by the success of the preemptive strike and overcertain of the possibility of realizing it a second time, and thus ignored the difficulties involved.

While the armor corps ignored the danger of the Sagger missile, the IAF placed the anti-aircraft missile threat in the center of its combat preparations. The fact that three years of preparations did not prevent the accumulation of failures, some of which may have been linked to the surprise element, supports the argument that even if a warning signal is given in time—and in this case the warning light was flashing for three years—technological and doctrinal answers to certain threats still remain elusive.[41] This example also illustrates the problem of overdependence on a warfighting doctrine that, in the course of force planning, is planned according to the approach that precisely fills in the information gaps on the enemy but ignores critical friction factors independent of the enemy, such as weather conditions and strategic decisions at the political level.

The IAF's failure in dealing with the SAM layout had fatal ramifications on the ground forces who were left without close air support. This resulted in failure at the operational level to prevent the Egyptians and Syrians from making gains on the ground in the beginning of the war. The ineffectiveness of the air force against the missile layout also influenced the decision makers at the strategic level, convincing them that Israel was incapable of realizing the doctrine for a lightning victory. The unforgettable statement by Moshe Dayan, the minister of defense, about "the destruction of the Third Temple" in connection with the ground forces on the Golan Heights may be linked to his realization of the gap between the IAF's wartime mission and its difficulty in achieving it. In my view, the air force's surprise was felt at the operational-strategic level.

RECOVERY FROM THE SURPRISE

The recovery from the surprise occurred while implementing two strata of the flexibility concept: cognitive flexibility (that made improvisation possible) and lesson learning and application, especially at the squadron level. In addition, there were technical improvements that were carried out in real time and in conjunction with the defense industry.

Cognitive Flexibility and Lesson Learning

Cognitive flexibility was expressed in the "private" fighting methods that many squadrons developed. According to Giora Rom, the commander of the 115th Skyhawk Squadron:

> What the captains, majors and flight leaders basically did was to design an entirely new fighting doctrine on the ruins of the instructions that proved abortive and

on the basis of the new reality that we had to find a solution for . . . We entered the war employing a very low-altitude approach to the missiles, using the ground as cover, and ended the war by attacking the missiles from 25,000 feet—the exact opposite. We changed the strike mode, the whole preplanned strike against the airfields, and without mentioning individual stories, Abudi (the head of IAF's historical branch during the war) recalled such terms as "pop-up" and "delayed pop-up" (two types of runs by fighter aircraft), the squadrons employed the "delayed pop-up" every time we attacked the canal, which was like swimming in a pool of anti-aircraft fire, and we had to find a way to get above the anti-aircraft weapons while dealing with the missile problem. But we devised this [method] on the fourth day of the war. It reflected how we attacked the canal and Golan Heights. The whole war was about creating a new warfighting doctrine, and I think the air force can be rather proud of itself for having accomplished this.[42]

In an interview, Rom relates two cases that occurred one day after the next in which the attack methods were re-modified in real time. In one case the fighting method that was ordered by air force's headquarters was changed in flight during a sortie.[43]

In another case, one of the pilots from the squadron recalls: "We were briefed to fly at a high altitude in order to stay clear of AA [anti-aircraft] fire. It was assumed that the missile sites had been knocked out; and since many planes had been hit by AA fire and shoulder-held missiles, it was decided to switch to the high-altitude attack method."[44]

Another illustration of doctrinal improvisation and inter-unit cooperation was the case of the Mirage interceptor squadrons:

> Following joint debriefings between the 110th and 113th Squadrons, a combat doctrine was formulated for operating in areas bristling with anti-aircraft weapons and missiles: "quick kills" or "snacks" as Bahrab (a pilot) termed them. Following the lessons learned in anti-aircraft identification and evasion . . . the squadron could only try to develop new combat techniques for dealing with this worrisome reality. This was done in two ways. First, I made sure that each plane received an up-to-date map of the SAMs . . . Second, we began the attack approach several miles from the target, where the air controller told us to begin in order to put a distance between ourselves and the AA fire and keep at an altitude that permitted evasive maneuvering. If there's only AA fire without the SAMs—then fly above it . . . These tactics had to be invented and refined in the midst of the fighting.[45]

The 69th Phantom Bomber Squadron managed to "speed learn" during the war. "Immediately after the rough opening round, the squadron was careful to carry out debriefings on a daily basis and hold lesson learning sessions between sorties . . . The high level of organization and manpower that developed during the war definitely proved itself in the field."[46]

Even before the war, the 107th Phantom Squadron, under the command of Iftach Spector, invented and refined an attack method different from the "meticulously preplanned operating procedure." Spector claims that the IAF's operational plans "were too cumbersome, complex, and rigid, and seemed bent on directing the battle by bureaucratic means." Spector supported the "hunters" technique that gave small flight formations the freedom to lure missile fire onto decoy aircraft and then swoop down on the exposed batteries with low-altitude attacks. This approach depended on a pilot's improvisational savvy and the Phantom's ability to maneuver quickly. At the start of the war, the 107th was the only squadron using this technique.[47]

The following illustrates Spector's view on initiative and improvisation during the war:

> Toldano [the squadron commander in the book *A Dream in Blue and Black*] thought while lowering the wheels for a landing that we mustn't wait until "the big machine" is revealed and lets us know that there's a glitch. If you're thrown into the jungle, you have to smell it out and feel it out for yourself, perceive things and their consequences, and act in time before the little stories sprout up and become one big mess.[48]

The lessons that the squadrons learned were not conveyed on a regular basis to other squadrons, apparently due to an overload in the communications system, which had to be free to issue orders, assign missions, and receive mission reports.[49]

These are examples of the various improvisations made in five squadrons that flew three different kinds of aircraft. Amos Amir, a former Phantom squadron commander who was at air force headquarters in the Yom Kippur War prior to his promotion to head of the IAF operations branch, discusses improvisations in this period: "Local initiatives were developed in the bases for attacking in small, 'flexible' formations that resulted in a number of successes as well as failures, without any significant effect on the air war's progress. I followed this unprecedented process with great apprehension."[50]

Although the cumulative effect of improvisation did not have a serious impact on the air war, there was a great deal of flexibility at the unit level.

IAF headquarters displayed little flexibility compared to the squadrons. According to Elhanan Oren, "One of the reasons for the IAF's relatively poor showing in the Yom Kippur War . . . was headquarters' slowness in lesson learning and conveying the lessons to the squadrons."[51]

Although IAF headquarters modified its warfighting doctrine—based in part on the lessons learned in the squadrons—it did so slowly and in a rather limited fashion. One change was in control methods. On October 9 the air force changed its system of target allocation. Instead of attacking individual tanks in coordination with ground units through the forward command posts, four attack sectors were designated: the Golan Heights, Port Said, Dewar Su'ar-Kantara, and Kabrit-Suez. One squadron was responsible for each area.[52] Abudi termed this "the contract system"—"that is, we divided a large area like the Golan Heights . . . into secondary areas that were 'contracted' out to an IAF base or wing according to its ability to perform sorties that day."[53] Another improvisation that evolved into a new method was the "delayed pop-up" run.[54]

On October 10 the air force experimented during an operational attack against the SAM layout at Port Said. This test helped improve offensive tactics against dense missile layouts.[55]

The missile challenge on the northern front was solved mainly because the Syrians shot off their entire supply of SA-6s and did not receive any replacements. This meant that the Israeli pilots could deal successfully with the older missiles.[56] Only four Syrian missile batteries were destroyed during the war.[57]

The last attempt to employ the "meticulously preplanned operating procedure" system against SAM batteries in the northern sector of the Suez Canal on October 18 ("Operation Cracker [Mefatzeach] 22") came at a heavy cost: six batteries destroyed; six Phantoms downed. The lessons from "Operation Dugman 5" were of no avail in this case. After the attack, Spector's previously mentioned hunters method (whose effectiveness was proven by his squadron on a number of sorties) became the official modus operandi. Within four days, the IAF and IDF ground forces operating on the western bank of the canal destroyed the Egyptian missile batteries.[58] Although IAF headquarters seems to have used the knowledge acquired by the squadrons, it failed to act quickly in adopting and circulating it and delayed introducing changes in tactics that were found to be problematic. However, it did not prevent the squadrons from improvising and redefining their methods.

Military-Industrial Cooperation

RAFAEL deciphered the SA-6 missile data during the fighting. Bonen elaborates on the way this was accomplished. On October 8 and 9, a team sent to the Golan Heights found missile fragments and homing seekers. Later a burnt homing seeker was discovered that served as the main source of research. On October 18 after exhaustive labor, a team of experts drew up an initial report of the missile with detailed information on its guidance systems and radio frequency. The report was immediately conveyed to the IAF's head of equipment wing at headquarters.[59] Although application of the missile's data was realized after the war, the case nonetheless illustrates rapid, effective cooperation between the IDF and the military industry.

Bergman states that during the war, a young female scientist at RAFAEL quickly developed a way to jam the enemy batteries' radar system. The device was installed on IAF planes during the war and prevented losses.[60] Another offshoot of the army-military industry cooperation was technological improvisation against shoulder-fired SA-7 missiles:

> The engineers in the technical layout had to improvise and modify during the war. Many Skyhawks had their tails hit by infrared missiles that locked onto the gas trails emitted from the engine, which caused serious damage to the plane's body and steering system. While the fighting was still in progress, IAF and industrial engineers designed, developed, and produced a sleeve that lengthened the after-burner, which put the aircraft's heat signature at a greater distance from the body of the plane, thus reducing the number of missile hits.[61]

Another area of on-the-spot creativity was aircraft maintenance. The basic assumption was based on Six-Day War concepts, namely, that all of the IAF's planes would fly together and then be serviced together. This method proved impractical because of the unexpected length of the war:

> During the war we decided to switch to a maintenance system similar to that used by civilian airline companies—every time a plane lands, certain checks and repairs are carried out allowing it to take off, so that in rotation, after a number of flights, all checks and repairs are completed—and none of the planes remains on the ground for too long of a time.... This is not the IAF's standard operating procedure. Only minor checks are made between flights so that the planes are not "pried open" each time—but during the war, there was no alternative since the goal was to have the greatest number of aircraft in the air.[62]

Benny Peled, the commander of the IAF, put to verse the air force's ability to improvise when encountering a surprise:

> We have two ways of playing a musical instrument
> One, according to prescribed notes,
> But sometimes the notes are missing,
> And if there aren't any notes, then we play without them.
> And this is what we did,
> And although the strings snapped, the melody produced was sweet,
> And resounded everywhere.[63]

THE SOURCE OF IMPROVISATION AND LESSON LEARNING CAPABILITY

Spector's lessons from the Yom Kippur War illustrate cognitive and command flexibility and the culture of lesson learning that enabled the air force to recover from the missile surprise:

Combat Methods
As commander of a squadron you have a lot of influence on the shape of the fighting, whether you choose certain fighting techniques or support a particular method . . .

Determine a clear position as quickly as possible, especially for rapidly changing situations. Introduce an "operational forum" consisting of leading officers and look for fighting methods that will answer tomorrow's challenges. Naturally the methods will be developed according to their results in combat . . .

Planning Versus Performance
As in air warfare, so in every field of combat, do not overplan; detailed planning will lead to failure because it overlooks the simplicity principle and because of "friction" in combat . . .

Briefing the Troops to Expect the Unexpected
At the start of a war (or a day), it is absurd to have a detailed briefing designed to answer every problem liable to occur in the field . . . It must be perfectly understood by the troops that war is an extremely fluid situation, and surprises can be expected . . .

We saw that men, who did not understand the fluid nature of warfare, were completely caught by surprise in situations they had not been briefed on; it was as though they suddenly forgot everything they had learned. Therefore, make it clear to your people, first and foremost, that they have to expect the unexpected.

A "Living" System and Corrections During Combat

The squadron is judged in combat by its ability to self-correct while engaged in missions and not repeat mistakes. This requires knowledge (precise, up-to-the-minute reports), thinking, and planning ... In this area—"the small ring"—was the main expression of the squadron as "a fighting organization." Do not shun criticism, analysis, and immediate correction in the small ring. Do not blindly follow instructions and the force of habit ... Beware of "indifference to change." If you believe that a change has to be made—then change the plan immediately and don't wait till the end of the day ...[64]

Gordon claims that initiative and creativity are the basic principles of aerial warfare and that the IAF has kept to this principle over the years. As evidence he presents the ideas of Colonel Ya'akov Nevo (Yak) on the concept of aerial warfare. In 1958 Yak formulated the doctrine that became an ironclad asset in the IAF for years. The manual emphasized the importance of initiative and agility (flexibility).[65]

Giora Rom, the aforementioned commander of a Skyhawk squadron in the Yom Kippur War, states that the flexibility demonstrated in the war was the result of two elements in the air force culture that Ezer Weizman inculcated when he was IAF commander between 1958 and 1966. The first was openness to express views that deviated from official doctrine, regardless of seniority or rank.[66] According to Weizman:

> In whatever role I was assigned in the air force, I nurtured freedom of expression. I abhorred officers who wanted to hasten their career advancement by blindly agreeing to my views. I could never stomach yes-men. The air force needs a thinking officer who develops his own tools for independent analysis and uses them openly for assessing things . . . I wanted to have an ongoing polemic with those who disagreed with me, a real confrontation, out in the open, in which those who thought differently had an equal opportunity to justify their evaluations. I frequently utilized the "commander's evening" to encourage officers and pilots to argue with me.[67]

The second element in the air force culture, according to Rom, was the leveling of partitions among systems elements. Pilots, maintenance crews, administrative staff, and technical experts worked without formal partitions and in the spirit of cooperation. Rom gives the example of the switch from separate mess halls, where personnel had eaten according to sectoral affiliation, to a common dining room. This change enhanced intergroup connectivity and resulted in fast and efficient cooperation in finding solutions to operational and technical problems throughout the war.[68]

Another type of flexibility that I touch on only briefly was the transition to ground-air cooperation as a solution to the SAM challenge. Colonel Y. and Major Y. write that the basic flaws in the IDF's warfighting doctrine before the Yom Kippur War was its ignoring the potential of an integrated, air-ground solution to the SAM problem and its relying exclusively on the air force. After crossing the canal, the armored forces began raiding the Egyptian missile bases, thus enabling the air force to fly with relative freedom and provide close air support to the ground troops. According to Colonel Y. and Major Y., this is a classic example of combined arms combat in which the air force (the original weapon) was helped by the armored corps (the assisting weapon) to respond to the SAMs (the opposing weapon). This is similar to armor being helped by combat engineers in the removal of anti-tank mines.[69] Believing that the air force was the sole solution to the SAMs violated the principles of diversity and redundancy. The increase in diversity enabled the employment of combined, mutually beneficial combat in which armor made up for the air force's shortcomings and vice versa. The operational challenge of the SAM batteries, which had originally been given an exclusive solution, was solved by employing an alternative means—armor raids.

In conclusion, the air force recovered with limited success from the SAM surprise. It managed to minimize the damage but failed to completely overcome the surprise. Recovery was based primarily on cognitive flexibility and the culture of lesson learning in the combat squadrons. Speed of response was extremely fast in the squadrons, but relatively slower in IAF headquarters. The increased organizational flexibility in the form of combined air-ground operations also helped meet the challenge of the surprise.

10 THE SLOW BRITISH RECOVERY FROM THE GERMAN ARMOR AND ANTI-TANK TACTICS

THE FIFTH TEST CASE deals with warfare in the Western Desert between 1941 and 1942—a battlefield with features foreign to both sides. The ability to respond to this challenge proves the importance of conceptual flexibility as an essential answer to battlefield fluctuations. The German doctrine (see above) was noted for its flexibility and the British doctrine for its dogmatism. The low level of British cognitive and command flexibility may be attributed largely to dogmatism. The British also suffered from stunted organizational flexibility that resulted in a serious gap in weapons diversity.

THE ORIGIN AND INFLUENCE OF THE SURPRISE

The novelty of desert warfare to both sides spawned many innovations in techno-tactics and tactics. One case in point was the open terrain that enabled anti-tank weapons to be used at maximum range.

These weapons, however, were not designated anti-tank guns but German 88-mm anti-aircraft and British anti-aircraft 3.7-inch guns. The 88s, for example, could penetrate the front of the American-made Sherman tanks (in service with the British army) at ranges of over 2700 meters, whereas the long-barreled, 50-mm Panzer III tank shells penetrated at only 900 meters, and the long-barreled 75-mm Panzer IV gun could knock out a Sherman at 2200 meters.[1]

The Germans made wide use of their anti-aircraft guns, not according to their original purpose but by transforming them into "miracle weapons" in the armored combat of the Western Desert. The reasons for this, I contend, stemmed basically from the flexibility of the German doctrine. In this case special permission was given to employ weapons not according to their original designation, but for battlefield exigencies. Section 812 of the *Truppenführung* states: "The use of anti-aircraft guns in ground operations reduces their effectiveness in their main task and must be limited to exceptional circumstances. Their use in this manner is justifiable only for close-range defensive fire against combat vehicles."² The seeds of flexibility that had been sown in the nineteenth century came to fruition in a flexible doctrine that encouraged the education and promotion of commanders who displayed initiative and creative thinking and that enabled the German army to improvise in the constantly changing conditions of the battlefield.

Handel claims that the British were shocked by the 88s even though the weapon had demonstrated its anti-tank capability in France in 1940. He further states that due to an information gap, the British continued to underestimate German anti-tank weapon capability to penetrate armor until late 1941.³ Bungay describes the combat technique of integrating anti-tank weapons with armor in the battles of El Alamein:

> However, one German tactical innovation was peculiar to the desert war . . . They perfected the technique of using their anti-tank guns offensively as well as defensively. When German tank units met British armour, the German anti-tank guns would rapidly dig in, forming a screen behind their tanks. The German tanks would withdraw behind their guns, and the British tanks, much as their Anglo-Saxon forbears had done at the battle of Hastings, would move forward in a charge, only to be cut down, not by Norman cavalry, but by the German anti-tank guns. The guns were barely visible, and so their impact was often erroneously ascribed to the tanks, which were [visible]. When advancing, German tanks and anti-tank guns would leap-frog each other, moving forward and screening each other in turn.⁴

In effect, the German surprise was conceptual, that is, the employment of anti-tank weapons in conjunction with armor was what surprised the British. Pinning the blame on technological inferiority is inaccurate.

According to the Israeli military historian Eado Hecht:

> Given that the brunt of the fighting and decision belonged to armor, both sides had great respect for tank and anti-tank technology and the warfighting doctrine at the basis of their [techno-tactical and tactical] employment. The British believed that the Germans held an advantage in both realms. The supposed technological superiority of German armor and anti-tank weapons attained a mythical status in the British army during the war, and is repeatedly referred to in postwar memoirs and unit histories. But the truth is that the Germans' advantage lay only in the area of the operating concept, whereas in technology, a balance between the advantages and disadvantages of weapons

on both sides was probably reached. Occasionally this balance was upset for brief spells, in favor of one side or the other with the entry of a new tank or the upgrading of an older model—until the opposing army acquired a new tank or retrofitted an older one.[5]

In any case the psychological impact of the 88-mm gun on the British troops was devastating. "A psychiatrist investigating the effects of different weapons on troops discovered that men willing to charge the lethal MG 34 machine gun were terrified by the 88 . . . When it did not appear, the fear that it might led experienced British tank crews to exercise caution."[6]

In this example, the source of the surprise was not the German attempt to cause a surprise, but the fact that the British attributed their armor's vulnerability to the wrong weapon.

The effect of the surprise was felt on the operational-strategic level. During Operation Battleaxe (June 1941), whose objective was to defeat Rommel and return Cyrenaica to British hands, the Germans employed a mere thirteen 88s[7] that, according to British General Frank Messervy (who seems to downplay the effect of trouble-plagued British planning), were a central factor in the operation's failure.[8] Churchill relates that all Britain hoped for Rommel's defeat, and the operation's miscarriage "was to me a most bitter blow."[9] After the failure, the commander of British forces in North Africa General Archibald Wavell was replaced by General Claude Auchinleck.

In November 1941 Auchinleck launched a more ambitious attack ("Operation Crusader"). This time the Germans used scores of 88s and 50-mm guns.[10] Liddell Hart states that the British "direct approach" ran smack into an antitank ambush of carefully concealed 88s. "As a result the British lost not only their strategic advantage but much of their numerical superiority in tanks. The Eighth Army was thrown off its balance, psychologically as well as physically."[11] However, since the size of the 88s made them difficult to camouflage, much of the damage that the British incurred was undoubtedly caused by the smaller and easier to conceal 50-mm guns.

THE SLOW RECOVERY FROM THE GERMAN SURPRISE

Recovery took over a year—from the first encounter with Rommel in late March 1941 until the Second Battle of El Alamein in October–November 1942—and in the end it was British quantity that defeated the Germans. The main reasons for the slowness of recovery were the low levels of cognitive, command, and organizational flexibility—the latter realized in the form of problematic interbranch cooperation and the lack of diversity in anti-tank and antipersonnel weapons.

Low-Level Cognitive and Command Flexibility

A good example of low-level cognitive flexibility is the 3.7-inch AA gun. Rommel's intelligence officer Friedrich Wilhelm von Mellenthin claims that the British 3.7-inch gun was a very powerful gun but was employed solely against aircraft[12] because of the British unwillingness to use it as the Germans did with their anti-aircraft weapon—the 88.[13] The British also employed their anti-tank weapons only as a defensive measure. ". . . they failed to make adequate use of their powerful field artillery, which should have been taught to eliminate our anti-tank guns."[14] The British conservative thinking is best described by a saying of a British prisoner of war to Rommel's personal aid Heinz Schmidt: "In my opinion . . . it is unfair to use flak [the 88-mm gun] against our tanks."[15]

Wolf Heckmann describes how the British developed the idea of applying the 3.7-inch (94-mm) gun for anti-tank purposes. He claims that using the gun in ground warfare was the brainchild of a group of young British officers who convinced their superiors and General Wavell himself, the supreme commander of the British forces, to give it a try. Wavell ordered the gun to be tested in the 1940–1941 winter campaign against the Italians. The task fell on Brigadier McIntyre in the Battle of Tobruk, and the guns proved accurate at 2500 meters. The upshot is characteristic of British dogmatic thinking: "The traditionalists in the officer corps naturally raised quite a hullaballoo about the heretics who had taken the gun away from the role ordained for it by God; and McIntyre was given the ugly name "Mad Mac."[16]

Michael Carver, a junior officer in WW II and future British chief of staff, claims that after the Eighth Army Headquarters finally convinced the general staff to allocate its 3.7-inch guns as anti-tank weapons,

> [o]ne would have thought that their long-delayed arrival would have been welcome, but their reception in the Knightsbridge "Box" was cool . . . It was described by Brigadier Lyndon Bolton, then an experienced field regiment commander: "One evening we were informed that four 3.7-inch heavy anti-aircraft guns were waiting outside the Box to be added to our defenses and to be used in a ground role against enemy tanks. No doubt this was the result of the success that the Germans were achieving with their 88-mm guns. Imagine the conster-

nation that the prospect of the new arrivals aroused! In this congested space there was no chance of concealing such enormous weapons, so only two were accepted... Alas! This was not time to experiment with such weapons... This must have been the first, or nearly the first, effort to use these guns in a ground role, and it must be admitted that it was not a great success..." One wonders if German gunners made the same sort of complaints when the first 88-mms were used as anti-tank guns.[17]

The anti-aircraft guns were eventually needed against the Luftwaffe and duly transferred from North Africa to the Greek Campaign. When the supply flow improved, Wavell was relieved from command. In the meantime McIntyre took ill and left the desert, and the idea petered out. Heckmann states that: "Nothing was ever done; in the very theater of war, where the range and penetration of guns was decisive, the standard British weapon reminded the absurd two-pounder. It is virtually impossible to estimate what rivers of blood this failure of the imagination cost the British."[18] This is an example of the way a dogmatic mindset impedes improvisation and thinking outside the box.

Norman Dixon, in his book *On the Psychology of Military Incompetence*, severely criticizes the dogmatic thinking of the British military. He analyzes, inter alia, the British failure to respond to Japanese tank employment in the Malayan jungles and rubber plantations in December 1941.

The Japanese employed light tanks in the Malaya jungles.[19] But since the British had not taken this contingency into account, the 11th Indian Division was caught by surprise at Jitra. A small Japanese advance party with a single tank company was able to subdue an entire division in one day and overtake its strongly established defense positions.[20] There were other reasons for the debacle, but denying the possibility of a tank attack from the jungles and not meeting its challenge undoubtedly played a large part in the British rout.

The British War Ministry's prewar instructions for dealing with enemy tank attacks failed to reach the units. In late November it was discovered that this stack of instructions from London, from the time of its arrival in Malaya, had remained ensconced in a storeroom at headquarters. Staff[21] and field officers[22] ignored suggestions to prepare anti-tank defenses on the main roads and train the troops in coping with enemy tanks. "Experts," brought over to lecture on the island's defense concept, swore that no army—let alone the Japanese army—could advance through the impenetrable jungles of the Malaya

Peninsula because it was impossible for tanks to negotiate, and the Japanese military structure was a primitive mechanism that need not be taken seriously.[23] An intelligence expert who had warned the defenders of Singapore in mid-1941 about the Imperial Japanese Army's capability "was officially pronounced pro-Japanese and defeatist."[24]

Dixon claims that the guardians of Singapore who had been lax for so long could not allow their negligence to be exposed; therefore, they concocted a two-prong cover up: they supplied their forces with fabricated information and prevented important information from reaching the population, alleging that it was liable to cause widespread panic.[25] In this instance, British inability to respond flexibly to change seems to have stemmed from a dogmatic mental attitude at various levels: from senior commanders, who convinced themselves of the jungle's impenetrability, to field officers who believed they could deal with the tanks despite unsuitable preparations.

What was the source of this dogmatism? Bungay claims that British doctrine tried to solve the problem of battlefield chaos by a system of centralized C2. Detailed, all-encompassing instructions were issued from higher up. Only in special instances were junior officers allowed to deviate from their orders and, even then, only after receiving permission from their superiors. "Initiative was effectively banned. Training emphasized obedience, which was inculcated by drill."[26]

Elizabeth Kier investigated the development of the French and British doctrines in the interwar period and found that the centralized system was based on the British prewar defense doctrine that called for meticulous preparations. According to this concept, maneuvering demanded superior firepower. Therefore, "[b]attles would be precisely planned, carefully timed, dominated by firepower, and methodically executed."[27]

Murray's study of the British land army's military effectiveness in WW II reveals that its tactical approach precluded the tactical and operational systems' efficient adaptation to battlefield changes. Murray found the source of the problem lay in its inchoate doctrine and ineffective training, the main causes of the substandard ability to adapt to changing situations. "The Staff College at Camberly issued staff solutions to all exercises and, as one attendee noted, 'discussion was very much frowned on after lectures.'"[28] Another officer recalled: "As for expressing an opinion which differed from the general point of view, that was almost unheard of . . . It would have been considered very bad manners not to agree with the senior officer."[29]

The German assessment of Britain's fighting ability includes many references to British inflexibility.[30] Rommel's view of the British officer cadre in the Western Desert was that "most senior British officers have a certain tendency to think along established lines."[31] He also states that "their [British officers] unwieldy and rigidly methodical technique of command, their over-systematic issuing of orders down to the last detail, leaving little latitude to junior commanders, and their poor adaptability to the changing course of the battle were also much to blame for the British failures."[32]

On the British operational conduct of the desert war, von Mellenthin noted: "In general the British method of making war is slow, rigid, and methodical."[33] On British training, one British junior officer admitted that "[i]t was pretty unimaginative, all the things that we had learned to do at battle school. A straight forward infantry bash."[34] Millett and colleagues found that the British army in both world wars provides examples of relative ineptitude at achieving tactical surprise and exploiting success.[35] Regarding British tactical effectiveness in WW II, Murray writes:

> Certainly the system did not encourage the flexibility of mind and the willingness to take initiative that are essential if one hopes to do more than break into enemy positions. For exploitation of fleeting battlefield advantages, one needs NCOs and junior officers who possess initiative and drive. Those qualities were not often enough in evidence in the Second World War.[36]

It seems, then, that the basically dogmatic doctrine that curbed initiative and innovation impeded British ability to make non-conventional use of their weapons.

Low Organizational Flexibility

A basic solution to anti-tank weapons is combined arms. Combined arms uses infantry and artillery to compensate for the weakness of a tank against anti-tank weapons in an open area, just as the Israelis did to recover from a similar surprise. The British army possessed unit diversity, which is necessary for combined arms, but it lacked the ability to apply combined arms tactics in combat. The core of the problem lay in the fact that the British forces had not been trained in combined arms, and this stemmed from a doctrine based on the approaches of Fuller, Broad, Hobart, and others that favored the almost completely independent use of tanks rather than their integration in combined arms. Carver has this to say on British tactical concepts: "Our real weakness was the failure to develop tactics for concentrated attack employing tanks, artillery, and infantry in depth on a narrow front."[37]

Without a central guideline in doctrine and training, a variety of approaches prevailed in the British units. These approaches were developed and promoted by individuals rather than within a unified, clearly-defined framework. The regimental system, whose main advantage lay in the development of unit solidarity, encouraged autonomy but reduced the ability to cooperate at the tactical level.[38]

William Jackson, an expert on the battles in the Western Desert, found that a psychological snag—the "clan spirit"—hampered the British and rendered them slower than the Germans in coming up with solutions to tactical problems. This spirit imbued the British regiments and corps with immense pride, but at the same time inculcated the sense that each branch could perform its job without recourse to others. In the Western Desert this problem was exacerbated since all of the tank units were regiments from the British Isles, while the units made up of Commonwealth troops had no armor formations.[39] The lack of joint exercises created a feeling of estrangement among infantry, armor, and artillery units and made it difficult for one branch to appreciate the other's problems.[40] Carver claims that the inability to carry out tactical concentration should be sought, inter alia, in ". . . parochialism of the peace-time British Army between the wars, in which cavalry, tanks, infantry, gunners, engineers and others saw little of each other in England. . . . Opportunities for all arms to work and train together, and for their tactical problems to be considered as one, were rare and fleeting."[41]

Even if some officers did realize the importance of developing interbranch cooperation, they were thwarted by the organizational-technological gap that held sway in the British army and prevented cooperation from being carried out. The British armor expert Percy Hobart admits that he was fed up with the attempts to combine the forces of the Seventh Armored Division with infantry and artillery, since the latter lacked the requisite radio equipment and mobility.[42]

The communications shortcomings stemmed from the British reluctance to depend on radio because of its unreliability (and the fear of eavesdropping) and budgetary limitations that retarded the development of tactical communications systems. The gap in armored mobility or anti-tank weapons per se was well-known (see below) and left the British infantry extremely vulnerable. As a consequence, British tankers regarded the infantrymen disparagingly, as troops that had to be protected rather than as comrades in combined arms. House observed that this was not surprising, since many British armor units reverted to the interwar concept of attacking in which armor trained as an independent tank force assigned to traditional "cavalry-like" missions.[43] This concept played into the hands of German combined arms groups (see above).[44]

Thus, unit diversity, even if physically present in an army, is not itself a guarantee for combined arms combat. An applied doctrine that emphasizes integration and according to which units are trained, equipped, and familiarized with one another, is sine qua non (in addition to the physical presence of the units).

Another level of organizational flexibility is that of weapons diversity. The British made an enormous effort to surmount their deficiencies in desert warfare in general and armor versus anti-tank weapons in particular. Their limited assortment of anti-tank and anti-personnel weapons was one of the reasons for the slow recovery from German obstacles.

A modern tank gun is capable of firing a wide variety of shells in various operational exigencies, especially penetrating enemy armor (the solution to which is high velocity, armor-piercing shells), and hitting enemy infantry and anti-tank squads and soft targets that require wide shrapnel-scattering, high-explosive shells. This was not the reality before or at the start of WW II. The Germans realized that an armored force requires two operational solutions; therefore, they developed diverse munitions for their tanks: the Panzer IV had a short, squat gun that fired a heavy, high-explosive shell mainly against infantry and was suited less for armor versus armor battles; the Panzer III tank, on the other hand, with its relatively long gun, fired a high-velocity, armor-piercing shell.[45]

The British 40-mm (two pound) gun mounted on most of the British tanks (Wellingtons, Matildas, and Crusaders) was also used as an anti-tank weapon but had a relatively low armor-penetrating capability when fired at the frontal plating of the German tanks. Moreover, the gun discharged only armor-piercing shells; that is, it was unable to fire high-explosive or smoke shells.[46]

The drawbacks in the organization of the British force meant that in the initial British-German clashes in the Western Desert in early 1942, the British two-pound gun on the Crusader tank barely dented the Panzer IIIs even at ranges of one thousand meters, whereas the German tank's 50-mm gun penetrated the Crusader's entire twenty-three millimeters of frontal armor.[47]

Another basic gap lay in anti-tank weaponry. The German 88s easily destroyed British tanks at nearly two kilometers—twice the effective range of most British tank guns.[48] The German anti-tank guns were often effectively camouflaged and capable of damaging British armor, while the latter lacked the necessary range and type of ammunition for dealing with them. In his book *Tobruk*, Carver observed: "It was certainly not realized by the British, as it was by the Germans, that the 88-mm gun, and perhaps the new 50-mm anti-tank gun also, had been the principle cause of their defeat."[49]

In addition to the anti-tank gun's advantage in range, the range of the Panzer IV's high-explosive shell was also much greater than that of the British tanks. Although its armor penetrating effectiveness was low, the difference in ranges meant that "the British tanks and any infantry near them had to endure a prolonged high-explosive pounding before they could get into range—high explosives which killed infantrymen and was also likely to damage tracks and suspension of the Crusader."[50] It should be specially noted that since the British were preoccupied with avoiding these shells, the German anti-tank guns were able to "sneak into" comfortable, close-range firing positions. When the British tanks began taking hits, they reported being fired on by enemy tanks at long ranges (due to the mistaken impression that the German tanks had an advantage in their armor-penetration range) when in effect it was the enemy's anti-tank guns that inflicted most of the damage.

One solution to the German anti-tank nemesis was the six-pound, anti-tank gun; however, it was a case of "too late and too little." In May 1942 there were only one hundred such guns in British service.[51] To overcome the gap, the British employed the twenty-five–pound artillery cannon as an anti-tank weapon; however, in the intermediate period, they desperately needed a weapon capable of destroying enemy tanks at ranges of one to two thousand meters. They possessed such a gun: the 3.7-inch anti-aircraft gun. This weapon was superior in almost every way to the German 88 though it was larger and clumsier to operate. For various reasons (see above), the British chose not to use it for closing the weapons gap in anti-tank warfare. A similar gap existed in artillery. The ranges of British artillery pieces were almost ten kilometers less than those of the German cannons. British artillery inferiority was offset by the long-range American cannons.[52]

The capability gap in the structure of the British force (high-explosive shells, tank ranges compared to German anti-tank guns, and artillery ranges) stemmed from a lack of weapons diversity. This imbalance contributed to a situation in which German armor tactics, which Rommel repeatedly applied to surprise the British, were left without a response. The British suffered from major C2 problems and a basic shortage in combined arms fighting skills, but the gap in force organization, which may be defined as limited weapons diversity in ammunition and range in anti-tank warfare, undoubtedly influenced the slow pace of response to the surprise of the German tanks and anti-tank tactics.

The solution to the diversity problem in the Western Desert developed sluggishly. It came in the form of closing the weapons gap with the arrival of American equipment as part of Uncle Sam's aid program. The Grant tank, for example, which had a 75-mm gun and frontal armored plating 50 mm thick, which surprised the Germans and caused heavy damage to their tanks, demonstrated a high degree of survivability against assorted anti-tank weapons.[53] The Grants began arriving only in the spring of 1942, more than one year after the German force invaded in North Africa. The 75-mm gun had one major limitation though—it was mounted on the side of the hull because it could not be fitted onto the turret. Still, it was the first opportunity the British had to fire high-explosive shells from a tank and deal more effectively than in the past with German anti-tank guns.[54]

Later, in the Battle of El-Alamein, the American Sherman, with its turret-mounted, 75-mm gun and high-explosive shells, was used for the first time. "It brought the British tank crews superiority over the ubiquitous Mark III and equality—or perhaps a little more—with the Mark IV. It brought the British infantry the comfort of effective high-explosive support, and it brought the German 88s the new fear of being pounded and destroyed at long range."[55] The reestablishment of diversity, despite the slowness of the process, enabled the British to meet the challenge of German anti-tank warfare more successfully.

The Problem of Lesson Learning
During WW II, the approach to lesson learning appears to have been an outgrowth of the flawed culture that was the hallmark of the interwar period. Between the wars, the British army had little inclination and slight interest in learning lessons from its field exercises. It also lacked an effective mechanism for disseminating combat experience to its units.[56] Only in 1932 was a committee established by the army for learning the lessons of WW I,[57] but the conclusions were heavily censored as the authors of the reports dared to criticize their nation's military capabilities.[58] Montgomery-Massingbierd, the British chief of staff, suppressed any conclusions that reproached the conduct of the British army.[59]

Despite the vast amount of material the High Command in England received from the East Mediterranean theater, it is debatable whether the battle reports and tactical lessons were seriously reviewed, especially since the High Command did not consider itself obligated to deal with tactical problems at the battalion level.[60] Any division preparing for combat for the first time had to figure out almost "everything" on its own.[61] Millett and colleagues contend that the British apparently failed to develop a mechanism for conveying the hard-won, desert-combat experience to the training facilities in Britain.[62] Hecht states that "the British needed two and half years of bitter experience—from the German invasion of France in May 1940 to the end of 1942—in order to internalize the necessary lessons and shift to combined arms warfare."[63]

In conclusion, this case illustrates extremely slow recovery from technological and doctrinal surprise. The brunt of the surprise came not from the anti-tank guns per se but from the manner of their application. The failure to recover quickly stemmed from rigidity in several areas: conceptual-doctrinal, cognitive-command, and organizational. The unwillingness to learn and the lack of tools for self-improvement precluded the rapid redressing of shortcomings. British ineffectiveness allowed technological and doctrinal surprise at the tactical level to precipitate a series of failures at the operational-strategic level.

11 THE SLOW SOVIET RECOVERY FROM THE SURPRISE OF LOW-INTENSITY CONFLICT IN AFGHANISTAN

THE SIXTH TEST CASE deals with the slow recovery of the Soviets from the surprise of the nature of the conflict in Afghanistan. The Soviet Union invaded Afghanistan in December 1979, and its army fought a low-intensity conflict (LIC)[1] between a conventional (in the meaning of regular forces and regular war fighting methods) state army and the guerilla force of an insurgency movement for over nine years (until February 1989). Scholars on the War in Afghanistan Lester Grau and Michael Gress have noted: "The war in Afghanistan gave the Soviet forces their first significant experience in the preparation and conduct of operations and combat against irregular guerrilla formations on mountain-desert terrain."[2] This chapter describes the Soviet recovery from the surprise. I contend that this case illustrates conceptual surprise: a conventional army's inability to subdue an unconventional enemy in large-scale operations. According to Stephen Blank, the Soviets were caught by surprise at the tactical and strategic level by the nature of the enemy and the fighting, a surprise that seriously hurt the army's operational effectiveness.[3]

The Soviet recovery was relatively slow and involved elements of conceptual and doctrinal flexibility and, especially, organizational flexibility. The low level of cognitive and command flexibility that was characteristic of the Soviet army retarded faster and more significant recovery.

Unlike the other chapters in this book, the current chapter deals with a prolonged, asymmetric LIC, one that the Soviets viewed as a "small war." As such, it is trickier to put a finger on the exact time when the surprised party—in this case, the Soviet army—recognized the gap between its expectations of the fighting and the reality that it faced. More than hours or days were to pass before the Soviets became aware of the gap. Unlike the other surprises described in this book, it developed over months and years. Also, the responses did not evolve evenly; instead, solutions to various problems can be identified that emerged in different phases of the war. The following discussion concentrates on the first years of the war (1979–1984) during which the Soviets introduced the majority of their adjustments to the new situation.

THE SURPRISE: SOURCES AND INFLUENCE

The Soviet surprise contained elements of concept, doctrine, organization, technology, training, and military education. The heart of the problem was that the Soviets prepared their forces for fighting in a nuclear scenario or high-intensity conflict (HIC), in open terrain, against European or Chinese state armies. Most of the elements of force structure that were based on this concept proved irrelevant in the Afghanistan theater.

The first element of the Soviet surprise stemmed from the attempt to decide the conflict with large-scale campaigns. The war began with Soviet battlefield success, in effect a rerun of the Prague takeover in 1968, and included airborne troops, the seizure of strategic facilities throughout the country, and a rapid mechanized and armored move along the main transportation axes. The combined effort enabled the Soviets to gain control of the Afghani population, economy, and infrastructure. The success of the operation and the perception that its influence would be decisive reflect the Soviets' erroneous belief in an unsuitable doctrine and the type of force required and their misunderstanding of the nature of the conflict (see below).

Guerillas, by their nature, exploit a regular army's weakness. Thus, the Mujahideen made a supreme effort to avoid direct contact with Soviet military power. They struck at typical weak points, such as supply convoys, and made frequent use of tactical ambushes to create a local advantage over the enemy, all the while conserving their strength by immediately breaking off contact after an attack and melting into the surrounding countryside.

The war in Afghanistan was characterized by Soviet search-and-destroy missions and clashes between small forces on both sides. This feature intensified in the mountain fighting when low-echelon, small-size units often attained a decisive advantage by making a surprise maneuver on the enemy's flanks or seizing a dominant position. To meet this challenge, the Soviets needed independent, operationally flexible forces and commanders at the tactical level. Soviet military leaders were aware of this, as an article written by Colonel Y. Pavlov in 1980 shows: "It was the skilled actions of small sub-units which often decided the outcome of combat in the mountains."[4] The problem was that in the years prior to the war, the Soviet war concept stressed the operational level but undervalued the importance of tactical action. Grau and Gress state that:

> The Red Army was victorious in WW II through their mastery of the operational art based on the skillful employment of large armies, fronts, and groups of fronts . . . The Soviets expected to fight WW III operationally. The Soviet Army incorporated tactical predictability that allowed it operational flexibility. Its structure, weaponry, tactics, and training all supported this operational focus . . . When the Soviets invaded Afghanistan, they soon found themselves fighting the Mujahideen. Naturally, the Soviets tried to defeat the Mujahideen with large, sweeping operations. Instead of victory, the operations used a lot of fuel, covered a lot of territory, and accomplished little. Still, the first years of the war were marked by a series of successive operations, more appropriate for the Northern European plain than the rugged mountains of Afghanistan . . . The best-executed Soviet operations were the invasion, "Magistral," and the withdrawal.[5]

Grau adds that the operational flexibility required for operational maneuver actions demands a certain degree of tactical rigidity and discipline. Battle drills were the basis for tactics at the squad and company level. The nature of the fighting in Afghanistan precluded successful operational-sized actions; instead, it focused on tactical activity and demanded tactical flexibility.[6] Soviet tactics were simple: they were designed for quick application by regular army troops and reservists without disrupting the development of the operational battle. Initiative at the tactical level received no incentive since it was liable to interfere with the timing of the entire operation.[7] Thus, surprise in this field was conceptual and doctrinal. The development of the tactical level in the modus operandi and especially in the education of junior officers encountered difficulties (see below).

The second element of the surprise was an outgrowth of the first—the Soviet force structure that had prepared for conventional fighting. The Afghan enemy—the Mujahideen—employed, as stated, guerilla tactics, which included the avoidance of direct clashes with Soviet firepower. Therefore, the basic Soviet techniques in HIC—the massive use of artillery, air strikes, and armored attacks in the enemy's depth—proved unsuitable. The Mujahideen refused to remain in permanent layouts and wait for Soviet artillery bombardments or armored attacks; instead, they abandoned the fire zone without a fight and returned later when the "coast was clear."

Grau discusses the unsuitability of the initial composition of Soviet forces and the steps taken to correct it:

> The Soviet 40th Army was outfitted for war on rolling plains with NATO or China. The 40th Army brought its full complement of tanks, air defense artillery, chemical protection units and all the other paraphernalia for conventional war against a modern mechanized force. Soon, the Soviets began sending home tank and air defense regiments and brigades and replacing them with more infantry.[8]

Tactically, the Soviets were surprised by the nature of the war (no clear frontlines, missile-defended zones, clear-cut targets)—all of which were the trademarks of Soviet tactical training.[9]

The way the mechanized infantry soldier was armed epitomizes the scenario according to which the Soviet force was built and its influence on problems at the lower tactics level. The Soviet warfare concept held that the mechanized infantryman should not fight at a distance of more than two hundred meters from his vehicle. His uniform, weapon, and equipment reflected this concept. Overweight gear, camouflage designed for a European environment, boots unsuited for mountain climbing, and even non-waterproofed sleeping bags—all of which caused serious problems when the mechanized infantry force had to leave its vehicles and advance intermediate distances, especially in mountainous terrain.[10]

The encounter with the reality of fighting a guerilla war and an enemy on whom artillery and tanks had relatively little effect seems to have caught the Soviets by surprise and forced them to introduce basic changes in the composition and application of their forces (see below).

Another element in the surprise stemmed from the nature of the country and the climate. The mountainous topography and the unpredictably inclement weather also surprised the Soviets. In addition to the difficulties in employing large-scale armored and mechanized forces, because of limited passage in the mountain areas and the need to offset this with relatively large numbers of light infantry, the Soviets also encountered problems in the use of military aircraft and artillery, complicated combat engineering challenges, and communications glitches due to equipment requiring a line of sight or operable only at short distances. For example, the engineering equipment designed for open areas proved inapplicable in the mountains. The mountain terrain with its abrupt weather changes rendered chemical warfare nearly irrelevant.[11] Soviet logistical elements entered the war unprepared for supporting combat operations in Afghanistan's demanding topography.

What was the source of the surprise and its degree of influence? The surprise may be defined as "self-surprise," since the Soviets were familiar with the nature of guerilla warfare and the country's geographical and climatic conditions. According to Blank, the Soviets made meticulous preparations for operations in the Afghan theater by studying its topography, ethnic divides, economy, and geography, and especially its transportation and economic infrastructure.[12] Grau and Gress acknowledge that the Soviets' faulty preparation for combat in Afghanistan raises a number of questions, given their considerable experience in the struggle against the Basmachi (who rebelled against Bolshevik expansion) in Central Asia between 1918 and 1933, since it seems like this knowledge was forgotten. The Soviets also seem to have forgotten what they learned from the experience of their partisan units against the Wehrmacht in WW II, their crushing of the anti-Soviet uprisings in the Baltic and Ukrainian regions after WW II, and their knowledge of other armies' struggles against guerilla forces, such as the Americans in Vietnam.[13] Nevertheless, they suppressed this vast body of experience when developing their combat doctrine, preferring to focus on the Red Army's massive, conventional campaigns against the Germans. The upshot was that the Soviet army entered Afghanistan without a doctrine for quelling insurgency and guerilla activity.[14]

Another possible reason for the conceptual surprise is that despite the Soviets' familiarity with Afghanistan and guerilla warfare, they had not really planned on fighting the Mujahideen. Therefore, they were surprised by the type of military activity facing them. Grau and Gress noted that the initial Soviet force numbered 52,000 troops. "This was considered sufficient to guarantee the viability of Afghanistan. It was thought that the Soviet forces would not have to fight during the invasion and subsequent stationing of Soviet forces. It was felt that the mere presence of Soviet forces would serve to 'sober up' the Mujahideen. Soviet military assistance would primarily be moral support to the DRA (Democratic Republic of Afghanistan)."[15] The Soviets planned to leave the fighting in the hands of the Afghani army. The inability to perceive the nature of the military challenge may have been the source of the conceptual surprise—regardless of whether it was rooted in faulty intelligence assessments or some other conceptual deficiency. This may explain the initial surprise but not why recovery took so long (see below).

As for the impact of the surprise, the Soviet army's entanglement in Afghanistan took a heavy toll in human lives, in the army's fighting spirit, and in the public's faith in the military system. Opinion is greatly divided over the question of whether the war's results influenced the breakup of the Soviet Union. Some claim that the war had a major effect on the dismantling of the Soviet Empire,[16] whereas others avow that its effect was minimal.[17] Whatever the case, a faster recovery would have undoubtedly lowered the cost of the war, though such quantification is elusive. Most scholars claim that even if the Soviet Union had adequately prepared for guerilla fighting in Afghanistan, it still would have lost the war because of the lack of troops. This was due to the unwillingness of the Soviet leadership to risk more losses (according to estimates, there were 26,000 Soviets killed in action [KIA]).[18]

The introduction of American Stinger anti-aircraft missiles into the Afghanistan theater in 1986 was a major technological surprise for the Soviets and certainly influenced their ability to recover. The American missiles had a devastating effect on the attack helicopters and armored assault helicopters that had been playing a key role in the Soviet response to the guerilla challenge.

THE SLOW RECOVERY FROM SURPRISE

Most scholars point to the Soviets' learning ability and adaptability, but at the same time they note its slow pace and limited effectiveness. Blank perceived that the Soviets underwent a ceaseless learning process that began shortly after their invasion of Afghanistan. "Having found itself fighting an unforeseen war there, the Soviet command began adapting its forces in Afghanistan to the requirements of the theater as it came to understand them."[19] Cordesman and Wagner write that the war was characterized by innovation and change. The Soviet forces that entered the combat theater, improperly trained and with equipment unsuited for counter-guerrilla operations, gradually improved between 1979 and 1984 and developed tactics and skills that threatened to quash Afghani resistance in 1985 and 1986.[20]

Most scholars, however, agree that the Soviet army adapted too slowly to the new fighting conditions. Thus, Luttwak asserts that the Soviets displayed "surprising slowness" in adapting to counterinsurgency warfare in Afghanistan.[21] Recovery from the surprise occurred at an uneven pace and with various degrees of success.

Slow Conceptual and Doctrinal Change

The Soviets had to shed the conceptual, doctrinal, and tactical notions they devised for fighting the Europeans or Chinese. Eventually they developed new concepts for non-linear war management that required a redefinition of the traditional conceptualization of fighting in echelons and the reorganization of layouts and units (with an emphasis on flexibility and survivability). During the 1980s, the Soviets modified their concept of the theater-strategic offensive by drawing up new concepts for cutting back the echelons at each level and revising their air units' employment concept. Inter alia, they tested new organizational structures at the corps, brigade, and combined-combat teams' battalion level. They also tested concepts in flexible logistics and adopted new techniques such as the armored group (Bronegruppa).[22]

The overall concept of the fighting emerged slowly. During the first years of the occupation, the Soviets applied the direct approach of large-scale, usually division-sized, attacks that included lengthy air strikes, preparatory bombardments, and attacks by armored columns with large amounts of supporting fire. Beginning in 1980–1981, modifications were introduced, with an increased emphasis on airborne assault troops, but they failed to have a significant impact on the effectiveness of the wide-scale conventional attacks.[23]

In 1983 after three years of fighting, a new approach can be observed in the Soviet combat operations. The first element is the removal of mechanized forces from direct contact with the Mujahideen. The mechanized troops were employed in defensive activities, guard duty, punitive actions (such as crop burning, destroying irrigation systems), and so forth that were designed to reduce the local population's ability to support the guerillas and eventually force the locals to leave the area.

The second element is the use of designated infantry forces against the guerrillas. These forces were originally air assault troops, paratroopers, and reconnaissance and special forces. Although they had been operating since the beginning of the war, it was from this stage forward that they bore the brunt of direct offensive activity against the Mujahideen. Mark Urban points to the emphasis on airborne troops in Afghanistan: air assault troops numbered 15 percent of all Soviet forces of this type, whereas mechanized forces, only 2.5 percent.[24] These forces engaged in assignments that can be defined as tactical rather than operational, such as long-range reconnaissance missions, night ambushes, helicopter raids, and so forth. These missions were highly effective as compared to the conventional activities in the first years of the fighting. But it should be noted that the designated forces, too, could have adapted themselves better to anti-guerrilla fighting and that they continued to rely "too much" on technological superiority and "too little" on tactical superiority.[25]

Some of the changes in the general concept took place in the use of the Soviet helicopter force. Like the Americans in Vietnam, the Soviets discovered the advantage of helicopters in quelling insurgents. In this case, too, the pace of change was relatively slow. The massive use of the Mi-24 (Hind) helicopter as a "flying tank" began relatively early in the campaign as a solution to ground limitations (movement and firepower), but until 1983 it was still used in standard attacks. Only when major changes were introduced in the Soviet air assault concept and emphasis shifted from ground action to vertical flanking moves was the helicopter's potential finally realized.

In 1984 the helicopters began to play a significant role in the war against the Mujahideen. For all practical purposes, the Soviets used the air force as an alternative to mechanized forces. The problem lay in overreliance on the solution, whose effectiveness was seriously reduced with the entry of shoulder-held, one-man-operated Stinger missiles in 1986. The Stinger's technological surprise critically impaired the Soviets' new modus operandi.[26] Robert Cassidy writes that the use of helicopters did not enable the Soviet forces to free themselves from conventional conflict thinking and make the conceptual leap to a force structure truly designed for anti-guerrilla fighting. But the use of helicopters definitely enhanced the Soviets' ability to transfer the battle to areas under Mujahideen control.[27]

At the tactical level, modifications in the general concept came at a faster pace. The new modus operandi that the Soviets adopted included: raids, blocks and searches, ambushes, and convoy escort. The raid was a modification of the Soviet forward formation engagement in a conventional battle. The use of this method began immediately after the invasion of Afghanistan. "Cordons and searches" did not appear in Soviet military training manuals as a combat technique, and for this reason the Soviet forces regarded it as a new tactic. Actually it had been employed after WW II mainly by the People's Commissariat for Internal Affairs (NKVD) forces in the forests and marshes of the eastern Soviet Union. The technique was immediately applied on a wide scale in Afghanistan.

The ambush, as an independent tactical action, was also employed extensively by the Soviets in Afghanistan in all forms of combat. The country's vast area, the rapid troop mobility, and the need to bring equipment from the Soviet Union, led to the massive use of truck convoys and the development of ways of protecting them and the logistic routes. A static layout of security posts and inspection points was established that functioned both defensively and as a solution to maintenance requirements and medical treatment. An electronic control system was built for supervising movement and other purposes. None of the units in the Soviet expeditionary force had trained in this type of activity before the war.[28]

The Bronegruppa was an innovation at the tactical level. The idea was to use mechanized infantry for infantry activity and armored fighting vehicles (AFVs) as a firebase for assistance and as a reserve. To realize the concept, the commanders of the mechanized force had to overcome their resistance to having the infantrymen distance themselves from their vehicles. But many cases called for precisely this separation because the terrain did not permit AFVs, *amphibious* tracked Bronevaya Maschina Piekhotas (BMPs), airborne Boyevaya Mashina Desantas (BMDs), and multi-wheeled Bronetransportyors (BTRs) to escort the infantry forces in close formation. On the other hand, the separation enabled the force commander to keep the vehicles in reserve for independently attacking the enemies' flanks, blocking possible enemy trails, serving as a mobile firebase for advancing foot soldiers, transporting supplies and ammunition to forces that had infiltrated earlier on foot, escorting convoys, and so forth.[29] The Bronegruppa included four to five AFVs and tanks.

Another area of change was in the development of mountain warfare. After failing in the employment of armored and mechanized forces in mountainous terrain in 1980, the Soviets reevaluated their doctrine. The result was a reduction in the role of armor in the Afghani theater and an emphasis on mechanized infantry activity, air assaults, special forces, helicopters, and anti-aircraft cannon that enabled rapid, direct-line fire on the mountain slopes. The entry of armored combat vehicles with improved mobility, such as the BTR-70 and BTR-80, and variants of the BMP and BMD, enabled a greater degree of off-road movement. Greater use was made of light artillery: 160-mm mortars and 76-mm mountain cannons.[30]

Another type of adaptation was based on the Soviets' awareness that mountain fighting, where forces operated with relative independence in limited areas, required the integration of support elements at low levels because of the difficulty in assisting combat troops in a more concentrated manner. The solution was to integrate support units at the platoon and company level that included engineer squads or teams, mortar crews, grenade launchers, radio operators, scouts, and forward-observation officers. The innovation was not the allocation per se of support forces to the maneuvering troops, but the low echelon at which this allocation was carried out.[31] These changes gradually increased the combat troops' effectiveness in mountain areas.

John Sray states that despite the development of the army's understanding of mountain warfare and the solutions it entailed, the Soviets failed to implement them during the war.[32]

Changes were also introduced in the methods of applying artillery. Soviet artillery had a high standing in the overall war concept. Artillery officers based their firing systems on covering wide areas with a specific number of shells for a specific target or particular area in order to neutralize or destroy it. The system required time, large amounts of ammunition, and a static enemy; in short, it was not suited to conditions in Afghanistan. The artillery officers were loath to forego this concept and its irrelevant terminology. A new system was developed during the war that enabled different fire assignments to be carried out simultaneously from a number of separate fire-direction centers in the same battalion.[33] As a rule, the artillerymen adapted themselves relatively slowly to the conditions of the Afghani battlefield.

Limited Organizational and Technological Flexibility

In this field two features are discernable: changes in the general arrangement of the forces, primarily the addition of infantry rather than forces that had been developed mainly for wars against regular armies; and changes in the size of the fighting formation and the degree of integration among the forces in the formation.

Basic changes in force structure were made relatively quickly when the 40th Army was reinforced with a division of mechanized infantry and two independent infantry brigades in the first half of 1980.[34] In mid-1980 the following units were shipped out of Afghanistan: the 40th Army's artillery and air defense brigades, one armored regiment, and several Frog surface-to-surface missile battalions—altogether about ten thousand troops. In 1981 an air assault brigade and regiment of attack helicopters joined the fighting forces.[35] The number of tanks in Afghanistan dropped from one thousand in the beginning of the fighting to about three hundred in late 1980. Between June 1980 and 1981, the number of helicopters rose from approximately sixty to nearly three hundred.[36] All of these changes took place within a relatively short time and illustrate the Soviets' perception of the new nature of the fighting.

During the war the Soviets tested various types of unit formations. These included independent mechanized infantry brigades that employed heliborne assaults for the first time,[37] mechanized infantry battalions that functioned as independent taskforces, and others.[38]

The rigidity of the original organizational structure was due to divisional headquarters' centralized control over combat support elements, such as artillery, engineers, and communications, and army headquarters' control over the air force. This type of organization left battalion commanders without permission to employ support units in emergencies; instead, they had to go through regimental, divisional, and sometimes even army headquarters before the green light was given.[39] The Soviets learned from their experience in the first year of the war and reinforced their mechanized infantry battalions with firepower when they were sent on independent combat operations. Reinforcements often consisted of up to a company of tanks, one or two artillery batteries, a platoon of mobile anti-aircraft cannons, a platoon of combat engineers, and one or two squads of chemical warfare troops. Sometimes they received support from fighter planes, attack helicopters, or armored assault helicopters.[40] This type of unit became the basic formation for troops operating against the Mujahideen. When engaged in mountain fighting, forces and weapons were integrated at the lowest levels. Most of the existing equipment remained without technological adaptations, but the older weapons were replaced with newer ones.[41]

The Failure to Develop Cognitive and Command Flexibility

In the field of cognitive and C2 flexibility, the Soviets were not very successful. They found it difficult to change traditional patterns of action that were based on centralized command, which inhibited initiative and laid great stress on detailed planning and adherence to plans. An Afghani general who deserted to the Mujahideen described his former comrades as lacking "initiative" and "addicted" to fighting "according to the book" and mechanically following "battle formulas" in various situations.[42]

Soviet military literature from the war period included within the military leadership's weaknesses in Afghanistan: lack of initiative, low-standard ability to exploit support elements, and the distrust that prevailed among the echelons.[43] The Soviets realized the need to decentralize decision making in combat, especially in mountain terrain that was so unlike the large-scale campaigns on sweeping battlefields. After the generals recognized that the traditional Soviet solutions to the problem of control—more radio equipment and relay stations—were incompatible with the conditions of Afghanistan, the role of junior officers in mountain warfare began to be reviewed. From these discussions began the gradual increase in authority of senior NCOs and junior officers so they could take greater initiative in combat situations. By 1984 there were signs that the Soviet army was implementing its conclusions on the decentralization of command in its forces in Europe. Another area of flexibility was that of granting field commanders the authority to order close air support without the requirement to go through higher echelon headquarters.[44]

According to Scott McMichael, the main conclusion from the Soviets' inability to develop the flexibility required for tactical operations in Afghanistan was that the typical mechanized infantry commander was incapable of adjusting to a rapidly changing tactical situation. The special tasks he was assigned were beyond his capabilities due to his conventional military education and training[45] in rapid armor maneuvering as part of large formations.[46]

Blank states that even after Afghanistan, Soviet training continues to emphasize "following the rules and procedures" as the standard for quality performance. The lack of creative thinking and independent judgment and decision making still characterize the Soviet officer cadre despite the introduction of reforms intended to ameliorate the situation. Detailed orders to the lower echelons leave them only with the technical performance and reproduce a generation of commanders trained in a centralized command environment.[47] McMichael adds that there is little evidence to show that the lessons from Afghanistan have precipitated a significant improvement in exercises that foster initiative, decentralized decision making in the lower ranks, or imagination in planning and training.[48] It seems, then, that in the area of command cognitive skills, the Soviets fall far short of flexibility.

Regarding the ability to learn while engaged in combat, Cassidy writes that the dissemination of knowledge—acquired in blood by units of the 40th Army—was poorly carried out. Many units gained experience in the Afghani theater but rarely conveyed it to other units.[49]

Why was recovery from surprise so problematic? First, the Soviets seem to have implemented organizational changes relatively quickly in their initial response to the surprise. A similar success occurred in their rapid organizational changes after Operation Barbarossa (see Ch. 4).

But the Soviets applied the other flexibility strata only slowly or not at all. The conceptual and doctrinal shift from large-scale armored and mechanized operations to relatively limited-scale, heliborne, air assault operations required three years. As for command-cognitive changes, the Soviets failed to develop a viable response to battlefield surprise. One possible explanation is the military and civilian culture of obedience and fear of authority—traits that cannot be changed by issuing an order "from higher up" or through training. The difficulty in achieving cognitive and command flexibility has more complex roots.

One source is the conceptual failure to distinguish between HIC against a conventional enemy and LIC against irregulars. The Soviets, Blank avows, were unable to perceive the conceptual gap between what they regarded as "local war" and "theater war." Thus, they included the Arab-Israeli wars, Vietnam, and the Falkland and Afghanistan wars in one group without discerning their distinction from a large "theater war," especially regarding the relevancy of warfare principles or employment of weapons systems. The war in Vietnam illustrates this problem. The Soviets believed that the lessons from Vietnam were universal axioms for future confrontations. However, they underestimated the importance of local war conditions and their influence on the lessons learned, and they overemphasized the so-called "big lesson": communist forces are "destined" to win every conflict regardless of specific conditions or local limitations.[50] This ideological mindset seems to have been the key stumbling block to a genuine change of doctrine in the first years of the Afghanistan War.

On the other hand, Theo Farrell notes that armies tend to adhere to norms of conventional war even if they have to prepare for LIC because these norms remain at the basis of the military organization's professional identity. Therefore, an army will make every effort not to deviate from them.[51] Cassidy claims that the Soviet army was aware of the changes that had to be introduced to succeed in Afghanistan, but it was mired in the cultural inertia of adhering to the blueprint for the "big war."[52] This problem may be especially acute in the armies of the superpowers, which must constantly maintain the ability to meet the challenge of the rival superpower even when engaged in a LIC. According to Grau and Gress (2002), even though a major conventional war does not appear on the horizon and even though the Russians were involved in LIC conflicts in Chechnya, Ossetia, Dagestan, Georgia SSR, and Tajikistan, Russian officers still spend their time learning large-scale battles in military academies.[53]

Another explanation for the resistance to change may be the organization's internal need to present WW III as a scenario that justifies force building and the army's outlay of enormous sums in preparing for it. In the same context, Deborah Avant has noted the Americans' difficulty in planning and adapting to combat in Vietnam.[54]

In summary, despite the large number of adaptations and modifications that the Soviets introduced, they still fell far short of having a decisive influence on the war. Grau and Gress conclude: "There is ample evidence that the Soviets never really understood their enemy or the neighboring country in which they were fighting" and that "the Russian army is unwilling or unable to apply the lessons of this conflict to their present situation."[55]

12 THE FRENCH FAILURE TO RECOVER FROM THE SURPRISE OF THE GERMAN BLITZKRIEG

ON THE MORNING OF MAY 10, 1940, the German army launched an offensive against France and overran the country within a few weeks. One of the main causes of the debacle was deploying a major, qualitative part of the French army and the British Expeditionary Force to implement the French plan for dealing with the German army on the plains of Belgium. The concentrated German effort in the Ardennes Forest and rapid breakthrough into northern France thwarted the French army from realizing its plans. Another cause of the failure to recover from the speed and momentum of the blitzkrieg was the French lack of flexibility.

THE SURPRISE: SOURCES AND INFLUENCE

The French army failed due to various weak points. May's book on the fall of France, *Strange Victory*, states that in 1940, and for a long time afterward, the accepted explanations for France's collapse were: unconditional German military superiority, poor French and British leadership, and unwillingness to put up a fight. According to May, scholars now realize that none of these facts are valid or the main reason for the debacle.[1]

In his book *Strange Defeat*, the French (Jewish) officer Marc Bloch contends that the main causes of the fall of France were exaggerated self-confidence in the French army's superiority; overemphasis on weapons that were supposed to reduce losses (e.g., bunkers and firepower) but that resulted in the army's slow response and limited initiative; and stodginess in French (as well as British and Belgian) military decision making.

May acknowledges these charges and adds that Germany won despite its military inferiority and that the French and British failed to exploit their superiority. He further claims that the Germans identified the psychological and procedural weak points that Bloch mentioned[2] and exploited them to the hilt. He quotes Hitler, who, on October 25, 1939, in a meeting on the plan for the invasion of France, made it clear to his generals that the operation must be a fast and deep penetration that forces the enemy "to operate and to act quickly, something that does not come easily to the systematic French or to the ponderous English."[3] Thus, as May sees it, Hitler had the insight that was the key to German success: exploitation of the Allies' procedural weaknesses. This chapter develops May's thesis and describes how French shortcomings influenced France's ability to recover from the surprise.

I maintain that the primary consequence of the French warfighting concept and combat doctrine (see below) was the French army's inability to recover from the surprise of the form of German fighting—the blitzkrieg—and especially from its momentum and speed. Liddell Hart has this to say on the way the French dealt with the speed and momentum of the blitzkrieg:

> The issue turned on the time-factor at stage after stage. French countermovements were repeatedly thrown out of gear because their timing was too slow to catch up with the changing situations ... The French, trained in the slow-motion methods of World War I, were mentally unfitted to cope with the new tempo, and it caused a spreading paralysis among them. The vital weakness of the French lay, not in quantity nor in quality of equipment, but in their theory.[4]

Bloch sums up the French military failure:

> Our leaders, or those who acted for them, were incapable of thinking in terms of a *new war*... The ruling idea of the Germans in the conduct of this war was speed. We, on the other hand, did our thinking in terms of yesterday or the day before. Worse still: faced by the undisputed evidence of Germany's new tactics, we ignored, or wholly failed to understand, the quickened rhythm of the times.[5] Our own rate of progress was too slow and our minds were too inelastic for us ever to admit the possibility that the enemy might move with the speed which he actually achieved.[6]

In his book *Blitzkrieg—From the Rise of Hitler to the Fall of Dunkirk*, Len Deighton writes: "It was the time factor that surprised and defeated the French. Not only did the French deploy slowly but they did not believe that the Germans could move any faster."[7]

Fuller described the German application of the military doctrine in France as the tactics of speed that included: maintaining the momentum of the initial attack until the objective is reached; integrating weapons of all branches so that the shock force is concentrated at the point of impact; disrupting the enemy's command. The goal of the tactics is more psychological than physical, namely, to undermine the enemy's will and throw his organization into turmoil.[8] The German Panzer divisions—the leading mobile force in the blitzkrieg—were built on a combined arms integration of units (armor, infantry, artillery, armored engineers, and anti-aircraft) and received close air support for reconnaissance and bombing. Branch integration and close air assistance decisively contributed to German superiority in battle.[9]

The French army's surprise did not come from German weapons (about which they had full knowledge) but from the Germans' method of employing them. The French army's problem in force employment stemmed from its inability to meet the challenge of a fighting method that was not methodical like its own. The biggest surprise to the French was the momentum and mobility of the German forces, whose combat doctrine emphasized the immediate adaptation of plans to circumstances and the exploitation of success by field commanders.

According to Fuller: "The attack had been so sudden and so overwhelming that the French Command did not understand what was happening. They did not realize that, having penetrated their enemy's front, the Germans would thrust straight ahead with their armored and motorized troops. It would seem that they expected a pause, a build-up and then a further attack."[10]

French and British POWs repeatedly admitted to experiencing total shock at the lightning assault of German armor units: "Opponents, who had been fighting in Belgium and had been beaten, wanted to regroup and defend, but they were surprised by the thrust of the Panzer divisions."[11]

On the night of May 16, Rommel's armored division advanced across the French fortification line. "The French were clearly taken by surprise, first that Rommel should have broken through the frontier fortifications with such speed, and secondly that, against all the rules, he should be continuing the advance by night."[12] These are just a few examples of the French tactical surprise from what may be categorized as doctrinal surprise. Later it will be shown how French forces failed to recover from this type of surprise because of their dogmatic mindset, overly rigid C2, and lack of communications equipment for relaying information.

The French surprise by the blitzkrieg can be classified as self-surprise. At first glance the blitzkrieg appears to have been secretly developed by German military geniuses before the war: German tank formations were superior to those of France in armament and quantity, and German force organization and applied doctrine caught the world by surprise. A closer look reveals that only the last statement is correct.

The origins of deep armor penetration, whose aim is to undermine the enemy's consciousness and defeat him seem to be a combination of the German infiltration doctrine, which was developed and employed at the end of WW I, and British thinking and testing of armored forces, especially in the first part of the interwar period. Major-General J. F. C. Fuller, the father of "Plan 1919" for applying armored masses in WW I, was scrupulously studied by Guderian and German armor officers. Fuller, a fascist in his twilight years, visited Germany several times during the Nazi regime, where his politically tainted writings were widely read.[13] Liddell Hart, too, adopted the concept of armor warfare from Fuller in 1922 and included Fuller's ideas in many of his works. The Germans were influenced by Liddell Hart's books and articles, although some scholars claim that this was less than generally assumed.[14] Whatever the case, British thinking had a decisive impact on the establishment of the German Panzer corps and development of its doctrine.[15]

The interwar period witnessed all of the industrialized nations engaged in polemics, testing, and the construction of armored forces at various levels. The period is also characterized by the transfer of much information between Germany, Britain, France, and the Soviet Union on these matters.[16]

Azar Gat, a scholar of military theory, believes that the British, French, Russian, and German military leaders were all aware of the potential of the new weapon for restoring the ability to advance and maneuver and the key role it would play in the coming war. Opinion was divided not over the primacy of the tank but over how it would be employed. The military planners could not say whether the development of mines and anti-tank guns would stop the tank, just as the machine gun had stopped the cavalry and infantry. During the 1930s, when all of the armies demanded large tank forces, conservatives and radicals differed over the establishment of designated armored formations that would function as concentrated, mobile fists or having the tank provide close support to infantry formations as it did in WW I.

Gat states that "lacking battle experience except for the First World War it was impossible to predict how armor would be used most effectively and no responsible military authority would put all the eggs in one basket." Therefore, all of the armies of the great powers decided on the "golden mean" in the 1930s, that is, building armor divisions while apportioning tanks for close infantry support. This approach also meant dividing the production effort between fast "cruiser" tanks for in-depth armor offensives and heavier, slower tanks for infantry assistance.[17]

The French Army and the Employment of Armored Maneuver Forces

The French had four thousand tanks in WW I but never used them in concentrated form for a breakthrough. The French view of the coming war emphasized firepower and a strict command system that made maximum use of firepower and that subordinated tanks to the infantry, mainly for close assistance.

Charles de Gaulle's book *The Army of the Future* (published in 1934) championed the use of mobile armored forces, but the idea fell on deaf ears in France's conservative military establishment, which determined that despite the use of tanks and airplanes, tactical warfare remained essentially the same as it had always been. In 1935 the French defense minister rejected parliament's proposal to establish an armored corps.[18] After Germany occupied the Rhineland in March 1936, France began upgrading its army. In September it decided to establish three mechanized and two armored divisions. In November the minister of defense ordered a study of the employment of tanks organized in armored divisions. The general staff, however, did not follow through.

The first testing of a heavy armored force was supposed to take place in March and October 1938 but was postponed due to the annexation of Austria and the Munich Crisis.[19] An analysis of French intelligence shows that it managed to attain substantial information on German military innovations. The latest weapons were described in great detail and sometimes photographed. Their tactical employment was analyzed—from armored cars to road construction equipment. However, the information seems to have been processed to fit the French defense concept, and the intelligence branches pooled their assessments to conform to the expectations of their superiors.[20]

At the same time as the ideas of force structure crystallized, the French began manufacturing advanced tanks. In 1935 heavy Model B tanks were being produced, and in 1936 the Somua, a fast, medium tank with greater firepower, was being manufactured. In 1937 the 47-mm anti-tank gun began rolling off the assembly lines.[21] Despite these late arrivals, the dominating concept on the use of tanks remained conservative and regarded the tank's main role as infantry support.

The French, like the Soviets, also misinterpreted the lessons from armor employment in the Spanish Civil War. In 1937 the French minister of defense stated that near Madrid "tanks lay 'pierced like sieves.'" He added "be reassured . . . our fortified works are sufficiently equipped to halt a sudden attack even on Sunday . . ." In conversations with a British general, the French chief of staff Gamelin avowed that German tanks in Spain were "inadequately protected, fit only for the scrap-heap." In his memoirs, Gamelin regrets that he had not "demanded in 1935 the creation of an independent armoured force. . . . Had I known that the war would not break out before 1939, I would certainly have tackled the problem."[22]

According to Robert Young, lessons from the success of defensive weapons, such as anti-aircraft and anti-tank guns, were overemphasized, whereas lessons from the successful application of tanks and aircraft were considered irrelevant because defense preparations had been shoddy, and the circumstances surrounding the war in Spain unique: in other words, everything was "doctored" to fit the French defense doctrine.[23]

French planners analyzed Germany's growing Panzer branch and drew the mistaken conclusion based on their own tests with armor. Since the French tests and those of the Germans (according to French intelligence reports) with independent tank formations failed to reach conclusive results, the French planners assumed that the Germans had drawn similar conclusions. That is, tank units had to be integrated into the infantry until greater numbers of newer tanks were produced and the mechanized infantry was sufficiently large to follow in their tracks since this would enable the formation of independent tank units. The German generals agreed with this analysis to be correct—but Hitler did not. Under his pressure, the Germans embarked on war against France with fewer tanks than the French had, and only a few of which were of superior quality, but with a doctrine that paralyzed the French army.[24]

The German conquest of Poland did not lead to any change in the French concept of armor employment. The collapse of the Polish army seemed irrelevant to France's situation. French intelligence was duly impressed by German mobility and an intelligence report, based on the testimony of French military attaches in Poland and the debriefing of six hundred men who fought or witnessed the fighting in Poland, was circulated among senior French commanders and explicitly indicated all of the German combat elements.[25] Nevertheless, French intelligence maintained that the absence of border fortifications, incomplete mobilization of the attacked party, and overwhelming German military superiority were the factors that rendered the lessons from the Polish invasion irrelevant for war against France.[26] The French armored force developed slowly and only began making headway just prior to the war. The first two armored divisions became operational in January 1940 and the third in April.[27]

On May 10 when France was invaded, the French were quantitatively superior to the Germans in weapons—tanks, planes, and artillery pieces. Taking all of the Allies weapons into consideration, Allied superiority was greater.[28] The armament of the French tanks was technically superior to that of the Germans. The number of new heavy Model B and Somua tanks with thick armor plating was greater than the number of Panzer IIIs and Panzer IVs. The Model B had heavier armor and superior guns to any German tank.[29] The main problem with the French tanks was that 80 percent of them lacked radio equipment, which naturally had a pejorative influence on control over them.

Little need be said about the results of the blitzkrieg doctrine on the conquest of France in 1940. Despite all that has been described—the development of the concept's roots in Britain; the Germans' meticulous study of the cumulative British knowledge; German-Soviet cooperation; the prewar, unclassified publication of German military literature on armor operations; and even the Germans' manifest use of blitzkrieg tactics while engaged in the conquest of Poland—nothing was able to save France and the British Expeditionary Force from defeat in May 1940. Despite the quantitative and qualitative advantage of its tanks, the French army was routed by the German organization and combat doctrine that made better use of its smaller and technically inferior armored force. For a long time the French army had been aware of the enemy's doctrine and organization, but it failed to take the necessary steps. Therefore, the surprise in 1940 may be classified as self-surprise. It was not a case of the Germans successfully concealing a new warfighting doctrine from the enemy.

How great was the surprise? It is difficult to quantify. Surprise resulted from the blitzkrieg's momentum and speed and from the fatal mistake of deploying the cream of the French and British forces in Belgium immediately at the outbreak of war and exposing the Ardennes sector to enemy penetration. Therefore, I will not discuss the strategic significance of the surprise, stating at the outset that its influence was felt at the tactical and operational levels.

What obstructed the French army from recovery?

THE INABILITY TO RECOVER FROM SURPRISE

A large part of the blame for French inflexibility may be placed on the conceptual-doctrinal element, which can be identified on two levels. The first is inflexibility of the "first order," that is, defensive fixation that caused doctrinal one-dimensional thinking. A second element of the first order is the "methodical battle" (*la bataille conduite*). This fighting method (see below) required a centralized C2 method that reduced the cognitive flexibility of the French junior command. Of no less importance, it included the new weapons (especially tanks) in the existing method that was relatively static, thereby stifling the development of dynamic thinking that could have been derived from maneuver.

Inflexibility of the "second order" was the suppression of thinking that deviated from the official military doctrine, and whose expression was intolerance with views and ideas that did not correlate with the official view. In addition to the mindset just described, this situation also hampered the ability to develop an alternative concept from which insights could be drawn when the dominant concept proved inadequate.

Doctrinal Dogmatism

The source of the dogmatism in defense thinking and in developing the offensive methodical battle is identical. It was based on the belief that the development of firepower, as a result of developments in weapons that modern armies had been equipped with between the wars, would transform the battlefield into a more lethal environment than in WW I.[30] The words of General Philippe Petain, the commander of French forces in WW I, that "firepower kills," concisely reflects this principle.[31] Movement on the battlefield without massive fire support appeared suicidal. The French solution was to make maximum use of the advantages of defensive fire and rely on firepower for the attack. The solution also called for the development of a fighting method based on limited movement and tight control suited to the pace of artillery activity. Overreliance on firepower, according to Bloch, stemmed from France's experience in developing doctrines and weapons that would save lives in the coming war.

In addition, the French erred by projecting their way of thinking onto the Germans. They believed that the Germans shared their view of the extreme importance of firepower in an offensive maneuver and that the German army's manner of operations was based on this understanding. The French discerned close similarity between their concept of force employment on the future battlefield and the German concept.[32]

As war approached, one-dimensional, conceptual-doctrinal thinking, expressed as the cult of the defensive, grew out of the cult of the offensive that had characterized French thinking before WW I.

There are two schools of thought on the factors that influenced the development of the French defensive concept. Posen stresses the impact of international relations, claiming that France had to adapt itself to the British concept (France's ally) that was seen as vital to French defense.[33] On the other hand, Kier stresses the prevalence of the organizational culture in the French army that was influenced more by internal French politics than from the perception of external threat. In 1928, when compulsory service was shortened to only one year, the army's organizational culture led its commanders to realize that the troops could not be educated and trained for complex offensive operations in so short a time. In essence, the French political echelon "forced" the army to opt for the defense—and so it did. Furthermore, in the years preceding the war, the standing army, suspected of right-wing tendencies, had to deal with left-wing state leaders. The sense of threat induced the military leaders to eschew external criticism and avoid changes that might lead to such criticism.[34]

In his book *To Lose a Battle: France 1940*, Alistair Horne notes that the building of the Maginot Line closed a cycle of military thinking that began in 1870. The French lost the 1870 war because of their defensive posture and overdependence on fixed strongholds. In 1914 they succumbed to the cult of the offensive and almost lost the war. As 1940 approached, once again they sought security in a mantle of cement and steel. "Rapidly the Maginot Line came to be not just a component of strategy, but a way of life. Feeling secure behind it, like the lotus-eating mandarins of Cathay behind their great wall, the French Army allowed itself to atrophy, to lapse into desuetude."[35]

Its defensive doctrine notwithstanding, France was also a partner in alliances with other countries (e.g., Czechoslovakia and Poland) that obliged it to develop an offensive capability in the event that they were attacked by Germany and French assistance was required. Offensive capability meant that heavy tanks and artillery had to be developed and manufactured. The concept of force employment in an attack was based on a methodical, meticulously planned offensive borrowed from the late WW I model. Although the civilian authority demanded a stronger offensive capability in 1934, the military establishment viewed mobility more in the defensive context than the offensive one.[36] France's generals may have been cognizant of the need to hasten preparations for meeting the German challenge—which included plans for adding large numbers of tanks and planes to their arsenal, but "no attempt was made to alter the army's predominantly defensive plans for the use of these weapons. No major preparations for offensive operations of any kind were made."[37] Posen claims that the defensive doctrine that the French developed in the 1920s and early 1930s:

> ... put French civilians and soldiers in an unfavorable position to absorb improvisations in tank and aircraft design ... the armed forces accepted the new weaponry without "adapting" to it. ... Moreover, French officers were taught abstract principles of war in a "formalistic and routine" way. Instruction in the actual conduct of complex military operations seems to have been sparse.[38]

Doctrinal one-directional mindset generally causes armies to lose their ability to conduct the neglected form of war and in this way limits flexibility. In the French case, this refers to the ability to wage a counterattack. According to Posen, the French doctrine avoided "encounter battles" (clashes between maneuvering forces fighting from unprepared positions) that are characteristic of mechanized warfare. Encounter battles are high-risk engagements, violent, and generally costly.[39]

Eugenia Kiesling, an expert in the French army's prewar preparations, mentions the dissonance that developed in the armed forces. On the one hand the French convinced themselves that their defenses would stand up to a German attack; on the other hand, they also succumbed to the belief, as expressed in their doctrine, that victory is gained by an attack and that the methodical battle would enable them to penetrate German defenses.[40] The upshot was that French officers who had been educated in the superiority of firepower came to realize that to overcome the difficulties inherent in an attack, vast amounts of ammunition must be prepared, the fire curtain strictly controlled, and friendly fire avoided. In other words, the shape of this type of battle must be meticulously maintained. An exercise carried out in 1938 that simulated a breakthrough from the Siegfried Line demonstrated the enormous difficulty in carrying out such an attack. The ambiguous message regarding the need for offensive capability—in light of the horrific cost it entailed—penetrated all levels of the army and cemented defensive thinking.

Centralized Command and Control Doctrine

The origin of the second element of inflexibility was the French C2 doctrine that was derived from the concept and doctrine of the methodical battle. This doctrine emphasized action in stages and centralism and limited commanders' flexibility.

The fighting method had an impact on both the operational and tactical levels. Evidence of this appears in French military studies and doctrinal literature. In a 1938 lecture at the Center of Higher Military Studies, a French general graphically outlined a complex military operation consisting of five corps with fifteen divisions deployed across a sixty-kilometer area. The force, made up of six first-echelon divisions, attacked a fifteen-kilometer zone in which each division attacked a sector measuring two and a half kilometers. The depth of the target was no more than half the width of the attacking front. Since the French had implicit faith in the methodical battle and the inherent advantage of making optimal use of lethal fire, it was natural that the artillery dictated the rate and stages of the fighting at the lower tactical level. French doctrine assumed that the infantry would advance one or two thousand meters and then halt to allow the artillery to redirect its fire. Intermediate objectives were determined at distances of one to two kilometers to control the infantry's rate of advance and enable the artillery to provide assistance. After the infantry advanced four or five kilometers, the artillery units had to move forward.[41]

In 1955 a French colonel wrote an article entitled "Adaptation" that charged the French army with not knowing how to adapt itself to the pace that the enemy dictated. The fighting method of 1940, the article attested, was based on the principles of the 1916–1918 fighting method with minor adjustments:

> Rigidity of formations, of zones of action, and of axes; chronometric tyranny; equations of tonnages of fire per acre; extreme centralization . . . the horizon of the familiar terrain compartment . . . worship of the complete, written order; of the complete report at a fixed hour; and of the detailed drawing showing everything down to the humblest rifleman. . . . During the 20-year period an entire generation of officers was thus prepared in accordance with norms which were proper to siege warfare.[42]

The French believed that decisions in the methodical battle had to be made at the highest levels so that tight control over the action of a large number of subordinate units was preserved. The doctrine and organizational structure stressed the authority of the commander of the army group, army, and corps and left little flexibility to lower echelon commanders.[43] The French war minister Charles Huntziger admitted that "the French army had relied too much on textbook solutions before the war, and that future training exercises should emphasize having commanders solve unanticipated problems, making decisions, and issuing concise orders rapidly."[44]

The French system operated according to the principle of "pressure from above," that is, decisions were made at the highest levels and implemented via rigid centralism and blind obedience. This was the exact opposite of the German system of "pulling from below," in which the method of decentralized C2 called for initiative and flexibility on the part of officers.[45] Due to cognitive and C2 constraints, France's system limited the ability to respond to surprise stemming from the enemy's warfighting doctrine.

An analysis of the French use of artillery on the battlefield, and its comparison to that of the Germans, illustrates the influence of the methodical battle concept on the rigidity of the French C2 system and the mistaken notion that such a concept would also be applied by the enemy. It will be recalled that the French artillery doctrine emphasized concentration of resources, meticulous planning, and methodical battle in stages. These were the principles of the battle system that "shackled" the main force's movement to the artillery's rate of movement, the timetable for the employment of artillery fire, and the number of artillery shells required.

The Germans, on the other hand, opted for decentralization and rapid-response fire. Their doctrine drew from lessons learned toward the end of WW I when artillery units were attached to infantry divisions to increase flexibility in the application of fire. The German War Regulations of 1933 stated that an artillery battery or battalion may be attached to infantry units. Furthermore, the German doctrine featured an immediate artillery response to infantry needs.

A French analysis, published in August 1938, of the German 1937 artillery regulations noted the differences between the doctrines. The author observed that the German regulations made almost no mention of the technical and organizational characteristics of artillery and the methods of its employment. He also noted that the Germans wanted to instill in their officers the initiative for solving operational problems.

A difficulty can be identified in the application of cognitive flexibility, stemming from the rigid French artillery doctrine and the attempt to credit it to the Germans during a particularly critical event—Guderian's crossing of the Meuse River. Liddell Hart claims that although the Germans reached the Meuse quickly on the morning of May 13, the French did not expect them to attempt a river crossing without heavy artillery support. He cites General André Doumenc, the French chief of staff, who said: "Crediting our enemies with our own procedure, we had imagined that they would not attempt the passage of the Meuse until after they had brought up ample artillery: the five or six days necessary for that would have given us ample time to reinforce our dispositions."[46] It is interesting to note that the German general staff also believed that the Meuse could not be crossed at so early a stage.

Despite the ideal conditions for these "destructive-fire concentrations"—the pinnacle of firepower as it appeared in the "scientific" five-hundred-page manual *General Instructions for Artillery Fire*—French canons were limited to thirty to eighty shells per artillery piece. Why? General Pierre Grandsard, the commander of the French Tenth Corps, states that he wanted "to spare ammunition." In spite of all the evidence of large concentrations of enemy forces on the opposite bank, Grandsard was convinced that Guderian would not dare to cross the river that day. "The enemy would be unable to do anything for four to six days, as it would take them this long to bring up heavy artillery and ammunition and to position them."[47] However, as stated, German artillery was replaced by tactical air support, an innovation that took the French time to realize.[48]

The methodical battle doctrine emphasized firepower over maneuvering. The French interpreted maneuvering to mean moving units to gain an advantage in the employment of fire. The moving of units for other requirements was hardly mentioned. The doctrine stressed the physical destruction of troops and equipment to crush the enemy's will to fight and underemphasized maneuvering a unit into a position where its presence alone would influence the enemy troops' mental attitude. This limited view of maneuvering pervaded French doctrine. "From the French perspective, the task of gaining superior firepower was far more important than acquiring greater mobility or preparing to counter a more mobile opponent."[49]

The main exception to the preference for firepower was the development of mechanized infantry divisions. These units were formed so that French forces could move rapidly into Belgium before the Germans arrived, but they were not built for maneuver battles. This explains why the development of the French mechanized force contributed so little to enhancing the army's ability to respond to a mechanized offensive.[50]

In his book *The Seeds of Disaster*, which deals with the sources of the French debacle in WW II, Doughty claims that the overemphasis on firepower's importance, and the meticulously organized methodical battle that was a by-product of this emphasis, hampered the French from linking innovations in tactical mobility and the Germans' WW I infiltration concept to a perception of the modern maneuver battle.⁵¹ The concept of armor employment was integrated into the French doctrine so that armor was given the role of infantry support under the cloak of artillery support. This role, in effect, stymied the tanks' ability to maneuver quickly and independently.⁵² Given the limited, symmetrical view of the battlefield, the French were not trained to think of a hasty assembly and a long-range armored attack, and certainly not an operation supported by attack planes.⁵³

In the accounts of French generals after the rout in 1940, the inspector general of the infantry acknowledged that one of the lessons to be gained from the debacle was the importance of "developing the 'spirit of maneuver' in infantry commanders through command flexibility and by training the infantry to deal with unexpected situations in combat."

"Second order" inflexibility, which increased the negative influence of the methodical battle and cult of the defense on flexibility, overstated their necessity and stifled arguments in favor of the attack and mobile armor combat. Keeping these views out of military discussions contributed to the French army's inability to devise an effective answer to German mobile-armored fighting. The French surprise did not come from the tanks themselves but from the nature of modern combat: mobility, maneuvering, and speed based on decentralized command.

The French military concept subordinated the tank to infantry support. On principle there is nothing wrong with this, but it was done while arresting the possibility of making better use of this weapon system. The 1937 tank regulations, signed by Gamelin, intentionally forbade discussion on mechanized divisions and the exploitation of a mechanized success in a breakthrough. "Generally, there was a dogmatic subjugation of the tank to the infantry and an unwillingness to exploit any of its maneuver possibilities."⁵⁴

A lecturer at the French War College in 1930–1931, who was discussing the regulations for the tactical use of large units, declared: "This document, which has hardly more than a hundred pages, systematically compels every detail of execution." Another audience was told that French doctrine had established the need for preponderance of fire as "dogma," and since French doctrine was very near to being the "truth," it "should only be modified with the greatest care."⁵⁵

The French army's training reflected its prejudice against receiving negative feedback on the prevailing doctrine. French exercises were designed to inculcate the doctrine, thus strengthening the belief that there were no alternatives. Almost no learning took place because the general command was certain that it had all the answers. In other words, no feedback was required for lesson learning from combat exercises or actual fighting.[56] French military exercises and tests were designed to attain the desired results. "The approach was invariably successful—until the large-scale exercise that the Germans scripted for them in 1940."[57]

The French army's interest in the Spanish Civil War, which had served as a testing ground for the Germans and Soviets was minimal. "While both the Soviets and German military journals devoted enormous attention to the study of this conflict, *Revue militaire française* rarely covered the developments and when it did, it provided little analysis."[58]

Doughty claims that:

> As the years passed in the interwar period, the army became much more protective of its doctrine. The tendency of the military hierarchy to resist criticism was reinforced by its energetic opposition to the ideas of Lt. Col. (later general) Charles de Gaulle, which called for a return to the offensive, the abandonment of the continuous front, and the creation of a professional armored corps. Although causing a political uproar, his book *Vers l'armée de métier* found little or no sympathy among the military hierarchy. His ideas were attacked systematically by its top military leaders of the 1930s, and by every individual who occupied the war office from 1922 through May 1940.[59]

De Gaulle's book raised a furor in the French public mainly because the left-wing government opposed the idea of a professional army, which in turn led the military hierarchy to vehemently oppose new ideas. In 1935 the military leaders declared that only the supreme command was authorized to define doctrine.

Kier points out that Gamelin tightened control over military writing in 1935 by ordering that every publication receive permission before being sent off. Thus, the official view alone would be presented. One officer recalled that "everyone got the message; absolute silence reigned, until our eyes were opened in 1940." In 1934 De Gaulle was refused permission to publish an article in a military journal and, following his nationwide struggle to develop offensive armor tactics, his name was removed from the army's promotion roster. An analysis of over six hundred articles that were published in five leading military journals in the interwar period shows the dramatic changes in the tone of the articles: in the first part of the 1920s a lively debate ensued, but the late 1920s and 1930s were characterized by acknowledgment and support of the official view.[60] The suppression of discussion on the use of armor for mobile and offensive operations resulted in the French inability to absorb the German mobile combat method and respond accordingly:

> For these reasons the military, who received from the state no more than spasmodic and contradictory impulses, continued to defer to doctrine. The Army became stuck in a set of ideas which had had their heyday before the end of the First World War . . . Hence the concept of the fixed and continuous front dominated the strategy envisaged for a future action. Organization, doctrine, training, and armament derived from it directly.[61]

Thus, beyond the problem of a one-dimensional doctrine and the implications of the methodical battle on flexibility, French dogmatism, it has been shown, precluded a discussion on the soundness of the doctrine.

The Problem of Lesson Learning and Information Transfer

There are no indications of French lesson learning in the course of fighting, but even if there were, it is most unlikely that the French would have succeeded in relaying them because of their shoddy communications systems. This topic is discussed briefly.

May asserts that the French command feared that the Germans were tapping their communications lines. To limit the possibility of eavesdropping, the French reduced the number of wireless radios in their tanks and other vehicles and minimized the lines running between senior echelon headquarters.[62] According to Deighton:

> The military telephone service, being no better than the French civilian telephone service, messages were usually conveyed by motorcycle dispatch riders. There was no teleprinter communication between the HQs and the army commanders. At Gamelin's HQ there was not even a radio. Gamelin's usual way of communicating with Georges was to go to him by car. Questioned about the lack of radio, Gamelin said it might have revealed the location of his HQ. Questioned about the speed with which he could get orders to the front, Gamelin said that it generally took 48 hours.[63]

Expecting a centralized, controlled, staged battle, the French felt under no pressure to develop a flexible communications network. In this slow type of battle, commanders were expected to have sufficient time to use the outdated radio equipment. Communications between headquarters and its subordinate units' headquarters depended on the latter's immobility. After they redeployed, C2 ability became drastically limited.[64]

Murray contends that in May 1940 most of the French company commanders stayed in the bunkers, glued to the field phones connecting them to their superiors. German generals, on the other hand, were on the banks of the Meuse.[65] Obviously the French communications system frustrated the flexibility required to respond effectively to rapidly changing events on the battlefield.

On the German side, wireless communications in combat operations received top priority, derived from the willingness of German officers to alter their plans by the minute, versus the enemy's opposition[66] and also from Guderian's advocation of sophisticated radio equipment (Guderian began his military career as an officer in the signal corps).

In summary, the combination of the cult of the defense and the methodical battle concept suppressed cognitive and C2 flexibility. Inflexibility increased because of the aversion to candid discussion on these topics.

> By the late 1930s, French military doctrine had moved from the ideal of being the basis of military education to the unfortunate status of being an inflexible prescription. Although the hierarchy in precept viewed doctrine as something that should be applied with judgment and accepted the need for flexibility and improvement, doctrine in practice became progressively more rigid during the interwar years and approached the realm of a mandatory formula that had to be applied regardless of the circumstances.[67]

A conceptual crisis resulted from the gap between the French view of the centralized, controlled methodical battle and the sense of chaos and loss of control and effectiveness that emerged from the actual nature of the war. The French surprise, Colonel Chêne observes, was of the worst type—mental surprise.

> They (the French units) found themselves without orders, or with orders no longer applicable because of a change of situation when they were face to face with uncertainty or the unknown, with adjoining units yielding on their flanks, and the enemy attacking their rear . . . the feeling that nothing of what they had learned was proving of any value was overwhelming to them and, unprepared for the unforeseeable, they were unable to find the necessary resources for adapting themselves instantaneously to the new forms of combat.[68]

Conceptual and doctrinal rigidity that produced a one-dimensional doctrine and intolerance toward views that did not correspond to the official doctrine, together with the cognitive intransigence of the French officers and C2 system, obstructed the ability to recover. Inflexible radio mechanisms also contributed to this because of the communications disconnect between units and command echelons. The doctrinal surprise, early signs of which appeared at the tactical level, gradually developed into an operational surprise and then into a strategic surprise. This illustrates the danger in force planning that is based on meticulous speculation on the nature of the fighting in future wars.

SUMMARY AND CONCLUSIONS

THIS BOOK PROVIDES AN ANSWER to one of the most basic questions in military and security matters: how armies deal with technological and doctrinal surprise. It focuses on technological and doctrinal surprise—an area that has been treated only superficially until now—thus closing the gap in academic research that has concentrated on the way armies cope with strategic surprise.

At the outset I stated that the main solution to the problem of technological and doctrinal surprise is not to be found in predicting the contours of the future battlefield or in attempting to discover the enemy's preparations, but in recovery from surprise by employing the range of abilities that come under the heading of flexibility. These abilities offer the indispensable balance between conceptual elements (such as the art of war), physical elements (such as force organization and military technology), and moral elements (such as the commanders' cognitive skills).

My analysis of the way the German, Israeli, Soviet, British, and French armies met the challenge of technological and doctrinal surprise uncovered various levels of application (or non-application) of flexibility that culminated in success (or failure).

The application of these conclusions can be of assistance in handling similar problems in the future. How should a military force be built for dealing effectively with technological and doctrinal surprise? A number of processes can be derived from the theory and introduced into the military system to produce the desired result.

The conceptual-doctrinal dimension includes a few aspects. Anyone who reads an army's basic conceptual documents should realize that combat uncertainty cannot be dealt with by prediction and meticulous planning, but only on the battlefield itself. The meaning for the commander is that the burden of this challenge falls on his shoulders—not on the war planners and development engineers who labor in peacetime. A future warfare concept must reduce the attempt to make "accurate" predictions, concentrating instead on the outer limits of technical developments from which the next technological and doctrinal surprise is likely to emerge.

Another consequence is the status of the various intelligence levels. The first conclusion is the need to reduce the status of the intelligence element, termed strategic intelligence, as a prediction tool in general and for identifying a surprise attack in particular. The second conclusion is that the growing danger of technological surprise requires developing technological intelligence and using it while fully conscious of its limitations. Tactical-operational intelligence on the battlefield is not related to our discussion because it is not the type of intelligence upon which force planning is based. Tactical-operational intelligence is a field that is rapidly developing and will remain a key tool in the effort to identify the timing, location, and intensity of battlefield surprise and in the support of decisions made in the course of the fighting.

The warfighting doctrine of an army, whose aim is to cope with technological and doctrinal surprises successfully, should relate in equal proportion to all types of war (from limited to total), conflict intensity, and forms of war and warfare. A warfighting doctrine must make it clear that it is not a sacred cow and that it must be constantly modified according to the battlefield contingencies. The analysis of the Israeli reaction to the Sagger missiles and the French response to the blitzkrieg illustrates the price that is paid for making one particular form of war into a cult. A warfighting doctrine must be defined as an "open" warfighting doctrine that commanders use while continuously upgrading and modifying their modus operandi at the technical-tactical level. The application of the theory in officer education and training programs should emphasize the development of skills that enable the commander to cope with surprise in general and technological and doctrinal surprise in particular.

The organization of the army and the technology that it develops and acquires, according to the theory presented in this research, combines ongoing development of new capabilities and retention of existing ones. The removal of weapons or an entire layout from active service must be weighed against the need for diversity in the event of surprise. This dilemma is illustrated in the analysis of the IDF's acquisition of water-fording equipment prior to the Yom Kippur War. A recent example is the current status of the attack helicopter, after its problematic usage in the Iraq War.[1] Given a basic operational challenge, weapons diversity must be developed by designing a number of alternatives that enable a force to deal with a challenge, even when the enemy may neutralize some of the alternatives employed on the battlefield, or when the effectiveness of the alternatives proves to be overestimated. The military technology that armies develop and acquire should be changeability oriented, even at the expense of other capabilities such as lethality and mobility.

The C2 method should be decentralized and its use encouraged even in situations where centralized command could be employed. The development of cognitive flexibility should be the essence of military education and training and the main yardstick for commander evaluation. The importance of a decentralized C2 method in dealing with technological and doctrinal surprise is demonstrated in the book's analysis of events related to the German army's doctrine. The price of a centralized system is shown in the case of the French response to the blitzkrieg.

Lesson learning and its fast dissemination comprises a number of elements. The first is the culture of learning where self-improvement is a central motif in military activity. A learning culture encourages the reporting of mistakes despite the risk of punishment. A second element is learning while fighting. Combat debriefing officers must be attached to the fighting forces and allowed access to every event that appears as technological and doctrinal surprise. Units should be trained to identify such changes, develop an effective counterresponse, and inform their sister units and commanders. The German army exemplified this in WW II. Another element is technology that enables lessons to be swiftly conveyed. Today's advanced C2 systems must have this capability.

Force planning that is based mainly on flexibility is not easy to carry out, as we have seen in the application of the decentralized C2 method. The price for applying the theory presented here comes at several levels, including a long and complex training period (relative to the training required for a limited number of military activities); the danger of the system's unfocused orientation because of the surfeit of opinions; and the possibility of a situation in which some of the equipment that was developed and procured is not used on the battlefield, whereas the required equipment suffers from insufficient quantities.

Despite the expected "high" cost of flexibility, the cost of failing to deal with technological and doctrinal surprise probably will be much higher as armies with a low level of flexibility have learned. This, then, justifies the application of the book's theory as can be seen in recent conflicts.

In the Second Lebanon War (summer 2006), the IDF seemed to have been doctrinally surprised by the requirement to conduct major combat operations.

This surprise stemmed from the partial loss of its capability to carry out large formation maneuvers. According to the blue-ribbon Winograd Report:

> In the years preceding the war, IDF ground forces were engaged in long and arduous fighting against Palestinian terrorism, which prevented them to a large degree from undergoing the necessary training and preparation. The results were as expected. . . .[2] The period between 2000 and 2006 was characterized by a decline in training periods, number of maneuver exercises . . . which crippled the ability of the troops, and especially the commanders at the battalion, brigade, and divisional level. . . .[3] The IDF entered the Second Lebanon War after a long period of not having fought a real war. This explains why maneuvering exercises with relatively large formations had not been conducted, and why it was not required to preserve its ability to engage in prolonged fighting—as opposed to swift raids and pinpoint operations—for achieving difficult goals.[4]

The training cutback resulted in a loss of military proficiency at all levels, from tank driver to division commander. As Matt M. Matthews noted: "Years of counterinsurgency (COIN) operations had seriously diminished its conventional warfighting capabilities. . . . The IDF lost many of these perishable combat skills during its long years of COIN operations against the Palestinians."[5]

Another form of surprise was the ineffectualness of the firepower-based, stand-off operations against Hezbollah rocket launchers.[6] The IDF's new operational concept (2006) held that:

> [c]hanges in the role of fire from a support element to the main element in achieving a decision with additional maneuvering elements on the ground, air, and sea reduces the need for the following actions:
>
> A. Deep, large-scale ground maneuvers for advancing fire means for attacking targets and achieving in-depth effects.
>
> B. Massive takeover of enemy territories to provide the large assault echelon with the required areas for waging mobile decisive battles.[7]

Although the new operational concept, with its emphasis on low-intensity conflict and stand-off fire (mainly from the air and based on accurate intelligence), was not unconditional, it did contribute to the sense of doctrinal justification in expecting a decision from the air campaign alone, and avoiding a ground maneuver, although the conditions—according to IDF plans—demanded it [the ground maneuver].[8]

This concept resulted in two imbalances: one, between counterinsurgency capabilities and the ability to engage in major combat operations; and two, between stand-off fires and maneuvering capabilities. The imbalance was felt less in the area of weapons and more in training-based skills.

The IDF acted vigorously and determinedly to correct the mistakes and, in the winter of 2008–2009, demonstrated its improved capability in the limited-scale ground maneuver in Operation "Cast Lead."[9] The American army, involved in counterinsurgency operations for almost a decade, currently finds itself hard pressed to conduct combined arms training. These problems are similar to those of the IDF: losing the ability to carry out combined arms operations in what Frank Hoffman terms "hybrid war,"[10] that is, against well-trained and well-equipped terrorists and guerilla forces. "The U.S. Army, focused as it necessarily is on preparing soldiers and units for duty in Iraq and Afghanistan, might be approaching a condition similar to that of the Israelis before the 2006 Second Lebanon War: expert at COIN, but less prepared for sophisticated hybrid opponents."[11]

The culture of flexibility is the leitmotif that runs from the conceptualization and doctrine stratum (acceptance of uncertainty as a given condition and willingness to study the possibilities that might develop in wartime) to the stratum of understanding the need for constant learning as the key for dealing with unavoidable changes. What is the culture of flexibility based on? The following is a brief look at the main research approaches that try to explain the process of changing military doctrines (in most cases the process occurs in the interwar period). Their relevance for research will be discussed.

According to the realistic approach, doctrinal change can be explained by combining two theories. The first theory is the balance of forces, according to which the dominant factor in the decision for change will be the perception of an external threat. The second is organizational theory, according to which armies tend to adopt doctrines that preserve and increase their strength as organizations. Posen contends that these doctrines will be offensive and that the adoption of a doctrine with other characteristics happens only when an outside factor, such as the political echelon, intervenes. He also argues that organizational considerations exert a greater influence in peacetime; however, in periods of a clearly developing threat, like when close allies suffer obvious defeat on the battlefield, considerations based on the external threat will dominate.[12] This approach is of little value for our study since it offers no explanation regarding the source of flexibility or inflexibility, but only the army's motivating factor for change. In this study the motivating factor is obviously external: technological or doctrinal surprise. The greater the intensity of surprise, the more obvious the need is. This need takes priority over any other factor (e.g., internal organizational politics) that works against the changes required for recovery.

Kimberly Marten (Zisk) developed Posen's theory, claiming that three additional variables are needed to complete it. First, military professionals will introduce necessary changes in the doctrine without civilian interference when faced with threats from an enemy's new doctrine. Second, not all military officers follow the institution's organizational needs. Some value change. Third, some civilian intervention, instead of forcing changes on the military, will build coalitions with certain segments of the officer corps.[13] These considerations seem to clarify better peacetime "reactive innovation" to emerging threats; however, excluding the second variable, they are not very helpful in explaining the basic conditions required for developing battlefield flexibility. The question of an innovative officer cadre developing in a military institution that is driven by organizational interests is undoubtedly of great importance. As noted in Chapter 2, a military organization that can contain controversy (usually developed by innovative officers), will probably succeed in devising solutions to battlefield surprise.

Kier criticized Posen's conclusions, claiming that the choice of a defensive or offensive doctrine was influenced more by national/local political considerations and the nature of the army's organizational culture than by a sense of external threat and organizational desire to preserve and increase resources.[14] The cultural approach can be divided into several secondary approaches ranging from explanations of the army's organizational characteristics and their cultural ramifications to purely cultural explanations.

Avant's book, which deals with the influence of the political control mechanism on the military echelon, presents an approach where elements of organizational theory can be identified that partially explain an army's level of flexibility while engaged in fighting. As a positive example, the author notes the British Cabinet's unified control in appointing commanders and appropriating the military budget. This mechanism, according to Avant, resulted in the British officer cadre being traditionally attentive to the demands of the civilian sphere and able to adapt itself relatively quickly to the changing demands of the battlefield.[15]

The author describes the British success against communist guerillas in Malaya (1948–1960) by changing their fighting patterns from those for a general conflict to a limited one in a civilian environment. As a negative example, she brings the two-headed American system where the president has the authority to appoint commanders and Congress has the responsibility for apportioning the budget. During the Cold War, this led to a situation in which the American military services tended to emphasize force planning for WW III in Europe in order to receive heavier funding. Avant avows that the failure of the United States to adapt concepts geared for an all-out conflict in Europe to a limited one in Vietnam stemmed from the American officer cadre having been "indoctrinated" in the concept of the "big war" in Europe, and under no pressure from the civilian sphere to change its outlook and prepare for other types of conflicts. The president's ability to get his views across by replacing officials was limited because of the fear of criticism. As a result, the army inclined toward a professional approach that stressed abstract principles of warfighting, instead of making a basic examination of the wide range of operational problems it faced.[16]

Hitler stood at the pinnacle of the German unified military-civilian system, thus wielding the authority to swiftly replace commanders when he thought necessary. This type of mechanism seems to have frequently helped the Germans recover from surprise situations. On the other hand, a number of British generals were relieved from command in the Western Desert, but this recovery mechanism was of little avail against Rommel's modus operandi.

In explaining the British case, a different cultural approach may be more appropriate, one that focuses on the internal characteristics of the military culture. One would expect that a military system with a tradition of sharp, interbranch rivalry in peacetime would be hard pressed to come up with a combined branch solution to battlefield surprise. The British difficulty in recovering from the surprise of Rommel's armor and anti-tank fighting tactics (see above) had a lot to do with the British military culture that revered the regiment as a unit that remained separate and closed to the outside world and that maintained a separation among the ground forces branches. On the other hand, rivalry among branches, units, or even commanders augments the diversity of operational opinions and solutions from which responses to surprise can be derived, as in the case of German recovery from British chaff. In the context of our study, it seems that the more significant the surprise, the more that prewar organizational rivalry is reduced. Given a situation in which the army's survival as an organization and the physical survival of the state depend on a swift response, no serious internal opposition stands in the way of a move initiated by one element to get out of the crisis.

A third cultural approach places great importance on the "soft" characteristics of the system, known as "military culture" or an army's particular strategic culture. Murray states that military culture is made up of the ethos and professional habits based on experience and intellectual learning that bequeath the military organization a common perception of the nature of war. The influence of this culture on the military organization is the result of long-term factors that are hard to pinpoint and are usually hidden from the eye of the historian or member of the military. This culture is shaped by the national culture and other factors such as geography and historical consciousness that create a military style with particular national traits. The Germans, for example, participated in European wars from the geographical epicenter of the continent for centuries and as a result tended to disregard logistical problems. The Americans, on the other hand, who faced the challenge of immense distances (beginning in the Civil War) attributed great importance to logistics.

For our study, Murray argues that the willingness (or unwillingness) to learn lessons from the past is characteristic of the military culture and that this has a major impact on the ability to adapt to battlefield reality.[17] He also contends that air forces tend to undervalue lessons from the past—sometimes even from the recent past—as they prepare for the technology-saturated future. This caused the American Eighth Air Force enormous problems in developing long-range escort planes despite the huge number of bombers lost in deep penetrating raids over occupied Europe in the summer and fall of 1943.[18]

Theo Farrell and Terry Terriff, who discuss the ability of military organizations to change, claim that changes in military culture may be systematically introduced by commanding officers who believe that the army must be transformed or by the political leadership for the same reason. Another basic cause for change may come from an external crisis that influences the military culture, such as conceptual failure in time of war. This kind of pressure can undermine and alter the army's norms.[19] Emulating foreign armies, for example, is a type of cultural transformation that does not necessarily arise from a change in the external threat, but from the need to develop and strengthen the army's identity (e.g., along the lines of a Western or modern army).[20]

Farrell holds that the theory of cultural adaptation explains how armies adapt their modus operandi to changes in the threat in ways that avoid upsetting the cultural norms of combat. In our context, the relevant norm, according to Farrell, is the preservation of the army's characteristics that are geared to a conventional conflict. Armies are reluctant to deviate from these characteristics because they form the basis of their professional identity.[21]

According to Cassidy, military strategic culture is defined as the complex of beliefs, approaches, and values that shape the army's concept on the manner and timing of employing military means to realize strategic goals. He also states that the Soviets' difficulty in adapting to the nature of the fighting in Afghanistan stemmed from the military-strategic culture that developed over a long period in which Russia, and later the Soviet Union, cultivated its status as an empire. The military culture concentrated on planning for the "big war," a form of thinking that detracted from the development of flexible thinking. In Afghanistan this lacuna was expressed in the Soviets' inability to adapt to a new situation—low-intensity conflict.[22]

This military culture approach for understanding an army's ability to change seems far more relevant to our study than the other approaches discussed. The infrastructure of the Israeli army's flexibility (discussed previously) can be analyzed in the context of the Israeli national improvisation culture. It is interesting to note that sometimes a wide gap seems to exist between the military culture and the national culture that envelops and nourishes it. The German army, with its impressive flexibility, came from a civilian society where strict discipline and obedience to one's superiors was a basic attitude. In this case, an internal military culture seems to have been built on the doctrinal foundations of Clausewitz in a reverse trend from that of German civilian culture. In the Israeli case, the military culture of improvisation clearly rests on the national, civilian culture of improvisation. The question of the relationship between the military and national cultures in this context deserves a separate study.

Military culture, then, appears to be of importance in establishing an infrastructure for the culture of flexibility that is required for recovery from surprise. Given the difficulty that Murray found in identifying the characteristics of this culture, and given the fact that research dealing with military culture has a sociological-cultural perspective, I have focused this study on possible reasons for the influence of military culture factors on recoverability, without engaging in an in-depth analysis that probably would have overstepped the boundaries of this research.

APPENDIXES TO CHAPTER 1

Appendix A: Definitions and Demarcations of Military Surprise in the Technological-Doctrinal Context

Chapter 1 introduced Kam's system for classifying the causes and reasons for surprise and analyzed the definitions of technological and doctrinal surprise. Since such surprises lie only in the margins in this field of research, the following is a brief survey of definitions and methods of differentiation in the general field of surprise, emphasizing their relationship to technological and doctrinal surprise. The discussion begins with the main system of differentiation in research—levels of war. Differentiation according to psychological impact also appears in mainstream scholarly literature on military surprise. The systematic arrangement of surprise according to intentions or capabilities, and classification according to timing, introduces other perspectives of technological and doctrinal surprise.

The commonly used term in literature—strategic surprise—is taken from the field of war levels. Surprise can be classified according to the level of its occurrence or the level at which its result is obtained. Surprise can happen at or influence the tactical, operational, strategic, or grand-strategic level when there may be a dialectical relationship between the levels and contradictory composites, such as defensive strategy with offensive tactics.[1] Most scholars agree that at the strategic level, surprise has a top-to-bottom influence, and technological surprise has a bottom-to-top influence. The problem with this concept is that surprise that begins at a particular level has a different influence on other levels depending on numerous intervening factors. For example, the technological surprise in the development of the atom bomb occurred at the grand-strategic level. When Israeli armor units encountered Sagger missiles in Sinai in the opening days of the Yom Kippur War, the result of the surprise at the tactical level spelled failure at the operational level. A connection can be made between the ways for creating surprise and the levels of war: feints are generally made at the tactical level, whereas deception occurs mainly at the operative, strategic, and grand-strategic levels.[2] Ruse usually takes place at a low level.

A methodological problem arises in a dichotomous classification of surprises according to the level of their occurrence or influence. The result of the American use of the atom bomb had tangible implications on the grand-strategic level. However, the surprise element was the technological development itself, which remained absolutely secret until the device was dropped. If such a surprise is considered strategic, then it will be difficult to compare this case to the surprise the French experienced when gas was used for the first time in WW I. This, too, was a technological surprise, but it did not cause a strategic volte-face; and in fact, no one claims that it had the effect of a turnabout; however, on the practical level, the manner of the surprise partially resembles that of the atomic bomb.

Another factor that complicates the term "strategic surprise" is the relation of the level and success of the surprise to its implications at the strategic or grand-strategic level and not to the reality that it forced on the combatants in the battlefield. Some authors make the connection between the result of the surprise, while it is actually taking place, and various processes with other elements integrated into them that are external to the surprise itself. Harkabi held that "only a surprise that produces prodigious strategic results is worthy of being called a strategic surprise in the full meaning of the word. The surprise at Pearl Harbor provided the Japanese with only a temporary advantage. True, it was a huge advantage at first, but not enough to decide the war. A similar thing happened in Operation Barbarossa when the Soviet High Command was caught by surprise."[3] Furthermore, attributing the success (or non-success) of Operation Barbarossa's surprise to such factors as strategic depth, the home-front economy, and the might of Soviet production that was realized months and even years after the event, makes it seem doubtful whether surprise can be analyzed while it is happening, let alone conclusions drawn from it. Putting aside the results of the war on the eastern front (which ended in a Soviet victory), it is obvious that for a Soviet commander, Operation Barbarossa presented a surprise: the Germans' doctrine for employing armored units.

Another difficulty is the term "strategic" when it refers to weapons. An example of this confusion is the way the term was used in the First Gulf War (1991) in reference to relatively small, tactical aircraft like the F-117 (Nighthawk) when it attacked strategic targets, and the use of strategic bombers, like the B-52, for demolishing tactical targets such as troop concentrations in Kuwait.[4] It would be incorrect, then, to link the term technological-doctrinal surprise to one of the levels of war.

Handel claims that the difference between technological and strategic surprise is discernible and that there are two distinct types of surprise. Technological surprise can be avoided since it primarily involves the relatively simple procedure of analyzing capabilities, the enemy is generally aware of the likelihood of a surprise being sprung, change is usually slow, and surprise often includes a number of concepts shared by victim and aggressor alike. On the other hand, strategic surprise is far more difficult to prevent. Predicting it requires a complex analysis of both intentions and capabilities since it involves cultural barriers, conceptual and evaluative complexities, and the possibility of extremely rapid change.[5]

This distinction suffers from the problems that were outlined earlier: a dichotomy cannot be created between the two types of surprises. Technological surprise is clearly demarcated and objectively defined. Strategic surprise is basically subjective, thus open to interpretation. Their relationship is that of cause and effect since technological surprise and/or doctrinal surprise can produce strategic surprise. Furthermore, almost every element in Handel's definition of technological surprise can be disputed. Thus, the IDF did not share the Egyptian concept of relying on anti-tank weapons as an effective weapon for knocking out tanks. A one-month technological modification enabled the Japanese to attack Pearl Harbor with an airborne-launched torpedo; Israeli electronic warfare devices against the Soviet sea-to-sea missiles in the Yom Kippur War were in a class of their own—and unknown to Soviet weapons designers.

For this reason this book does not deal with the level at which the technological and doctrinal surprise was realized or the degree of its success. From the force planning perspective, technological and doctrinal surprise can occur at and influence all levels and demands suitable preparation.

Another aspect of surprise is its psychological effect. Zvi Lanir states that fundamental surprise has the capacity to undermine the enemy's worldview—the surprised party's grasp of reality. Situational surprise occurs in the surprised party's "logical" reality, creating a problem but staying within the boundaries of the familiar. The surprise opening gambits of the attack on Pearl Harbor, of Operation Barbarossa, and of the Yom Kippur War can be categorized as fundamental surprises. An example of situational surprise is the exact location of the Allies' D-Day landing site in 1944. But this definition has a methodological problem: a differentiation among the types of surprise cannot be made dichotomously. Employing psychological influence as a variable that defines the level of surprise is misleading due to the subjective ability of individuals and organizations to withstand the psychological pressure that created the surprise in the first place. Furthermore, attempting to link a given situation (such as the unexpected outbreak of war or the gap in combat techniques in relation to enemy weapons or doctrine) and the cognitive result is a formidable task.

Lanir claims that use of the Sagger missile was not a situational surprise (or any other type of surprise) given the intelligence the IDF had garnered prior to the war.[6] On this matter, Lanir appears to have erred. The important point is not the information that the IDF had on Sagger missiles, but its preparedness, or lack of, to cope with the threat (see Ch. 8). At this stage, suffice it to say that the Israeli army's competency in dealing with Soviet anti-tank missiles was extraordinarily defective. Lanir further states that the psychological crisis that overwhelmed the IDF officer cadre came not from the Sagger surprise, but from the gap between its belief in the Arabs' congenital military incompetence and the opposite reality it suddenly faced on the battlefield. On the other hand, the reverse may be true: the opening move of the war, which was considered a major factor in the surprise, had less of an effect on the results of the armored battles in the first days than the sudden encounter with the deadly Sagger missiles. When the initial shock wore off on the Suez Canal front, and IDF officers and soldiers reverted to standard battle procedure, their main problem lay at the technical-tactical level. Had they not been surprised by the way the missiles were applied against the tanks and aircraft, and by the weapons' capacity to frustrate air and armor operations, the overall effect of the surprise would undoubtedly have been much lower. It seems that Israeli armor commanders were so convinced of the "tank shock" they would impose on the enemy that their inability to thwart the anti-tank missiles was a "basic surprise" for them, whereas the sudden outbreak of war was a "situational surprise." An example of basic surprise resulting from the technological factor is the Israeli navy's introduction of electronic warfare in the Yom Kippur War, which broke the rules of the game in that arena. However, because of the complications in its typology, it is not used in the book.

Sometimes a distinction is made between surprise at the capabilities level and at the intentions level. Assessment of the enemy's capabilities and intentions is a basic element in surprise prediction. Intelligence warnings, based mainly on qualitative assessment, are supposed to prognosticate the enemy's intentions.

Capabilities assessment takes into account the technical capabilities and the number of enemy frontline weapons (tanks and planes), the number and proficiency of enemy units, enemy combat doctrines, training, C2, and other factors that influence the enemy's ability to implement military operations (e.g., distance from the border is a major factor in the enemy's ability to launch an offensive or advance to attack). Naturally this is primarily a quantitative evaluation.

Assessment of intentions tries to predict the enemy's plans to carry out military operations in three areas: basic, expected goals; plans for achieving the goals; and the decision to embark upon a particular action.[7] Predicting the enemy's plans, ideas, and decisions is a daunting task for a defense analyst. In fact, assessment of intentions is so difficult that some analysts suggest removing this area from intelligence work altogether.[8]

Military literature generally distinguishes between intelligence on enemy capabilities and military intelligence, on the one hand, and intelligence on intentions and state (strategic) intelligence, on the other. According to Lanir, this distinction is problematic for a number of reasons. Although assessment of ability is supposed to be simpler (since it is based on data), it is also prone to error. Two cases will illustrate this: German intelligence's lack of information on the number of Soviet divisions and tanks prior to Operation Barbarossa and Israel's faulty assessment in 2003 regarding Iraqi surface-to-surface missiles tipped with chemical warheads. In addition, a mix-up exists between capabilities-oriented intelligence used for solving "basic" problems (as Lanir defines them) and intelligence on enemy intentions used for solving "situational" problems.[9] Even if Lanir is correct, this point will not be pursued any further.

In general, there is a link between intelligence for discovering intentions and the effort to identify surprise attack, on the one hand, and intelligence on the enemy's capabilities and the attempt to identify technological or doctrinal surprise (or surprising strength), on the other. Surprise of time and place integrates intentions and capabilities, such as the enemy's ability to move his forces (e.g., France's misreading of Germany's capability to move a significant number of forces through the Ardennes in May 1940) and the speed to do so (a fighting method known as the blitzkrieg that was based on rapid advances that cause a surprise of time). Technological surprise, by its nature, lies squarely in the field of surprise of capabilities. Doctrinal surprise relies to a great extent on the assessment of the enemy's capabilities and integrates a considerable degree of assessment of intentions. Intentions in this case are not the enemy's decision to launch a surprise attack, but concepts of force employment that are often hard to predict or to detect from observation or listening to the enemy. Given the relative objectivity of this typology, it is used in the book whenever necessary.

Uri Bar-Josef, an Israeli intelligence analyst, believes that in cases of strategic surprise, the classification of time of surprise is cogent. In other words, a distinction must be made between surprises at the end of the interwar period, such as Pearl Harbor and the Yom Kippur War, and surprises during wartime, such as the June 1942 Battle of Midway in the Pacific Ocean and the American landing at Inchon in the Korean War. Bar-Josef argues that a comparison of cases in different categories is incorrect for a number of reasons. Entirely different psychological status exists between surprise at the outbreak of war, which is usually accompanied by shock, and surprise in the course of war. The tendency to avoid preventative steps before the outbreak of war, lest they lead to escalation, no longer exists once war has erupted. The victim goes through a learning process, especially after having been caught by surprise and suffering defeat at the outset of the war. If knowledge of the enemy's capabilities was missing before the hostilities broke out, it is acquired during the fighting.[10]

Bar-Josef's reasoning justifies the two categories in cases of strategic surprise—such as surprise attack—but in general dismisses them in instances of technological or doctrinal surprise. In cases of protracted conflict, far-reaching technological and doctrinal surprises can take place despite the victim's knowledge of Side B. There are many examples of this: the debut of tanks, nerve gas, and infiltration tactics in WW I; proximity fuses and chaff for radar interference in WW II—all of which were developed during the war. For research purposes, the distinction between surprise at the outbreak of war and surprise in the course of the fighting seems less valid in cases of weapons and doctrinal surprise.

This book employs the following limited definition for technological and doctrinal surprise: the use of weapons and combat doctrine that the victim did not anticipate and cannot obstruct with countermeasures during an engagement.

It is defined by the manner of the effect of the surprise on the enemy and is objective regarding the definitions derived from its effect on it. Surprise is not connected to the level of the war (a problematic relationship that has already been discussed), nor is it associated with its impact on the mind of the surprised party (since this involves the element of subjectivity), and nor is it influenced by the source of the surprise—the enemy's intention or the victim's failure to interpret available information. This is the definition of the process that causes the surprise, rather than of the result of the surprise, which is influenced by other factors that are difficult to weigh in. Technological and doctrinal surprises may result in tactical, operational, or strategic success or failure. This is why an attempt to critically examine surprise by the criterion of success is problematic. Nevertheless, to analyze the ability to cope with technological and doctrinal surprise, it is not enough that the ability to implement surprise was acquired. If surprise was not perpetrated, then, obviously, nothing happened that can be analyzed. The test cases illustrate a wide spectrum of attempts at executing surprise and various levels of success.

Appendix B: Exacerbation of the Problem—
The Growing Probability for Technological and Doctrinal Surprise

As mentioned in Chapter 1, my contention is that in view of the technological development in recent years, there is a greater likelihood for technological and doctrinal surprise than for a surprise attack. There are three strong arguments for this:

> The importance of the technological dimension has surpassed the operational dimension in grand strategy.
>
> The differential influence of technology on military layouts: the accelerated development of observation, tracking, and identification as opposed to the modest improvements in mobility. This has created a situation in which effective strategic warning is far more possible than in the past.
>
> Developments in numerous technological fields, especially since the 1960s, have produced a wide range of weapons that have increased the likelihood of technological and doctrinal surprise.

A methodological comment is necessary at this point: a difficulty exists in providing a detailed, empirical proof of the argument, especially regarding the second and third reasons. This is due to two main factors. The first is the reduction in the number of conventional wars in recent decades in which a strategic surprise attack was often the key element. This has made it difficult to demonstrate the success of early identification of a surprise attack. The second problem is the censorship that countries have placed on information regarding both the use of technological surprises and the ensuing damage. In most cases the blackout is still in effect. For example, the book by the air-warfare expert Alfred Price, *Instruments of Darkness,* which describes electronic warfare in WW II, was published only in 1967—two decades after the war was over. R. V. Jones, the father of British scientific intelligence in WW II, published his memoirs only in 1978. Another example is that a description of the Israeli air force's technological surprise against the Syrian surface-to-air missile batteries in the First Lebanon War (1982) has yet to be published by either side. Israeli intelligence officer Eado Hecht, who has written about relatively recent cases of technological deception, stated in 1999 that information on the effectiveness of the American-made, surface-to-air Stinger missiles in Afghanistan (introduced in 1986) is totally absent in literature.[11]

The Rise in Importance of Technology as a Dimension of Grand Strategy Compared to the Decline in Importance of the Operational Dimension
The rise in the importance of technological and doctrinal surprise is linked to the growing importance of technology on the battlefield and at the grand-strategy level. The British historian Michael Howard has studied the neglect of this dimension along with the social and logistic dimensions and claims that the source of this neglect lies in the technological parity between the armies that existed up until the time of Clausewitz. According to his view, the operational factor in war—especially the surprise attack—attained a supreme position despite the rise in importance of other dimensions. He further notes that technological development began to gain in importance in the American Civil War and the 1870 Franco-Prussian War. Howard contends that technological development may someday restore the operational dimension to the point of achieving an offensive or defensive decision.[12]

Fuller exaggerates the importance of the technological element[13] and according to Ben-Israel, "moving the strategic center of gravity backwards on the time scale is one of the main expressions of the technological age."[14] In other words, the greater the importance of technology, the more the battlefield goes from a clash between weapons operators to a clash between weapons developers. The conditions at the opening of war regarding weapons systems, especially naval and air weapons systems, seems to determine, more than in the past, the end result of the combat engagements. To balance the picture, as Stephan Biddle shows in his book *Military Power*, the place of technology in determining military success becomes secondary to the method of force employment. This is true when one side employs its force according to the "modern system"[15] while the adversary does not. If both sides employ their forces in the same basic manner, then the technological advantage is of great importance. Secondly, Biddle's measure for technological superiority relies on the ratio of modernization of airplanes and tanks between the adversaries. In many cases, this is not the right measure. In few of the cases analyzed in this book, technological differences between opposing sides came from using radar and counter-radar devices, anti-tank guns and missiles (not tanks), and surface-to-air missiles (not planes). It should also be noted that in "asymmetric" confrontations, the technologically weaker side tries to compensate for its inferiority by neutralizing the superior technology of the enemy by relatively simple means.

Another argument involves the link between the reduction in importance of military decision and the reduction in the likelihood of a surprise attack.

Kober[16] believes that operational, technological, social, and political changes have eroded the status of battlefield decision. The following is a summary of the main changes:

The possibility of deciding a war by an integrated maneuvering attack has become increasingly difficult because of the greater sophistication of the firepower compared to the maneuvering. This trend has strengthened defense. Strategic surprise, whose main expression is the surprise attack using ground forces,[17] is becoming increasingly complicated to implement. This is due in a large part to the growing importance and effectiveness of anti-tank and anti-aircraft weapons. Other factors that have influenced the decline in battlefield decision include the expansion of war from battlefronts to the civilian rear; the transfer of the center of gravity from military confrontation to an all-out social confrontation; and the growing difficulty in making a connection between military decision and victory.[18]

The strengthening of firepower and, in its wake, defense, has reduced, as stated, the ability to launch a surprise attack (strategic surprise) on the ground. On the other hand, this process has almost no effect on the ability to launch a technological and doctrinal surprise because today's technological elements exist to a similar degree in both defensive and offensive weapons systems. These types of surprise are applicable to the strong side, as well as the quantitatively weak side. The surprise that IDF tankers encountered in the Yom Kippur War came from defensive weapons—Sagger missiles. The extremely troublesome air battle against Arab anti-aircraft missiles in the Yom Kippur War also comes under the category of technological surprise from defensive weapons. The reason for this is not because of the rise of the technological factor, but because of the rise in its importance compared to the decline of the operational factor in the form of a surprise attack.

The connecting thread for the theoreticians dealing with this is the fact that the operational ("decisional") dimension is in decline vis-à-vis the technological dimension. The implication for the surprise phenomena is the rise in importance of technological and doctrinal surprise.

The Differential Influence of Technological Development on Various Military Layouts Reduces the Likelihood of a Strategic Surprise Attack
Modern scholars of the influence of technology on strategy are fully aware of the likely impact of technology on the surprise-warning equation. Given the problems inherent in the definition of strategic surprise, henceforth it is referred to by the more clearly defined term—"surprise attack." In the following discussion surprise attack includes the movement of ground and/or air forces and not only a missile attack as in the case of a surprise nuclear attack.

In his analysis of Clausewitz's thinking in the light of technological change, Handel drew the conclusion that because of development in military mobility, the warning time for a strategic surprise was reduced from weeks and months in the pre-industrial age of Clausewitz to hours and days in the age of air power, and then to hours or even minutes in the nuclear age. This means that the chances of a successful strategic surprise have increased.[19] Sherwin and Whaley show how the nations of the world implemented their understanding of this development by statistically analyzing the rising success of strategic surprises in the modern age.[20]

I claim that technological development in the last two decades has created a situation in which the probability of a successful strategic surprise is on a downward curve. The reason for this is the differential influence of technological development on various military layouts. The mobility capability of ground and air forces, and even ballistic missiles, are now inferior to the warning capability against the use of these means in a strategic attack.

In 1985 Simpkin wrote that:

> [c]ertainly the tempos made possible by fast tracks and the rotary wing greatly widen the scope for operational and tactical surprise. On the other hand, a distinguished school of thought holds that one particular advance, satellite surveillance, rules out strategic surprise. This question bears on the whole shape of future warfare.[21]

The comparison between the mobility of ground forces in WW II and the Iraq War indicates no significant improvement, though tank speed improved by 45 percent. The ratio between engine (horsepower) output to the weight of the tank improved by 85 percent (see Table A.1).

Michael O'Hanlon, an expert in technology and the battlefield of the future, claims that the coming decades will probably not witness a major breakthrough in the propulsion systems of armored combat vehicles. Even a two-fold economizing of gasoline intake will not significantly reduce the logistic "tails."[22] The airborne transportation of forces (as paratroops) was performed in a similar fashion in both WW II and the Iraq War.

Table A.1 The development of tank mobility 1945–2003

WW II – 1945			Iraq War – 2003			
Tank type[1]	Horsepower / Weight (tons)	Max. speed (km/h)	Tank type[2]	Horsepower / Weight (tons)	Max. speed (km/h)	Comparison 2003/1945
T-34	500/28=17.8	53	T-80U	1250/46=27.2	70	Horsepower/Weight = 1.53 Speed=1.32
Panther	594/44.8=13.2	54	Leopard-2	1500/62.3=24.1	72	Horsepower/Weight = 1.83 Speed = 1.33
Sherman	375/31=12	46	M1-A2	1500/69.5=21.6	67	Horsepower/Weight = 1.80 Speed = 1.45

1. Orgill, *The Tank*. Appendix of Hebrew translation.
2. "T80-U," "Leopard II," Wikipedia, http://www.wikipedia.org/wiki/Tank; "M1A2," http://www.army-technology.com/projects/abrams/.

On the other hand, a vast improvement has taken place in the field of electronic warning.[23] The reduction in the amount of time required for decision makers to obtain air reconnaissance photos between WW II and the first decade of the twenty-first century has been revolutionary—the difference is measured in thousands of percents. In WW II an enemy located far from the front was identified by air reconnaissance. The photos were of relatively low resolution and even in the best of circumstances, several hours would elapse from the moment the picture was taken until it reached the consumer. Today the consumer can view air photos in real time from satellites or unmanned aerial vehicles (UAVs).

The warning satellites in the 1991 Gulf War effectively identified in real time Scud missiles being launched toward Israel, thus enabling the activation of anti-missile interceptor systems. The Israeli Arrow system designed for ballistic missile interception has its own radar system with a range of hundreds of kilometers, thereby making it extremely independent. Military satellites, like the American spy satellite KH-11, have separation strength down to a few dozen centimeters. In mid-1998 the United States began operating three such satellites.[24] Developments in rapid communications enable information to be transferred directly from the satellite to the decision maker in the control center.

During the Iraq War, the JSTAR system demonstrated similar identification and communications capabilities at the operational and tactical level. American UAVs also proved their worth in tactical operations. Even assuming that communications technologies have a similar influence on mobility speed and firepower of Side A launching a surprise and on the rate that intelligence data reaches Side B (the victim), because Side A's need to advance is accompanied by friction, the victim can be warned almost in real time.

The basic problem with the uncertainty factor is that human beings evaluate the data as they have done throughout history and probably will continue to do long into the future. This is of crucial importance. Some scholars believe that information gathering,[25] regardless of its accuracy and speed, will remain ineffective as long as the analysts and decision makers suffer from human foibles and many other limitations (mentioned in Ch. 1). With this in mind, it should be noted that uncertainty over data has been significantly reduced both quantitatively and qualitatively in recent years, and information accessibility has dramatically increased. Decisions regarding the identification and response to strategic surprise are now based on data far more reliable than in the past.

Assuming that today's strategic surprise includes long-range fire and troop mobility, then technological development may be said to have reduced the chances of its success in recent years. This conclusion of current trends appears in Table A.2. Conversely, the same technological development also raises the likelihood of technological and doctrinal surprise (see below).

The abovementioned target acquisition and tracking capabilities are available today (and probably in the next decade, too) only to technologically and economically developed states. This means that strategic surprise can occur with the same degree of suc-

Table A.2 The differential influence of technology on maneuverability, firepower, and tracking abilities

The component in the military system	1860–1914	1939–1979	1991–
Means of moving troops	On foot, horseback	Motorized, airborne, airdropped	Motorized, airborne, airdropped. Slight improvement
Range of fire and accuracy	From hundreds of meters to a few kilometers	Bombers, rockets: hundreds of kilometers; area target	Ballistic missile, cruiser missile—thousands of kilometers; pinpoint targets
Information gathering technologies for warning	Observation, espionage, balloon observation (optical)	Listening, radar, air photography (optical, early radio waves)	Satellite, UAVs, radar for intercepting missiles, advanced SIGINT
Communications technologies	Runners, carrier pigeons, telegraph	Telephone, wireless (only audio). Narrowband communication (few channels)	Digital, high-speed C2 systems, links to intelligence layouts by audio and video. Immune, encoded, frequency hopping, broadband: optical fibers, and satellites. Personal mobile communications. Internet and intranet

cess as in the past when the surprised party lacks a warning capability based on satellite tracking. It has been pointed out that surprise attacks in the last decades are few and far between due to the general decline in the number of conventional wars. But surprises of this type still occur: the outbreak of the 1980 Iran-Iraq War when Iraq launched a surprise air attack similar to Israel's "Operation Focus";[26] Iraq's invasion of Kuwait in 1991; and to a certain degree, the simultaneously launched ground and air attack in the 2003 war in Iraq that caught Sadam Hussein unprepared. Even if the future proves that attempts to challenge the enemy with a surprise attack are on the wane, it is not because of a decline in the strategic value of the surprise attack, but because of the difficulty in achieving the mobility of large forces without detection by the enemy.

Technological Developments Increase the Likelihood of Technological and Doctrinal Surprise
Technological progress increases the possibility of launching a surprise. The arsenal of an army that wants to launch a surprise grows as technology develops.[27] This does not contradict what was just said regarding the ratio between strategic warning and mobility technologies since technological surprise can develop in both fields, as well as in many others not related to warning or mobility, such as information warfare or warhead efficiency.

It should be recalled that technological surprise is not necessarily an impressive technological breakthrough. The source of the surprise can be the enemy's use of old and familiar weapons in an unexpected way or circumstance.[28] This category includes the use of the Sagger anti-tank missile of the Yom Kippur War and the use of tanks and dive bombers in the blitzkrieg. The other side was familiar with the technical data but was surprised by the method of application and the powerful impact of the weapons on the fighting.

Data-gathering developments—especially in the field of visual intelligence (VISINT)—have enhanced strategic, operational, and tactical warning capabilities but have not improved the ability to identify technological innovations to the same degree.

Lanir noted that as technological development improved weapons systems, a parallel development began in tracking and data-gathering that became available to intelligence agencies so that the identification of new weapons and the ability to intercept and destroy them often neutralizes new technological capabilities. Lanir further observed that because of the time required for the development of new weapons—including basic and applied research, testing, production, experimental implementation, and the formation and training of units—a decade may pass before the weapons are fully operational. According to Lanir, leakage of a secret cannot be avoided over such a long time. He argues that the example of the development of the atomic bomb is irrelevant for weapons systems that are produced and circulated in mass quantity in the interwar period.[29] Mahnken writes that the broad array of American sophisticated airborne and space-based reconnaissance and surveillance platforms enhances the ability of its intelligence organizations to detect foreign military weapons testing, but does not eliminate the possibility of surprise.[30]

My contention is that despite improvements in information collecting and the problems of concealing the development of new weapons, technological advances increase the likelihood of technological surprise for the simple reason that the inflation of possibilities for employing the weapons has a direct impact on the ability to identify and provide a response to any new technology. One reason for this inflation is the reduction of the time required from the "birth" of a new device to its implementation.[31] It is accepted today that the life span of a "computer generation" is three years. One example of inflation is the air surveillance radar used for naval vessel identification, which was in its infancy at the outbreak of WW II. By 1943 over sixty designs of vessel-tracking, air-surveillance radar devices had been adopted or abandoned.[32] Bonen attributes great significance to two technological fields: micro-electronics (whose roots go back to the invention of the transistor in 1948) and missiles (that date back to wartime Germany).[33]

Basic changes on the battlefield can occur with the secret addition of electronic components into an existing system. This does not require introducing a new platform into the arena. The enemy will be hard-pressed to foresee specific developments in a particular technological field when old, familiar systems are being retrofitted. In addition to the increase in the variety of weapons, a number of trends have made detecting new technological weapons systems more difficult than in the past.

Advances in weapons systems technology has reduced the importance of large platforms (tanks, planes, and ships), whose production requires a complex industrial infrastructure generally found only in large states, and increased the importance of missiles, radar systems, night-vision equipment, and other systems that can be produced in small and middle-sized countries and easily transported from state to state. This basic change raises the likelihood of technological and doctrinal surprise for a number of reasons.

The concealment of technology-based weapons development and production is easier than the concealment of mass-produced weapons. Small teams of engineers in military-industry companies can now develop technologies capable of causing a major surprise after testing them in only a limited number of tests. Bonen recalls that Israel's Shafrir-2 air-to-air missile entered operational service after only nineteen trials. The first missiles to shoot down enemy planes in the summer of 1969 were left over from the development stage. The Python 4 missile underwent only eighteen tests. Individual contributions to weapons development can result in battlefield surprise. Cold indium-antimonide sensors developed by Professor Yitzhak Kidron and his team from Israel's Technion enabled Israel to produce the Python 3 air-to-air missile.[34] Limiting the number of people involved in a project and the amount of testing greatly diminishes the enemy's ability to identify technological surprise. Upgrading a known system by altering the frequencies of radar or guidance systems in missiles makes it almost impossible for the enemy to detect changes and plan for them. The introduction of electronic warfare systems into Israeli missile ships before the Yom Kippur War is such a case. Another issue is the increasing use of modeling and simulation by military organizations that limits the opportunity to observe field experiments of developing technologies or doctrines before their surprise introduction on the battlefield.[35]

As stated, technological surprise is possible even in a small country like Israel if weapons are developed independently so that secrets are not leaked. This capability did not exist when sophisticated weapons systems could be developed and manufactured only by small superpowers, which meant they were out of reach for small- and middle-sized states. Israel, for example, developed a surprise weapon—electronic warfare—against the Syrian and Egyptian fleets in 1973. Israel's technological devices that neutralized Syrian SA-6 missile systems in the Lebanon Valley in 1982 are another example of weapons system development that remained hidden from the enemy until their debut in combat.

For the same reasons it has become increasingly difficult to keep tabs on weapons systems from foreign sources. A country that concentrates its information-gathering efforts on enemy weapons development is sometimes at a loss to monitor weapons transactions, especially when several enemies are being tracked simultaneously. The growing number of weapons-producing states adds another headache for an intelligence organization. In the case of Israel, the Soviet Union, until its disintegration, had been the almost exclusive supplier of weapons to Syria. Following the breakup of the Soviet Union, Syria had to look for other sources. Its weapons deal with North Korea for example, via a third party, makes technological data-gathering on Syria a difficult task for Israel. The proliferation in weapons sales in recent years has exacerbated this problem.

Of even greater significance is the fact that advanced technologies, which in the past were under government control, are now accessible to small, independent organizations.[36] Examples of this are the Mujahideen's use of Stinger missiles in Afghanistan against Soviet helicopters and Hezbollah's use of sophisticated anti-tank missiles against Israeli forces during the Second Lebanon War. This trend also enables a technically inferior enemy to employ technological surprises and has intensified following the transition from high-intensity confrontations between states to low-intensity confrontations and terrorism, in which the belligerents are not always internationally recognized states. In recent years the IDF has been involved in limited confrontations with Palestinian groups that employed various technological surprises, such as improvised explosive devices (IEDs) capable of destroying tanks and tunnel systems hundreds of meters in length that were dug under Israeli strongholds and then detonated.

Another trend, linked to the diminishing importance of weapons of mass production, is the increasing value of certain highly sophisticated weapons, such as satellites, radar systems, and surface-to-air missile batteries. The minimal critical mass for the successful use of weapons is decreasing, making it increasingly difficult to identify new weapons systems before their employment and less likely that they fall into enemy hands. Leonhard claims that the rate of technological development will cause a continuous decline in the use of mass-produced weapons and their replacement with "prototype warfare." Armies will employ the latest weapons. Their number on the battlefield will be small because of a shortage of time to mass produce them. This kind of prototype warfare will cause a radical change in military organization and warfighting doctrines and will increase the appearance of technological surprises.[37]

There are many examples of technological surprises that used active or passive deception to enhance their effectiveness. Technological deception has three goals: tricking the enemy regarding Side A's technological capabilities; deluding the enemy regarding its own technological capabilities; and convincing the enemy that a particular system may or may not exist so as to force it to squander resources in developing a similar device or countermeasures.[38] A few examples from WW II will suffice: the German deception regarding the actual size of the warship *Bismark*; British deception regarding the effectiveness of the Foxer device for disrupting the Germans' acoustic torpedo; British deception regarding its centimeter radar for detecting German submarines; British concealment that it cracked the Enigma code. As technology develops, the likelihood of technological deception also increases.[39]

Given the rapid advances of technological development, it is much more difficult today to draw conclusions from the last war and employ them for the next one. Ben-Israel writes that "today we measure the lifespan of weapons systems in 'generations' of a decade at most."[40] The accelerated pace of change casts a pall of doubt on the likelihood of predicting the next generation of technological surprises based on the technology of the last confrontation. Expanding the war to technological areas such as satellite and information warfare (two of the four areas in the revolution in military affairs [RMA]) opens new possibilities for technological surprise. "The increasing use of information technology in military systems implies that the characteristics of individual weapon platforms may be less important than the way military organizations integrate weapons, sensors, and command and control systems."[41] Surprise in information warfare can occur even when the inferior side employs means for interfering with enemy information systems. Sobeleman insists that the Russian army acknowledges its inferiority to the United States army but perceives the imbalance as an advantage by acquiring the capability to attack America's soft underbelly—its information networks.[42] This concept heightens the likelihood of surprise on another front.

Given all of these developments we can easily understand today's difficulty in detecting technological innovations capable of being used in a surprise move.

NOTES

Introduction

1. Michael Howard, "Military Science in an Age of Peace," *Journal of Royal United Services Institute for Defence Studies* 119, no. 1 (1974): 7.

2. Michael I. Handel, "Technological Surprise in War," *Intelligence and National Security* 2, no. 1 (1987): 3.

3. Richard K. Betts, *Surprise Attack: Lessons for Defense Planning* (Washington, D.C.: Brookings, 1982), 111–118.

4. Thomas G. Mahnken, *Uncovering Ways of War: U.S. Intelligence and Foreign Military Innovation, 1918–1941* (Ithaca, N.Y.: Cornell University Press, 2002).

5. Eado Hecht, "Technological Military Deception in the Twentieth Century," *Ma'arachot* 364 (1999b): 30–43.

6. George H. Heilmeier, "Guarding Against Technological Surprise," *Air University Review* 27, no. 6 (1976): 2–7.

7. Williamson R. Murray and Allan R. Millett, *Military Innovation in the Interwar Period* (Cambridge, U.K.: Cambridge University Press, 1996). Other books that deal mainly with interwar innovation and reforms are David Johnson's *Fast Tanks and Heavy Bombers: Innovation in the U.S. Army, 1917–1945* (Ithaca, N.Y.: Cornell University Press, 1998); Harold R. Winton and David R. Mets, eds., *The Challenge of Change: Military Institutions and New Realities, 1918–1941* (Lincoln, Neb.: University Press of Nebraska, 2000); Theo Farrell and Terry Terriff, eds., *The Sources of Military Change: Culture, Politics, Technology* (London: Lynne Rienner, 2002).

8. Kimberly Marten Zisk, *Engaging the Enemy: Organizational Theory and Soviet Military Innovation, 1955–1991* (Princeton, N.J.: Princeton University Press, 1993).

9. Stephen P. Rosen, *Winning the Next War* (Ithaca, N.Y.: Cornell University Press, 1991) describes three case studies: the British Army and the tank, 1914–1918; U.S. Navy submarine warfare, 1942–1944; and the U.S. strategic bombing force, 1941–1945.

10. T. N. Dupuy, *A Genius for War: The German Army and General Staff 1807–1945* (New York: Military Book Club, 1977); Martin van Creveld, *Fighting Power: German and U.S. Army Performance, 1939–1945* (Westport, Conn.: Greenwood Press, 1982).

11. Williamson R. Murray, "Armored Warfare: The British, French and German

Experiences," in *Military Innovation in the Interwar Period*, ed. Williamson R. Murray and Allan R. Millett (Cambridge, U.K.: Cambridge University Press, 1996), 46.

12. Gilad Ariely, "Operational Knowledge Management in the Military," in *Encyclopedia of Knowledge Management*, ed. David Schwartz (Hershey, Pa.: Idea Group, 2006), 713–720.

Chapter 1

1. Avi Kober, "Theory, Doctrine and Planning in Military Force Design and Development," in *Quantity and Quality in Force Planning* [Hebrew], ed. Zvi Ofer and Avi Kober (Tel Aviv: Ma'arachot, 1985), 78, 89.

2. The United States Defense Department defines the term *force planning* as "the military departments' and services' responsibility for the creation and maintenance of military capabilities"; U.S. Department of Defense, Joint Publication 1–02, *Department of Defense Dictionary of Military and Associated Terms* (U.S. Department of Defense, 2001, updated 2007), 209.

Another term, *force development*, "starts with the operational capabilities desired of the Army as specified in national strategies and guidance such as the Defense Strategy, Strategic Planning Guidance (SPG), Joint Programming Guidance (JPG), the National Military Strategy (NMS), and the Army Vision as well as the needs of the Combatant Command commanders. It then determines Army doctrinal, organizational, training, materiel, leadership and education, personnel, and facility capabilities-based requirements, translates them into programs and structure, within allocated resources, to accomplish Army missions and functions." U.S. Army, *How the Army Runs: A Senior Leader Reference Handbook 2007–2008* (Carlisle, Pa.: U.S. Army War College, December 13, 2007), 45.

IDF Dictionary uses the phrase "force organization and structure process" and included in it: doctrine, organization and size of the forces, training and education, and equipping and maintaining the weapons systems and units; IDF Doctrine and Training Division, *IDF Dictionary* [Hebrew], 1–10 (1998), 74.

The British *Glossary of Joint and Multinational Terms and Definitions* defines *force generation* as "[t]he process of providing suitably trained and equipped forces, and their means of deployment, recovery and sustainment to meet all current and potential future tasks, within required readiness and preparation times." "United Kingdom Glossary of Joint and Multinational Terms and Definitions," *Joint Doctrine Publication* 0–01.1 (JDP 0–01.1): (Shrivenham: The Development, Concepts and Doctrine Centre, Ministry of Defence, Edition 7, June 2006), F-8.

Israeli scholars disagree over the areas included in force planning. Kober, for example, puts the force planning doctrine on three levels (tactical, operational, and strategic), with an emphasis on structure and organization, and removes combat doctrine from the field (Kober, "Theory, Doctrine and Planning in Military Force Design and Development," 90–91). Zigdon and Ohana discuss the preparation of the force for combat and expand the definition to include structure and organization, as well as elements of combat doctrine and training in Ya'akov Zigdon and Nisim Ohana, "Force Preparation—The Bridge That Links Force Design and Development to its Application," *Ma'arachot* 372 (2000): 52; whereas Issac Ben-Israel, a former head of the Israeli Directorate of Defense R&D, limits force planning to the field of technology, the development and procurement of arms, formation of new units, and absorption of new weapons in the existing order of battle (ORBAT). Isaac Ben-Israel, "The Theory of Relativity in Force Planning," *Ma'arachot* 352–353 (1997): 33.

3. Israel Tal, *National Security: The Few Against the Many* [Hebrew] (Tel Aviv: Dvir, 1996), 68–84.

4. Kober, "Theory, Doctrine and Planning in Military Force Design and Development," 91–92.

5. Predicting the security reality of the next decade is almost impossible. Although demographic and social factors are relatively easy to predict, less stable factors such as the economy and political coalitions influence this reality. In a reality where enemies can become friends, friends can become enemies, international alliances are fickle, or superpower patrons can step in or scale down their involvement—all of these scenarios have a powerful influence on force planning. Examples of this are the IDF's force planning against the "eastern front" (Iraqi forces reinforced Syria in the Yom Kippur War) and the front's disintegration after the Iraq War in 2003; Syria's reliance on lavish Soviet arms supplies and the subsequent termination of this conduit following the Soviet Union's economic crisis. The conflict type also defies prediction. The American army defeated Iraq relatively easily in 2003 but has been bogged down there until the current writing (2010) in a type of conflict that it was not adequately prepared for.

6. Force planning-related decisions based on the latest input are supposed to provide responses to the enemy's future capabilities. The enemy may increase in size, acquire weapons systems that it lacked when the decisions were made, or develop new weapons systems and combat doctrines. Following a confrontation, one side may derive lessons contrary to those drawn by the other side and embark upon a different vector of development. For example, consider the conclusions reached by Israel and Egypt regarding Israel's armor victory in 1967. The IDF decided to depend almost exclusively on the tank, whereas Egypt "jettisoned" the tank in favor of the anti-tank missile. A danger always exists that combat doctrine or weapons that were developed with great effort will wind up being inferior to the enemy's.

7. Yehezkel Dror, "The Quality of the Decision-Making Process as a Critical Factor in Force Design and Development," in *Quantity and Quality in Force Planning* [Hebrew], ed. Zvi Ofer and Avi Kober (Tel Aviv: Ma'arachot, 1985), 462; see also Isaac Ben-Israel, "The Logic of Military Lesson Learning," *Ma'arachot* 305 (1986): 28.

8. Thomas C. Schelling, *Arms and Influence* (New Haven, Conn.: Yale University Press, 1966), 274.

9. Uncertainty arises from the difficulty in predicting the application of weapons still on the drawing board, especially in cases of a decade or more (Ben-Israel, "The Theory of Relativity in Force Planning," 33). Israel commenced work on its Gabriel sea-to-sea missile in 1955. The project required an enormous investment and many years to develop the prototypes. In the Yom Kippur War, seventeen years after the beginning of the project, Israel finally reaped the benefits of its investment. "In 1955 nobody knew whether this weapon system would turn out [to be] successful" (Zvi Shor, "Military Strength in the Budget Clamp," in *Quantity and Quality in Force Planning* [Hebrew], ed. Zvi Ofer and Avi Kober [Tel Aviv: Ma'arachot, 1985], 306). Choosing the right technology out of a wide range of possibilities to answer future operational challenges is another dilemma facing force-planning planners. Zeev Bonen, a former director of the Israel Armaments Development Authority (RAFAEL), gives the example of acquiring the Shafrir 2 air-to-air missile prior to the 1969–1970 War of Attrition (Zeev Bonen, *RAFAEL* [Hebrew] [Tel Aviv: N.N.D. Media, 2003], 85–87). This problem is growing more complex as technological advances continue. The cost of sophisticated weapons systems is another factor that must be taken into consideration, as uncertainty also surrounds the size and stability of the budget from the time the decision is made to final production (Ibid., 112–113).

10. Rosen, *Winning the Next War*, 64–75. If the new weapons and doctrines have not been developed yet, then obviously exercises based on them cannot be carried out. Even if technical experiments with new weapons or simulated exercises based on a sample number of them succeed, this is not reliable criteria for determining their performance in a full-scale confrontation. Sometimes the surprise can arise, paradoxically, when the weapons or fighting doctrine succeeds beyond expectations, as illustrated in two cases from WW I—the British tank attack in Cambrai, France, in November 1917, and the German gas attack in the Second Battle of Ypres in April 1915. The spectrum of uncertainties is a constant factor facing those engaged in force planning.

11. Developing a long-term project requires vast resources that reduce short-term operational ability. Even if the supply of resources is great, concentrating on short-term developments enhances the military's operational readiness at any given time but generally diminishes the product's innovativeness. The development of innovations calls for much time and thought if the weapons are to have a significant impact on the battlefield. Given today's accelerated pace of technological development (see Appendix B), a military planner is under heavier pressure than in the past to decide on the direction force planning must take. Such pressure inclines toward conservatism, which reduces risk in decision making (Uri Dromi, "The Mutual Relationship Between Doctrine, Technology and Weapons Systems," in *Quantity and Quality in Force Planning* [Hebrew], ed. Zvi Ofer and Avi Kober [Tel Aviv: Ma'arachot, 1985], 442). In general, force planning integrates short- and long-term development simultaneously or long-term developments one after another. The right amount and pace of development is where the uncertainty factor comes into play.

12. For example, in Israel the problem of financial instability is best illustrated by fluctuations in the defense budget and debates over the Merkava IV tank project in recent years.

13. Ephraim Kam, *Surprise Attack: The Victim's Perspective* (Cambridge, Mass.: Harvard University Press, 2004), 12–22.

14. Barton Whaley, *Stratagem: Deception and Surprise in War* (Norwood, Mass: Artech House, 2007), 112.

15. Richard E. Simpkin, *Race to the Swift: Thoughts on Twenty-First Century Warfare* (London: Brassey's, 1985), 181.

16. Yehoshafat Harkabi, *War and Strategy* [Hebrew] (Tel Aviv: Ma'arachot, 1990), 493.

17. Whaley, *Stratagem*, 64–65.

18. Sometimes the principle appears with a different name. For example, the French term it "the law of the unexpected," and the Soviets refer to it as the principle of "surprise and deception" (Harkabi, *War and Strategy*, 496). The American army defines surprise as "strike the enemy at a time or place or in a manner for which he is unprepared" (U.S. Army FM 3-0, Operations [2008]). In the IDF the principle is called stratagem. "Stratagem is designed to surprise the enemy in order to throw him off balance and defeat him" (IDF Doctrine and Training Division, "IDF War Principles," 18). Germany's doctrine in WW II put excessive emphasis on the element of surprise (John Erickson, "Threat Identification and Strategic Appraisal by the Soviet Union, 1930–1941," in *Knowing One's Enemies: Intelligence Assessment Before the Two World Wars*, ed. Ernest R. May [Princeton, N.J.: Princeton University Press, 1984], 398). As stated, this principle has received various appellations in the world's armies.

19. Robert R. Leonhard, *The Principles of War for the Information Age* (Novato, Calif.: Presidio Press, 2000), 182–193.

20. For a detailed analysis of deception in national military doctrines of Britain, Germany, the United States, Russia, Italy, Japan, France, Israel, and China, see Whaley, *Stratagem*, 13–43.

21. Ronald G. Sherwin and Barton Whaley, "Understanding Strategic Deception: An Analysis of 93 Cases," in *Strategic Military Deception*, ed. Donald C. Daniel and Katherine L. Herbig (New York: Pergamon Press, 1982), 177–194. Their analysis shows that active deception was carried out in seventy-six out of ninety-three cases (82 percent). All seventy-six cases succeeded in surprising the enemy. In eleven of the seventeen remaining cases, concealment was employed, and surprise attained. In six cases surprise was not achieved.

22. Richards J. Heuer Jr., "Cognitive Factors in Deception and Counter-Deception," in *Strategic Military Deception*, 60.

23. Michael I. Handel, *Military Deception in Peace and War*, The Leonard Davis Institute for International Relations, Jerusalem Papers on Peace Problems 38 (Jerusalem:

Magnes, The Hebrew University, 1985), 20–21. Handel describes the Cold War "bomber gap" that was created by the Soviets' use of deception in their July 1955 air parade. By flying the same planes over the same air space several times, Western observers received the impression that the number of Soviet strategic bombers was four times greater than that of the Americans. The "missile gap" resulted from the 1957–1958 Soviet attempt to present a better staging of their ICBM program than what it was in reality.

24. Hecht, "Technological Military Deception in the Twentieth Century," 40, discusses American and British success in deception operations in WW II by concealing technological development up until the moment of its use. Examples of this: British chaff for disrupting radar and bombing German cities and the atom bomb (Handel, *Military Deception in Peace and War*, 24–25). Jones gives a case of active deception designed to conceal technological development after it was already in use: British devices for detecting German submarines in 1943. The device employed was centimeter radar, and the British let the Germans think that it was an infrared detector (R. V. Jones, *The Wizard War: British Scientific Intelligence, 1939–1946* [New York: Coward, McCann & Geoghegan, 1978], 320–321). Another example is the British deception regarding the effectiveness of the Germans' acoustic torpedo (Handel, "Technological Surprise in War," 16–17).

25. Handel, *Military Deception in Peace and War*, 37–38.

26. James Cary, *Tanks and Armor in Modern Warfare* (New York: Franklin Watts, 1966), 59.

27. Edward Luttwak, *Strategy: The Logic of War and Peace*, rev. and enlarged ed. (Cambridge, Mass.: Harvard University Press, 2001), 28–29. An example of this is the brilliant performance of Israeli armor and aircraft in the Six-Day War, and, following this, the IDF's obsession with offensive weapons while the enemy was devising a response to them in the form of anti-tank and anti-aircraft missiles. Another example of a paradoxical situation is the measures and countermeasures in electronic warfare that the combatants developed in WW II. After the British installed Monica radar in their bomber fleets to identify German fighters approaching from the rear, the Germans responded by equipping their fighter planes with the Flensburg receiver, which enabled them to pick up the radar beams from the bombers at a range of eighty kilometers, thus rendering the British aircraft easy targets (Jones, *The Wizard War*, 466).

28. Handel, "Technological Surprise in War," 6.

29. Other scholars treat the definition of technological surprise in much less detail. Betts refers to technical surprise as "weapons innovations" (Betts, *Surprise Attack*, 111). Azriel Lorber, an Israeli military technology expert, attributes technological surprise to the sudden appearance of new weapons on the battlefield. He divides technological surprises into two types: those employing weapons that the enemy was completely in the dark about and those in which the weapon was known before its debut, but its potential was not properly recognized (Azriel Lorber, *Science, Technology and the Battlefield* [Hebrew] [Tel Aviv: Kronenberg, 1997], 73–86). This book will not differentiate between

Lorber's types. Shmuel Gordon, an Israeli security analyst, has adopted Handel's definition (Shmuel Gordon, *The Bow of Paris* [Hebrew] [Tel Aviv: Sifriyat Hapoalim, 1997], 82).

30. Betts, *Surprise Attack*, 115.

31. Shimon Naveh, "Was the Blitzkrieg a Warfighting Doctrine? Part I," *Ma'arachot* 330 (1993): 28–43.

32. Betts, *Surprise Attack*, 114.

33. Dromi, "The Mutual Relationship Between Doctrine, Technology and Weapons Systems," 433.

34. For a comprehensive survey, see Harkabi, *War and Strategy*, 447–487.

35. Avi Kober, *Decision: Military Decision in Arab-Israeli Wars, 1948–1982* [Hebrew] (Tel Aviv: Ma'arachot, 2001), 103.

36. Carl Von Clausewitz, *On War* (Princeton, N.J.: Princeton University Press, 1976), 282. General Colmar von der Goltz, a Prussian military officer who had a strong influence on military thinking before WW I also accepted this concept (Leonhard, *The Principles of War for the Information Age*, 104).

37. B. H. Liddell Hart, *Thoughts on War* (London: Faber and Faber, 1944), 33.

38. Ibid., 68.

39. Allan R. Millett, "The United States Armed Forces in the Second World War," in *Military Effectiveness: Vol. 3: The Second World War*, ed. Allen R. Millett and Williamson R. Murray (Boston: Allen & Unwin, 1988), 72.

40. Williamson R. Murray, "British Military Effectiveness in the Second World War," in *Military Effectiveness*, vol. 3, 119; see also 129.

41. Leonhard, *The Principles of War for the Information Age*, 104–106.

42. Kober, *Decision*, 135.

43. Eli Yoffe, "China—Awakening Giant" [Hebrew], in *Quantity and Quality in Force Planning*, 139–140. See Allen R. Millett, Williamson R. Murray, and Kenneth H. Watman, "The Effectiveness of Military Organizations," in *Military Effectiveness: Vol. 1: The First World War*, ed. Allen R. Millett and Williamson R. Murray (Boston: Allen & Unwin, 1988), 26.

44. Luttwak, *Strategy*, 72.

45. The U.S. 1998 defense budget of 281 billion dollars was larger than the combined annual sum of all the members of NATO and Russia (Bill Owens, *Lifting the Fog of War* [Baltimore: Johns Hopkins University Press, 2001], 3–4).

46. Barry R. Posen, *The Sources of Military Doctrine: France, Britain, and Germany Between the World Wars* (Ithaca, N.Y.: Cornell University Press, 1984), 48.

47. Barry R. Posen, "Explaining Military Doctrine," in *The Use of Force: Military Power and International Politics*, ed. Robert J. Art and Kenneth N. Waltz (Lanham, Md.: Rowman & Littlefield, 1999), 24.

48. Posen, *The Sources of Military Doctrine*, 47–48, 58.

49. Kober, *Decision*, 130–140.

50. Ibid., 169.

51. Kober, "Theory, Doctrine and Planning in Military Force Design and Development," 259.

52. Posen, *The Sources of Military Doctrine*, 97.

53. The success of predicting the technological and doctrinal characteristics of the future battlefield is examined independently as much as possible from the intelligence factor. An extensive literature exists on prediction methods. The chapter, "A History of Thinking About the Future," in RAND Report's *Shaping the Next One Hundred Years*, details the advantages and disadvantages of various methods of thinking about the future. These methods include the Delphi analysis, model-based simulations, and scenario analysis. The author concludes that no method alone can substantiate definitive conclusions about the needs of long-term planning; a number of methods must be integrated (Robert J. Lempert, Steven W. Popper, and Steven C. Bankes, *Shaping the Next One Hundred Years: New Methods for Quantitative Long Term Policy Analysis* [RAND, MR-1626, 2003], 11–37). For details on the scenario method, see Peter Schwartz, *The Art of the Long View: Planning for the Future in an Uncertain World* (Chichester, U.K.: John Wiley, 1998), and James A. Dewar, *Assumption-Based Planning: A Tool for Reducing Avoidable Surprises*, RAND Studies in Policy Analysis (Cambridge, U.K.: Cambridge University Press, 2002), 215.

In his book, *The Rise and Fall of Strategic Planning*, Henry Mintzberg acknowledges that the failure of this approach, which began in the middle of the 1960s, lies in the misconception that strategic planning is the same as strategic thinking. The author claims that the basic failure of this approach stems from three false assumptions: that prediction is possible; that strategic planning can be separated from strategic activity and planners separated from operations; and that formal, mechanical systems or other types are the solution to the problem and can replace human intuition by tracking a labeling list for attaining the plans' result (Henry Mintzberg, "The Fall and Rise of Strategic Planning," *Harvard Business Review* 72, no. 1 [1994]: 107–114; Henry Mintzberg, *The Rise and Fall of Strategic Planning* [London: Prentice Hall, 2000], 221–321). Mintzberg's conclusion regarding the failure of introducing strategic thinking into a framework is reminiscent of the German army's conclusion in the interwar period. According to the German Unit Command Regulations, because of the complexity of the battlefield and constantly changing events, combat cannot be condensed into a number of rules. The human attempt to arrange and "control" reality seems to have resulted in the adoption of similar paradigms and a list of rules in military science and strategic planning.

Another aspect of the problem is the adaptation of various prediction methods to appropriate time spans. The British Prime Minister's Strategy Unit proposes three distinct time spans and adapting different prediction methods to each of them. A one-to-three–year period is recommended for analyzing quantitative trends and extrapolating the data to the near future. Qualitative trends analysis, which focuses on social, organizational, and commercial matters and the Delphi method for a five-to-ten–year period, are used for identifying "meta-trends." For a ten-year period, the British suggest using the scenario and "wild card" methods for events deemed highly unlikely to occur; however, if they do occur, they will have an enormous impact (*Strategy Survival Guide*, [British Cabinet Office. Prime Minister's Strategy Unit, Ver. 2.1., May 2004]).

Hugh Courtney and colleagues suggest another track for dealing with uncertainty: defining four levels of uncertainty and adapting to each level the most suitable methods of analysis and prediction (Hugh Courtney, Jane Kirkland, and Patrick Viguerie, "Strategy for Uncertainty," *Harvard Business Review* 75, no. 6 [1997]).

A fourth perspective is the method of military planning. There are various analytical approaches to the challenges that characterize military planning bodies (for details, see: *The Use of Scenarios in Long-Term Defence Planning* [10.4.2007], http://www.plausiblefutures.com.cparticle55074–6691a.html)—each with its own implications for uncertainty. Two basic approaches can be identified. One is the conservative approach that includes methods such as incremental planning, which gradually improves what already exists, and planning that focuses on answers to the gaps identified in past lessons. This approach is useful, for example, in a drawn-out confrontation between armies when no observable window of opportunity exists in which proven, but outdated, weapons can cease to be used, and when new weapons, whose effectiveness is still undetermined, can be acquired. A second approach to planning is bolder. It tries to dissociate from the present and leapfrog into the future by adopting weapons or warfighting doctrines still untested on the battlefield. Such planning methods are "technological-superiority-oriented" force planning and "capability-based" force planning (see Paul K. Davis, "Uncertainty-Sensitive Planning," in *New Challenges, New Tools for Decision Making*, ed. Stuart E. Johnson, Martin C. Libicki, and Gregory F. Treverton, [RAND, MR 1576, 2003]). The distinction between these methods is constrained and is employed only heuristically. In practice, a number of approaches are integrated in force planning. The most relevant ones for predicting the future battlefield are technologies-oriented, capability-oriented, and scenario-oriented planning. One of the most frequently applied methods in force planning is scenario-based planning, which analyzes cases where military force is expected to be employed. The scenarios are generally built on assumptions about the battle theater, the nature of the enemy, and the planner's force. By means of an analysis of the army's ability (in the scenario), based on various methods (from academic discussion to war games and large-scale field exercises), conclusions are reached regarding the required force planning for a particular scenario. The advantage of the method is that it includes concepts, technologies, and threats in various alternative frameworks of a predicted future. It enables discussion to proceed on a large range of solutions of hypothetical problems without determining a particular technology or ability.

54. These limitations are apparent to those implementing the plethora of methodologies that aim to forecast or have foresight into the future, as can be seen in the choice

of narrative of "futures." Whereas the common methodological approaches include variations on extrapolating from the past, various futures methodologies aim to identify trends and portray potential, alternative futures. Such methodologies (e.g., scenario planning) are derived from a social-constructionist approach, which considers a view into the future as something that actually helps shape it. This approach is different than attempts to predict the future and thus is not the focal point of this book.

55. Michael I. Handel, "Clausewitz in the Age of Technology," *The Journal of Strategic Studies* 9, no. 2–3 (1986), 73.

56. Bush was believed to have unparalleled means at his disposal for predicting future weapons. In 1949 he published a book on the "coming" war with the Soviet Union. His predictions were wide off the mark. Despite the expected developments in jet planes and radar, he ignored the helicopter (already in operational use by both sides) and air-to-sea missiles that Germany had employed in WW II. He claimed that jet fighters would be incapable of engaging in dogfights because of their supersonic speeds. He shrugged off the possibility of intercontinental ballistic missiles, certain they would incinerate in flight through the earth's atmosphere and have very limited accuracy (Vannevar Bush, *Modern Arms and Free Man: A Discussion of the Role of Science in Preserving Democracy* [New York: Simon & Schuster, 1949], 50–51, 86–87). This is another instance where military technology defies prediction despite available information on the latest developments. Heilmeier points to a 1937 study entitled "Technological Trends and National Policy" that failed to foresee the helicopter, jet engine, radar, rocket-powered missiles, and other weapons (see Heilmeier, "Guarding Against Technological Surprise"). The latter three technologies were operational in WW II.

57. In February 1943 when the German scientific community got hold of British radar using a ten-centimeter wavelength, it was nonplussed. Until then German scientists believed that the development of radar at such a wavelength was impossible (Alfred Price, *Instruments of Darkness: The History of Electronic Warfare* [London: Macdonald & Jane's, 1977], 137).

58. Before WW II, many German commanders did not believe that a tank force larger than a division was operable either practically, technically, or doctrinally. Murray claims that the development of armor was met with skepticism on the part of senior officers. General Otto von Stulpnagel brought to the attention of Heinz Guderian, the leading advocate of armor fighting before the war and a senior commander during it, that Guderian's ideas were utopian and that large armor formations were militarily impossible and impractical. General Gerd von Rundstedt, the commander of the Wehrmacht's Army Group A in the Ardennes in the Battle of France, is quoted as saying to Guderian—in the late 1930s at the conclusion of a major armor exercise—"All nonsense, my dear Guderian, all nonsense" (Williamson R. Murray, "Armored Warfare," 40–43). General Douglas Haig, commander of the British Expeditionary Force in WW I, said in 1925, " . . . I am all for using aeroplanes and tanks, but they are only accessories to

the man and the horse . . ." (quoted in Douglas Orgill, *The Tank: Studies in the Development and Use of a Weapon* [London: Heinemann, 1970], 85).

59. Prior to WW II, the British admiralty extended its full backing to a submarine detection device—Allied Submarine Detection Investigation Committee (ASDIC)—that could identify a vessel's movement at 1000 meters, in the belief that this would prevent enemy subs from attacking British ships. However, the assessment of the technological effectiveness was flawed and had a disastrous influence on British anti-submarine warfare in the first years of the war (Holger H. Herwig, "Innovation Ignored: The Submarine Problem—Germany, Britain and the United States, 1919–1939," in *Military Innovation in the Interwar Period*, 245–246).

60. For example, despite research on the performance of the Patriot missile (which was retrofitted from an anti-aircraft missile to an antimissile-missile by introducing a number of basic assumptions into its deployment doctrine), whose antimissile effectiveness was "proven," it failed to knock out a single Scud missile in the Gulf War. Another example from the same war was the mistaken American prediction on expected losses, which was based on an analysis of previous wars between a Western army and Arab armies (the Six-Day War and Yom Kippur War). The prediction was off by approximately 90 percent (Lorber, *Science, Technology and the Battlefield*, 60–61, 225). Biddle claims that the Americans overestimated their troops' expected losses in the Gulf War by a factor of two to ten or more (Stephan Biddle, *Military Power: Explaining Victory and Defeat in Modern Battle* [Princeton, N.J.: Princeton University Press, 2004], 17).

61. Williamson R. Murray, "Strategic Bombing: The British, American and German Experiences," in *Military Innovation in the Interwar Period*, 125; see also David Johnson, *Fast Tanks and Heavy Bombers: Innovation in the U.S. Army 1917–1945* (Ithaca, N.Y: Cornell University Press, 1998), 164–165.

Another example is the French combat exercise in 1932, in which an integrated maneuver of tanks and infantry went awry because of faulty troop management. French commanders drew the conclusion that independent tank action was ineffective and that armor should be kept in the general reserve force. The chief of staff Maurice Gamelin later claimed that he regretted the lessons that were derived from an exercise in which the tanks operated so poorly (Eugenia C. Kiesling, *Arming Against Hitler: France and the Limits of Military Planning* [Lawrence, Kans.: University Press of Kansas, 1996], 152–153).

62. Handel, "Clausewitz in the Age of Technology," 66.

63. Michael I. Handel, "Intelligence in Historical Perspective," in *Go Spy the Land: Military Intelligence in History*, ed. Keith Neilson and B. J. C. McKercher (Westport, Conn.: Praeger, 1992), 187–188.

64. Ibid., 189.

65. A few examples: the American knowledge gap regarding Japan's success in stabilizing the torpedo, so it could operate in the shallow waters of Pearl Harbor (a technological feat that was accomplished just prior to the attack); the German data lacuna on

British chaff used for radar jamming; the Israeli air force's ignorance of electronic data in SAM-6 (surface-to-air) missiles; and the Egyptian and Syrian navies' knowledge gap regarding Israel's electronic naval weapons in the Yom Kippur War.

66. Roberta Wohlstetter, *Pearl Harbor—Warning and Decision* (Stanford, Calif.: Stanford University Press, 1962).

67. An example of important data on force planning, which was of superb quality but was interpreted as a German ruse, was the Oslo Report on German plans for weapons development at the outset of WW II (R. V. Jones, *Reflections on Intelligence* [London: Mandarin, 1989], 265–268).

68. Zvi Lanir, *Fundamental Surprise: The National Intelligence Crisis* [Hebrew] (Tel Aviv: Center of Strategic Studies, 1983), 126–127, 146–148.

69. John Ferris and Michael I. Handel, "Clausewitz, Intelligence, Uncertainty and the Art of Command and Military Operations," *Intelligence and National Security* 10 (1995): 48–50. See also Heuer, "Cognitive Factors in Deception and Counter-Deception," 61; Richards J. Heuer Jr., *Psychology of Intelligence Analysis* (C.I.A: Center for the Study of Intelligence, 1999), 57–58.

70. Aharon Levran, "Surprise and Warning: Reflections on Basic Questions," *Ma'arachot* 276–277 (1980): 18.

71. A. W. Marshall, *Problems of Estimating Military Power* (RAND, P-3417, 1966), 15–19.

72. Kam, *Surprise Attack*, 64–72.

73. The difficulty in capabilities assessment is not finding the number of enemy weapons, troops, or paths leading to the front but estimating the enemy's ability to render them operationally effective via organization, warfighting doctrines, and training programs. A trickier problem lies in identifying innovations in one or more of the abovementioned areas (Ashley J. Tellis, Janice Bially, Christopher Layne, and Melissa McPherson, *Measuring National Power in the Postindustrial Age* [RAND, MR-1110–A, 2000], 149–157.) Assessing "quality" factors is a very complicated task. For example, France's difficulty prior to 1940 in assessing the Germans' fighting power, to a great extent, was based on C2 doctrine and organization. These areas defied prediction. According to the American historian Ernest May, if the 1940 data had been fed into a modern computer program, the conclusion would have been a French and British victory (Ernest R. May, *Strange Victory: Hitler's Conquest of France* [New York: Hill & Wang, 2000], 6).

74. So far this issue has been discussed from the force-planning perspective regarding Side A building a warfighting doctrine. Now the discussion focuses on the attempt to identify Side B's (the enemy's) warfighting doctrine. The problem is exacerbated when new weapons are used that have not been tested under battlefield conditions so that prediction of their doctrinal application is all the more elusive. The Yom Kippur War was the first time anti-tank missiles were employed on the battlefield by regular armies.

Added to this was Egypt's method of operation, which created a lacuna regarding the missile's impact potential. The blitzkrieg in France was the first time that massive armor forces were used against a modern army. Concentrated carrier air strikes were first tested in battle at Pearl Harbor. Doctrinal prediction has been gaining in importance as technological advances are made more rapidly than in the past.

75. Mahnken gives the example of the failure of American naval intelligence in detecting the Japanese innovative carrier-aviation doctrine, partially at least because such a doctrine was non-existent in the U.S. Navy (Mahnken, *Uncovering Ways of War*, 72–85).

76. Britain's negative impression of the Japanese army's performance against the Chinese army resulted in an underestimation of the Japanese airplane and pilot. A similar case is the IDF's assessment of the Arab armies after the Six-Day War. Given the failure of the first British tank-supported attacks in WW I, the Germans concluded that armor would have little influence on the battlefield. The Battle of Cambrai proved how wrong their assessment was.

77. This was the case of the Allies' artificial harbors built for the Normandy invasion or "dam busting" bombs designed for the special mission against the Ruhr dams (Handel, "Technological Surprise in War," 14–15).

78. The Americans miscalculated the number of Japanese planes, training level of Japanese pilots, and range of the Zero fighters (Wohlstetter, *Pearl Harbor*, 337). The Germans underestimated the number of Soviet tanks (David Kahn, *Hitler's Spies: German Military Intelligence in World War II* [New York: Macmillan, 1978], 457–458) and divisions (Betts, *Surprise Attack*, 131).

79. Adir Pridor, "The Ability to Measure Military Power," in *Quantity and Quality in Force Planning*, 54–69.

80. Biddle, *Military Power*, 18.

81. Klaus Knorr, *Military Power and Potential* (Lexington, Mass.: Heath Lexington Books, 1970), 22–23.

82. Michael I. Handel, *Perception, Deception and Surprise: The Case of the Yom Kippur War*. The Leonard Davis Institute for International Relations, Jerusalem Papers on Peace Problems 19 (Jerusalem: Magnes Press, The Hebrew University, 1976), 21.

83. Another example is the French overestimation of the place of firepower on the battlefield after WW I. This caused them to overestimate the superiority of defense against an attack. Their assessment of the Germans' ability remained unchanged even after the latest German aircraft and tanks rendered it outdated in the 1930s. Another example is the Americans' miscalculation of the Japanese fleet's capabilities (Wohlstetter, *Pearl Harbor*, 337–338, 369–370). According to Mahnken, Britain's sense of technological superiority led it to lose sight of German radar developments despite British advances in the same field. In reality the Germans had surpassed the British in many areas of radar technology (Thomas G. Mahnken, "Uncovering Foreign Military Innovation," *The Journal of Strategic Studies* 22, no. 4 [1999]: 28).

84. Donald C. Watt, "British Intelligence and the Coming of the Second World War in Europe," in *Knowing One's Enemies*, 258–259.

85. Wesley K. Wark, *The Ultimate Enemy: British Intelligence and Nazi Germany, 1933–1939* (Ithaca, N.Y.: Cornell University Press, 1985), 66.

86. Watt, "British Intelligence and the Coming of the Second World War in Europe," 269. Another example of overestimation is the Americans' view of the ballistic missile gap with the Soviet Union in 1958 (Handel, *Military Deception in Peace and War*, 20–21).

87. Gideon Hoshen, "Intelligence in Weapons Development," in *Quantity and Quality in Force Planning* [Hebrew], ed. Zvi Ofer and Avi Kober (Tel Aviv: Ma'arachot, 1987), 529–533.

88. May, *Knowing One's Enemies*, 510.

89. Mahnken, *Uncovering Ways of War*, 84.

90. Ibid., 168–176.

91. Mahnken, "Uncovering Foreign Military Innovation," 28–29.

92. May, *Knowing One's Enemies*, 534.

93. Dixon deals with individual shortcomings, such as an authoritative personality, that are frequently characterized by limited thinking within rigid formulas and an inflexible mindset (Norman Dixon, *On the Psychology of Military Incompetence* [New York: Basic Books, 1976]). Bar-Joseph and Kruglanski studied the influence of cognitive enclosure on the functioning of people engaged in intelligence work (Uri Bar-Joseph and Arie W. Kruglanski, "Intelligence Failure and the Need for Cognitive Closure: On the Psychology of the Yom Kippur Surprise," *Political Psychology* 24, no.1 [2003]: 81–89.

94. Tversky and Kahenman identified three judgmental strategies employed in probability assessment and value prediction: availability, representation, anchoring, and adjustment (A. Tversky and D. Kahenman, "Judgment Under Uncertainty: Heuristics and Biases," *Science* 185 [1974]: 1124–1131). An example of availability bias in force planning is the French underestimation of air power as it was employed in the Spanish Civil War (Robert J. Young, "French Military Intelligence and Nazi Germany, 1938–1939," in *Knowing One's Enemies*, 301–302). Heuer is convinced that availability bias in the minds of intelligence analysts tends to lead them to see deception in places where there is none (Heuer, "Cognitive Factors in Deception and Counter-Deception," 48–49).

Another type of bias comes from the pressure that is created in conflict situations during decision making. Janis and Mann examined a variety of decision-making blunders and concluded that the failure of Admiral Kimmel, the American naval commandant at Pearl Harbor, in preparing for the surprise attack stemmed from "bolstering" defensive avoidance (Irving L. Janis and Leon Mann, *Decision Making: A Psychological Analysis of Conflict, Choice, and Commitment* [New York: The Free Press, 1977], 120–129). Cognitive dissonance is the primary expression of defensive avoidance. Dixon suggests that one of the reasons for France's capitulation in 1940 was Chief of Staff Maurice

Gamelin's suppression of "bothersome" data on Germany's ability and intentions. Norman Dixon, *Our Worst Enemy* (London: Jonathan Cape, 1987), 37–46.

95. Robert Jervis, "Hypothesis and Misperception," *World Politics* 20, no. 3 (1968): 454–479; Wohlstetter, *Pearl Harbor*, 56. The psychological basis of conceptual inertia lies in the human need for consistency and stability (Robert Jervis, *Perception and Misperception in International Politics* [Princeton, N.J.: Princeton University Press, 1976], 117–142). A basic problem is the absorption of contradictory information into existing information. Interpretation of new information is influenced not only by the existing concepts of the recipient, but also by expectations that guide him when he inquires into the new information. Many studies have been carried out in this field, inter alia: Jervis, 1968; Heuer, *Psychology of Intelligence Analysis*, 1999; and Kam, *Surprise Attack*, 2004.

96. Janis's book, *Victims of Groupthink* (Irving L. Janis, *Victims of Groupthink* [Boston: Houghton Mifflin, 1972]), presents a detailed analysis of the symptoms of group thinking, which include, inter alia, the illusion of invulnerability; direct pressure on any member who expresses cogent arguments against the stereotypes, misconceptions, or group commitments; self-censorship of divergences from group consensus; and the collective delusion.

97. Samuel Huntington, *The Soldier and the State* (New York: Random House, 1957), 59–60; Azar Gat, "Weapons, Warfighting Doctrine and Basic Organization," *Ma'arachot* 278 (1981): 51. According to Liddell Hart: "If a soldier advocates any new idea of real importance he builds up such a wall of obstruction—compounded of resentment, suspicion and inertia—that the idea only succeeds at the sacrifice of himself; as the wall finally yields to the pressure on the new idea it falls and crushes him" (Dixon, *On the Psychology of Military Incompetence*, 114).

98. On struggle and rivalry among German intelligence organizations in WW II, see Michael Geyer, "National Socialist Germany: The Politics of Information," in *Knowing One's Enemies*, 310–311. According to Watt, the competition between the British air force, admiralty, and land army for resources resulted in the rejection or distortion of information that the Luftwaffe was not concentrating its efforts on building its strategic bombing capability but on providing air support to the ground forces. The Royal Air Force (RAF) feared that such information would offset the importance it placed on deterring Germany by constructing a mighty bomber force (Watt, "British Intelligence and the Coming of the Second World War in Europe," 267–268).

99. The influence of this concept on doctrinal surprise occurred in British Malaya when the "weak" Japanese army advanced swiftly through the "impassable" jungles of Malaya with tanks (James Leasor, *Singapore: The Battle That Changed the World* [London: Hodder & Stoughton, 1968], 116–117). In the same context is the British and American refusal to acknowledge the superiority of the Japanese Zero fighter. According to Lowe: "It is difficult to escape the impression that Japan was underestimated because of unconscious or semi-conscious feelings of racial superiority and because Japan had

not fought a major war against occidental power since 1904–1905" (Peter Lowe, "Great Britain's Assessment of Japan Before the Outbreak of the Pacific War," in *Knowing One's Enemies*, 474).

100. Mahnken, *Uncovering Ways of War*, 9–11.

101. Eyal explains that the quality of intelligence depends not only on content, the method of transmission, and timing, but also on the level of education, open-mindedness, and skill of the person digesting the material. "Intelligence will never be better than the use made of it." Eyal lists four "paradoxes" for the consumer of intelligence [decision maker] (Reuven Eyal, "The Intelligence Consumer's Four Paradoxes," [Hebrew] in *Intelligence and National Security*, 475–485). If the leadership is too involved in the intelligence work, it is liable to shunt it off the strictly "scientific" path (Yehoshafat Harkabi, "Complications Between Military Intelligence and the Leader" [Hebrew] in *Intelligence and National Security*, 441). The opposite case is if the decision maker is detached and overreliant on intelligence sources (Shlomo Gazit, "Intelligence Estimates and the Decision Maker" [Hebrew] in *Intelligence and National Security*, 456–457).

102. Dewar, *Assumption-Based Planning*, 189–191.

103. Jervis, an expert in concept distortion, recommends a number of cautionary measures for intelligence analysts (Jervis, "Hypothesis and Misperception," 462–466; see also Jervis, *Perception and Misperception in International Politics*, 409–424). Janis offers solutions to the problem of groupthink (Janis, *Victims of Groupthink*, 207–224). Janis and Mann present various approaches for coping with defensive avoidance to improve the quality of decision making (Janis and Mann, *Decision Making*, 367–404). Bar-Joseph and Kruglanski provide solutions to the problem of cognitive enclosure (Bar-Joseph and Kruglanski, "Intelligence Failure and the Need for Cognitive Closure," 91–92), while Ben-Porat suggests upgrading intelligence analysts' training (Yoel Ben-Porat, "Intelligence Estimates, Why Do They Collapse?" [Hebrew] in *Intelligence and National Security*, 224).

104. For suggestions for methodological improvements in intelligence work, see Lanir (Lanir, *Fundamental Surprise*, 170–185). Ben-Israel studied the methodology of intelligence work and suggests employing refutation rather than verification when testing an assumption. Isaac Ben-Israel, *The Philosophy of Military Intelligence* [Hebrew] (Tel Aviv: Defense Ministry, 1999), 53–58, 104–111. Ben-Zvi dealt with strategic warnings prior to a surprise attack and recommends that intelligence analysts focus on identification of tactical warning signs separate from assessing strategic assumptions (intentions). Abraham Ben-Zvi, "Hindsight and Foresight: A Conceptual Framework for the Analysis of Surprise Attack," *World Politics* 28 (1976): 394–395. Hoshen proposes sounding a warning when a threat is being created, even if the data is not foolproof, and focusing on the "expected damage level" if the weapons become operational (Hoshen, "Intelligence in Weapons Development," 533). According to Heuer, the solution lies in cognitive tools and methods, such as constructing research problems in a way that enables the intelli-

gence analyst to dissect and separate the problems, so he can simultaneously observe all of the factors taken into consideration and analyze competing assumptions in a detailed, eight-stage process (Heuer, *Psychology of Intelligence Analysis*, 85–109).

105. Lanir, *Fundamental Surprise*, 181–185. The Agranat Commission (Israel's official investigation commission on the Yom Kippur War) recommended creating organizational pluralism by strengthening the research branch of the foreign ministry and the research capability of the Mossad (Shlomo Gazit, *Between Deterrence and Surprise: Responsibility for Formulating Israel's National Intelligence Assessment* [Hebrew] [Tel Aviv: Jaffee Center, University of Tel Aviv, 2003], 7). Pluralism cannot guarantee immunity from error. It may be assumed that constraints and external biases, such as a sense of superiority over the enemy, will have a similar effect on different organizations that present similar ideas. Pluralism exacerbates tension and rivalry between organizations. And, even if numerous evaluations are laid out, there is no guarantee that the decision maker will make the right choice (Kam, *Surprise Attack*, 225–228).

106. Mahnken, *Uncovering Ways of War*, 180–181. Mahnken suggests the improved use of military attaches and liaison officers and the nurturing of relationships with foreign professional military academies. These suggestions are limited to situations where future rivals are cooperating in military matters.

107. Ben-Porat, "Intelligence Estimates—Why Do They Collapse?" 243–249. Gazit presents systems that include improving the training of decision makers, attaching an "intelligence advisor" to the decision maker, and so forth (Gazit, "Intelligence Estimates and the Decision Maker"). These solutions may help solve the problem, but they still leave the decision maker exposed to conceptual distortion, cognitive biases, and so forth.

108. Yoel Ben-Porat, "Impossible Prediction: An Interview With Haim Lapid and Hagai Ben-Zvi," *Skira Hodshit* 36, no. 12 (1990): 52.

109. Alfred Thayer Mahan, *The Influence of Sea Power Upon History, 1660–1783* (London: Methuen, 1965), 9–10.

110. Handel, "Clausewitz in the Age of Technology," 72; see also Gordon, *The Bow of Paris*, 316.

111. Ben-Israel, "The Theory of Relativity in Force Planning," 36.

112. Eliot A. Cohen, "American Views of the Revolution in Military Affairs," in *Advanced Technology and Future War* (Ramat Gan: BESA Center, Bar-Ilan University, 1996): 3–18.

113. Rosen, *Winning the Next War*, 221–245.

114. Zeev Bonen, "Tactical Testing and Weapon Development," *Ma'arachot* 272 (1980): 50–51.

Part Two

1. Theo Farrell and Terry Terriff, "The Sources of Military Change," in *The Sources of Military Change: Culture, Politics, Technology*, 7–10.

Chapter 2

1. Posen, *The Sources of Military Doctrine*, 14.
2. "Land operations" (British Army: Directorate General Development and Doctrine, May 2005), 1.
3. Dromi, "The Mutual Relationship Between Doctrine, Technology and Weapons Systems," 442–443.
4. Michael McNerney, "Military Innovation During War: Paradox or Paradigm?" *Defense & Security Analysis* 21, no. 2 (2005): 205.
5. The concept of "forms of combat" is valid at the tactical, operational, and strategic levels. See Harkabi, *War and Strategy*, 407.
6. Liddell Hart, *Thoughts on War*, 297
7. Murray, "Armored Warfare," 22.
8. Dixon, *On the Psychology of Military Incompetence*, 112.
9. Ibid., 114. Bond also notes that Hobart's divorce "handicapped him for years and influenced his removal from his command in Egypt in 1939." Brian Bond, *British Military Policy Between the Two World Wars* (New York: Oxford University Press, 1980), 70–71.
10. Dixon, *On the Psychology of Military Incompetence*, 114. On Liddell Hart's place in prewar politics, see Harold Winton, "Tanks, Votes and Budgets: The Politics of Mechanization and Armored Warfare in Britain, 1919–1939," in *The Challenge of Change*, 74–107.
11. Ervin Rommel, *The Rommel Papers*, ed. B. H. Liddell Hart, trans. Paul Findlay (London: Collins, 1953), 520.
12. B. H. Liddell Hart, *The Other Side of the Hill*, enlarged and rev. ed. (London: Cassell, 1956), 65. According to Murray: "Montgomery-Massingberd was an out-and-out opponent of serious innovation"; Murray, "Armored Warfare," 22.
13. Elizabeth Kier, *Imagining War: French and British Military Doctrine Between the Wars* (Princeton, N.J.: Princeton University Press, 1997), 102.
14. Murray, "British Military Effectiveness in the Second World War," 126. Murray, "Armored Warfare," 29.
15. Bond, *British Military Policy Between the Two World Wars*, 62. The regimental system's shortcomings regarding innovation are described in ch. 2—"The Character and Ethos of the British Army."
16. Liddell Hart, *The Other Side of the Hill*, 122.
17. Heinz Guderian, *Achtung-Panzer! The Development of Tank Warfare*, trans. Christopher Duffy (London: Cassel, 1992), 13.
18. Azar Gat, *British Armour Theory and the Rise of the Panzer Arm: Revising the Revisionists* (London: Macmillan, 2000), 69. Corum notes that Guderian's self-serving memoirs overshadow other officers who promoted the mechanization of the German army and overemphasize the high command's opposition to him. James S. Corum, "A Comprehensive Approach to Change: Reform in the German Army in the Interwar Period," in *The Challenge of Change*, 57.

19. Rudolf Steiger, *Armour Tactics in the Second World War: Panzer Army Campaigns of 1939–1941 in German War Diaries*, trans. Martin Fry (New York: Berg, 1991), 2–3.

20. Murray, "Armored Warfare," 43.

21. John J. Mearsheimer, "Hitler and the Blitzkrieg Strategy," in *The Use of Force*, 140–141.

22. Matthew Cooper, *The German Army, 1933–1945: Its Political and Military Failure* (New York: Stein and Day, 1978), 149–158.

23. Dupuy, *A Genius for War*, 305.

24. Corum, "A Comprehensive Approach to Change: Reform in the German Army in the Interwar Period," 54–55.

25. Murray, "Armored Warfare," 47.

26. Rosen, *Winning the Next War*, 85. Douglas Macgregor, *Transformation Under Fire* (Westport, Conn.; London: Praeger, 2003), 242–243.

27. James S. Corum, *The Roots of Blitzkrieg: Hans Von Seeckt and German Military Reform* (Lawrence, Kans.: University Press of Kansas, 1992), 131.

28. Eado Hecht, *The "Operational Breakthrough" in German Military Thought 1870–1945* [Hebrew] (Tel Aviv: Ma'arachot, 1999), 120–121.

29. Gat, *British Armour Theory and the Rise of the Panzer Arm*, 63–66.

30. Hecht, *The "Operational Breakthrough" in German Military Thought 1870–1945*, 121.

31. Corum, *The Roots of Blitzkrieg*, 193.

32. Orgill, *The Tank*, 106–107.

33. John Erickson, *The Soviet High Command* (London: MacMillan, 1967), 258.

34. Charles de Gaulle, *The Complete War Memoirs of Charles de Gaulle*, vol. 1 (New York: Simon and Schuster, 1964), 15–16.

35. Heinz Guderian, *Panzer Leader*, trans. Constantine Fitzgibbon (London: Penguin Books, 1996), 20. Liddell Hart's name was added to the English version of the book on the urging of Liddell Hart himself.

36. Guderian, *Achtung-Panzer!* 167.

37. Dupuy, *A Genius for War*, 304–305.

38. Bond, *British Military Policy Between the Two World Wars*, 70.

39. Guderian, *Achtung-Panzer!* 7.

40. Donald E. Graves, "War Is an Art," review of *On the German Art of War—Truppenführung*, by Bruce Condell and David T. Zabecki, *The Army Training and Doctrine Bulletin* 5, no. 2 (2002): 71.

41. Bruce Condell and David T. Zabecki, *On the German Art of War—Truppenführung* (Boulder, Colo.: Lynne Rienner, 2001), xi.

42. Stephen Van Evera, *Causes of War: Power and the Roots of Conflict* (Ithaca, N.Y.: Cornell University Press, 1999), 194–197.

43. Jack Snyder, "The Cult of the Offensive," in *The Use of Force*, 113.

44. Azar Gat, *The Development of Military Thought: The Nineteenth Century* (New York: Oxford University Press, 1992), 114.

45. Ariel Levita, *Offense and Defense in Israeli Military Doctrine* [Hebrew] (Tel Aviv: Center of Strategic Studies, 1988), 128.

46. Ibid., 71.

47. Hanoch Bartov, *Daddo: 48 Years and 20 More Days* [Hebrew], 2 vol. (Tel Aviv: Ma'ariv, 1978), vol. 1: 74; see also Doron Almog, "Lessons From the Six-Day War as a Crisis in the Development of the Warfighting Doctrine," *Ma'arachot* 354 (1997): 4–5.

48. Dov Tamari, "The Yom Kippur War—A Question of Ignorance," *Zmanim* 84 (2003): 30.

49. Shimon Naveh, "The Cult of the Offensive Preemption and Future Challenges for Israeli Operational Thought," in *Between War and Peace: Dilemmas of Israeli Security*, ed. Efraim Karsh (London: Frank Cass, 1996), 175.

50. "Agranat Commission Report," *Investigating Commission on the Yom Kippur War, Third and Last Report*, vol. 4 [Hebrew] (Jerusalem, 1975), 1350–1351.

51. Ibid., 1439.

52. Charles W. Sanders Jr., *No Other Law: The French Army and the Doctrine of the Offensive* (RAND, P-7331, 1987), 14–15.

53. Eliyashiv Shimshi, *Storm in October* [Hebrew] (Tel Aviv, Ma'arachot, 1987), 32.

54. Almog, "Lessons From the Six-Day War as a Crisis in the Development of the Warfighting Doctrine," 5.

55. Aryeh Arad and Haim Laskov, *Looking Back: Soldiers Discuss the Yom Kippur War Battles They Participated In* [Hebrew] (Israel: Arad Family, 2003), 32–33; Shimshi, *Storm in October*, 18, 22, 23.

56. Eitan Haber and Zeev Shif, *Yom Kippur War Lexicon* [Hebrew] (Tel Aviv: Zmora-Beitan-Dvir, 2003), 164.

57. Azar Gat, "Ideology, Policy, Technology and Strategic Doctrine Between Two World Wars," *Ma'arachot* 390 (2003): 11–13; Posen, *The Sources of Military Doctrine*, 183.

58. Posen, *The Sources of Military Doctrine*, 218.

59. Jürgen Förster, "The Dynamics of *Volkegemeinschaft*: The Effectiveness of the German Military Establishment in the Second World War," in *Military Effectiveness*, vol. 3, 206.

60. Martin van Creveld, *Fighting Power: German and U.S. Army Performance, 1939–1945* (Westport, Conn.: Greenwood Press, 1982), 29.

61. Condell and Zabecki, *On the German Art of War—Truppenführung*, 24.

62. Shimon Naveh, *In Pursuit of Military Excellence: The Evolution of Operational Theory* (London: Frank Cass, 1997), 117.

63. Ibid.

64. Condell and Zabecki, *On the German Art of War—Truppenführung*, 284.

65. Steiger, *Armour Tactics in the Second World War*, 110–111.

66. Wilhelm Ritter von Leeb, "Defense," in *Roots of Strategy, bk. 3: 3 Military Classics* (Harrisburg, Pa.: Stackpole Books, 1991, 132–133.

67. Yehuda Wallach, Preface [in Hebrew] to Leeb, 1993, 12.

68. Robert M. Citino, *The Path to Blitzkrieg: Doctrine and Training in the German Army, 1920–1939* (Boulder, Colo.: Lynne Rienner, 1999), 236–237.

69. Guderian, *Achtung-Panzer!* 167–168.

70. Naveh, *In Pursuit of Military Excellence*, 137.

71. Mearsheimer, for example, refers to the blitzkrieg as a strategy (Mearsheimer, 1999, 130); Messenger, on the other hand, describes the evolution of the blitzkrieg "as a technique of war" (Charles Messenger, *The Blitzkrieg Story* [New York: Scribners, 1976], 7).

72. Rainer Kriebel, *Inside the Afrika Korps—The Crusader Battles, 1941–1942*, ed. Bruce Gudmundsson (London: Greenhill Books, 1999), 295.

73. F. M. von Senger Und Etterlin, *Neither Fear Nor Hope*, trans. George Malcolm (London: MacDonald, 1963), 233.

74. Ibid., 98.

75. Guderian, *Panzer Leader*, 259–260.

76. Steiger, *Armour Tactics in the Second World War*, 111–112.

77. Hans von Luck, *Panzer Commander: The Memoirs of Colonel Hans von Luck* (New York: Dell Publishing, 1989), 80–82.

78. Condell and Zabecki, *On the German Art of War—Truppenführung*, 7.

79. Rommel, *The Rommel Papers*, 138.

Chapter 3

1. Liddell Hart, *Thoughts on War*, 243.
2. Ibid., 245.
3. Ibid., 246.
4. Yitzhak Samuel, *Organizations: Features, Structures, Processes* [Hebrew] (Haifa: Haifa University Press, 1996), 342–343.
5. Simpkin, *Race to the Swift*, 136.
6. Handel, "Clausewitz in the Age of Technology," 71.
7. Kober, *Decision*, 85–86.
8. Handel, "Clausewitz in the Age of Technology," 71–72.
9. Naveh, *In Pursuit of Military Excellence*, 136.
10. J. Nazareth, *Dynamic Thinking for Effective Military Command* (New Delhi: Tata McGraw-Hill Publishing, 1977), 111.
11. Ibid., 112–113.
12. Anthony H. Cordesman, *The Iraq War: Strategy, Tactics, and Military Lessons* (London: Praeger, 2003), 183–184.
13. Luttwak, *Strategy: The Logic of War and Peace*, 192.

14. Cordesman, *The Iraq War*, 195.
15. Liddell Hart, *Thoughts on War*, 28–29.
16. Victor Suvorov, *Icebreaker: Who Started the Second World War?* trans. Thomas B. Beattie (London: Hamish Hamilton, 1990), 100–106.
17. Ibid., 82–99; Richard Overy, *Russia's War* (London: Penguin, 1997), 64–65.
18. Ibid., 68–89.
19. Beth M. Gerard, "Mistakes in Force Structure and Strategy on the Eve of the Great Patriotic War," *Journal of Soviet Military Studies* 4, no. 3 (1991): 479–482.
20. Ibid., 483.
21. Ibid., 472.
22. Almog, "Lessons From the Six-Day War as a Crisis in the Development of the Warfighting Doctrine," 7.
23. A. P. Wavell, *The Good Soldier* (London: MacMillan, 1948), 10.
24. Martin van Creveld, *Supplying War: Logistics From Wallenstein to Patton* (Cambridge: Cambridge University Press, 1977), 231. Harkabi (*War and Strategy*, 1990, 353) and Kober (*Decision*, 96) also describe the danger of imbalance in this area.
25. Van Creveld, *Supplying War*, 180.
26. Ibid., 151.
27. Posen, *The Sources of Military Doctrine*, 97.
28. Steiger, *Armour Tactics in the Second World War*, 113–128.
29. Naveh, *In Pursuit of Military Excellence*, 136.
30. Condell and Zabecki, *On the German Art of War—Truppenführung*, 9.
31. Kober, *Decision*, 294–295.
32. Ernst Mayr, *What Evolution Is* (New York: Basic Books, 2001), 201–203.
33. Fuller, *Armament and History*, 21–34.
34. Handel, "Clausewitz in the Age of Technology," 72–73.
35. Kober, *Decision*, 90–93.
36. Leonhard, *The Principles of War for the Information Age*, 73–74.
37. Luttwak, *Strategy: The Logic of War and Peace*, 34–38.
38. Rosen, *Winning the Next War*, 243–244.
39. Jonathan M. House, *Combined Arms Warfare in the Twentieth Century* (Lawrence, Kans.: University Press of Kansas, 2001), 4–5.
40. Robert R. Leonhard, *The Art of Maneuver: Maneuver-Warfare Theory and Air-Land Battle* (New York: Ballantine, 1991), 93–97.
41. Leonhard, *The Principles of War for the Information Age*, 70–72.
42. John H. Cushman, "Challenge and Response at the Operational and Tactical Levels, 1914–45," in *Military Effectiveness*, vol. 3, 332.
43. House, *Combined Arms Warfare in the Twentieth Century*, 164–165.
44. Martin Blumenson, *Kasserine Pass* (Boston: Riverside, 1967), 233–238.
45. Ibid., 243.

46. Johnson, *Fast Tanks and Heavy Bombers*, 189.
47. Blumenson, *Kasserine Pass*, 255.
48. John Shy, "First Battles in Retrospect," in *America's First Battles 1776–1965*, ed. Charles E. Heller and William A. Stofft (Lawrence, Kans.: University Press of Kansas, 1986), 334.
49. Martin Blumenson, "Kasserine Pass 30 January–22 February 1943," in *America's First Battles 1776–1965*, ed. Charles E. Heller and William A. Stofft (Lawrence, Kans.: University Press of Kansas, 1986), 226–265, 274–275.
50. Rick Atkinson, *An Army at Dawn: The War in North Africa, 1942–1943* (New York: Henry Holt, 2002), 385.
51. See Heilmeier, "Guarding Against Technological Surprise."
52. Luttwak, *Strategy: The Logic of War and Peace*, 29.
53. Price, *Instruments of Darkness*, 117.
54. Luttwak, *Strategy: The Logic of War and Peace*, 30.
55. Ibid., 31.
56. Ibid., 39.
57. Ibid., 39–40.
58. For a survey, see Ami Shamir, "The Combat Engineers Corps," in *IDF—A Military and Security Encyclopedia* [Hebrew] (Tel Aviv: Revivim and Sifriyat Ma'ariv, 1981), 109–125.
59. For details, see Shaul Nagar, "Our Water Crossing, Our Victory," *Shiryon* 19 (2003): 136–140.
60. For details, see Uri Gazit, "The 'Crocodile' Marches Past," *Shiryon* 11 (2001): 30–31.
61. Shamir, "The Combat Engineers Corps," 114.
62. Tamari, "The Yom Kippur War—A Question of Ignorance," 29.
63. Gazit, "The 'Crocodile' Marches Past," 30–31.
64. Price, *Instruments of Darkness*, 20–49; Jones, *The Wizard War*, 161–185.
65. R. V. Jones, *Future Conflict and New Technology* (Washington, D.C.: The Washington Papers, The Center for Strategic and International Studies, Georgetown University, 1981), 84.
66. Moshe Kress, *Operational Logistics* [Hebrew] (Tel Aviv: Ma'arachot, 2002), 83–84.
67. Rosen, *Winning the Next War*, 244.
68. Stanley Karnow, *Vietnam: A History* (New York: Penguin, 1984), 454, 540, 652.
69. Shmuel Gordon (*The Last Order of Knights: Modern Air Strategy* [Hebrew] [Tel Aviv: Ramot-Tel Aviv University Publications, 1998], 113) refers to Ezer Weizman (*On Eagles' Wings: The Personal Story of the Commanding Officer of the Israeli Air Force* [Hebrew] [Tel Aviv: Ma'ariv, 1975], 194–196).
70. Gordon, *The Last Order of Knights*, 115.
71. Jones, *Reflections on Intelligence*, 310–311.
72. Rommel, *The Rommel Papers*, 185–186.

73. Isaac Ben-Israel, "Technological Lessons," *Ma'arachot* 332 (1993): 13.
74. Eli Shvili, "Thinking About Countermeasures," *Ma'arachot* 355 (1998): 39.
75. James N. Constant, *Fundamentals of Strategic Weapons: Offense and Defense Systems* (The Hague: Martinus Nijhoff, 1981), 160.
76. Len Deighton, *Blitzkrieg: From the Rise of Hitler to the Fall of Dunkirk* (London: Jonathan Cape, 1979), 154.
77. Shlomo Na'aman and Roni Cohen, "Tank Armament—Past, Present and Future," *Ma'arachot* 247–248 (1975): 37–38.
78. Luttwak, *Strategy: The Logic of War and Peace*, 136.
79. Shvili, "Thinking About Countermeasures," 42.
80. Rosen, *Winning the Next War*, 244.
81. Ibid., 244–249.
82. Zeev Bonen, "Problems in Developing Military Layouts," *Ma'arachot* 245 (1975): 5–6. Macgregor suggests basically the same as part of his advocacy for U.S. Army right way of transformation (*Transformation Under Fire*, 278).

Chapter 4

1. Dan Sharon, "Surprise, Problem Solving and Commanders Education," *Ma'arachot* 278 (1981): 30–34.
2. Sun Tzu (400–320 B.C.) understood the commander's need to adapt quickly to unforeseen circumstances and revise plans according to changing conditions (Sun Tzu, in *Roots of Strategy*, vol. 1, ed. T. R. Phillips [Harrisburg, Pa.: Stackpole Books, 1985], 36–37). Machiavelli realized the importance of adapting to circumstances, noting that cognitive flexibility is not a trait common to most men because of their inability to alter personal characteristics and the psychological need to stay within the parameters of familiar patterns of action that have proved effective in the past. Machiavelli's solution was to replace the commander (Niccolo Machiavelli, *The Prince and Discourses* [New York: The Modern Library, 1950], 441–443). Clausewitz claimed that when political and military leaders develop the ability to identify the enemy's and their own errors, they have to be sufficiently flexible in order to change and often abandon their initial strategy (Michael I. Handel, *Masters of War: Classical Strategic Thought* [London: Frank Cass, 2001], 95). The need for flexibility, according to Clausewitz, is not the main element in force structure or the traits of the commander.
   Liddell Hart discussed flexibility at length from various angles. Regarding adaptability, he wrote that "[m]odern warfare demands an ever greater power of adjustment to new conditions; the effect of the higher direction may be paralysed unless the executants also acquire an increased degree of mental adaptability" (Liddell Hart, *Thoughts on War*, 105). He considered flexibility to be an essential skill of a commander (ibid., 59–60). Mao Tse-Tung, who dealt primarily with guerilla warfare, lauded flexibility as a trait or as a command skill and avowed that planning must adapt itself to the current of the war

(flowing with the change) and the level of the war (Mao Tse-Tung, *Six Essays on Military Affairs* [Peking: Foreign Languages Press, 1985], 294–295). Liddell Hart (*Thoughts on War*, 70) and Martin van Creveld (*Command in War* [Cambridge, Mass.: Harvard University Press, 1985], 270) noted the need of a flexible warfighting doctrine and C2 system.

3. Harkabi, *War and Strategy*, 490–493.

4. Flexibility—a war principle of the world's armies:

The term flexibility first appeared in 1920 in the British army's list of war principles under the principle of "mobility" (John I. Alger, *The Quest for Victory: The History of the Principles of War* [London: Greenwood, 1982], 240). After WW II Montgomery changed the "mobility" principle to the "flexibility" principle (ibid., 152)—the name that remains till today. The termination emphasizes the ability to respond quickly through such mental skills as "elasticity of thinking," speed in altering a decision, and dedication to the objective (British Army, *The Application of Force: An Introduction to British Army Doctrine and to the Conduct of Military Operations* [The British Army: DGD&D/18/34/66. Army Code No. 71622, 1998] 2-13–2-14). The armies of Canada, New Zealand, and Australia followed suit. In some cases "simplicity" appears with flexibility. For example, in 1973 the French army's list of principles contained "freedom of action" and the law of "simplicity and flexibility" (Alger, *The History of the Principles of War*, 152–153).

The German army has never drawn up a list of war principles because of the Clausewitzian concept that the art of war is too complex to be based on a number of guidelines. German military doctrine is rooted in the command concept that Moltke developed that favors mental flexibility and maximum independence of the commander. This approach still applies in Germany. The 1973 edition of the field manual on force command states, "Success is ensured only by the free action of the commanders within the framework of their missions. Creative, exacting, and critical thinking during maneuvers of all kinds will prepare military commanders of all ranks in peacetime for their combat assignments" (ibid., 154). The 1983 dictionary for military cadets at the German Command College defines "flexibility" as mental agility, the individual's ability to adapt, and the skill in responding flexibly in a tactical situation (The German Army, *Command and Staff College Dictionary* (Hamburg, 1983), 396). The term versatility is defined as multiple abilities, the capacity to reinvent oneself, speed, mobility, and adaptability (ibid., 942). These are the traits that commanders, units, weapons, and plans must possess. Flexibility does not appear as a war principle in the Soviet army. According to Leites, the Soviet command system stifled deviation from the original plan (Nathan Leites, *Soviet Style in War* [New York: Crane Russak, 1982], 179–180).

American war principles do not include flexibility but do consider "versatility" as one of the four principles of operations. Versatility is defined as "the ability of army forces to meet global, diverse mission requirements of full spectrum operations" (U.S. Army. FM 3–0. Operations [2001], 4-17–4-18). The term *versatility* does not appear in all armies. The American army's use of the term stems from the wide range of fighting scenarios that the military is likely to find itself engaged in. Frost suggests that the American army adopt flexibility as a "meta-principle," which stands by itself and at the same time brings "harmony" to the inherent tension that exists between the other war principles (Robert S. Frost, *The Growing Imperative to Adopt "Flexibility" as an American Principle of War* [Carlisle, Pa: Strategic Studies Institute, U.S. Army War College, 1999]). Dickerson, on the other hand, proposes that the American army adopt "adaptability" as the answer to future challenges (Brian Dickerson, "Adaptability: A New Principle of War," in *National Security Challenges for the 21st Century*, ed. Williamson R. Murray [Carlisle, Pa.: Strategic Studies Institute, U.S. Army War College, 2003]). The IDF has dealt "off and on" with flexibility over the years. Ben-Gurion elaborated on the subject of flexibility in his talks with the high command in 1951 regarding the need for initiative (David Ben-Gurion, *Army and Security* [Hebrew] [Tel Aviv: Ma'arachot, 1955]), 290). IDF commanders like Yigal Allon (*Contriving Warfare* [Hebrew] [Tel Aviv: Hakibbutz Hame'uchad, 1990], 157), Haim Laskov (*Military Leadership* [Hebrew] [Tel Aviv, Ma'arachot, 1985], 13), and Haim Bar-Lev ("A Lecture on War Principles at the IDF Command and Staff College," in *War Principles—Selected Readings* [IDF Command and Staff College Publications, 1 October 1982], 39) emphasized the importance of independence and flexibility. Van Creveld takes the IDF as an example of an army that from its beginning emphasized improvisation and creativity, abilities that came to fruition in the 1956 Sinai Campaign and the Six-Day War (van Creveld, *Command in War*, 194–203). From its inception the IDF officially adopted flexibility as a war principle (apparently from the British); however, over the years it gradually disappeared from the list of war principles (Meir Finkel, "Where Did Flexibility Go?" *Ma'arachot* 393 [2004]: 54–55).

5. Cohen and Gooch sought the reasons for failure in war and found that the inability to adapt to battlefield contingencies was one of the main causes. They blamed both the commanders and the command concept in the military organization (Eliot A. Cohen and John Gooch, *Military Misfortunes: The Anatomy of Failure in War* [New York: Vintage Books, 1991], 239–241). Another book, *America's First Battles 1776–1965* (John Shy, "First Battles in Retrospect," 330), put its finger on the link between the ability to cope with a new battlefield situation and the level of cognitive and command flexibility. In his book *Fighting Power* (1982), van Creveld provides a comprehensive analysis of the Wehrmacht's fighting capability against the American army in WW II. The book draws a connection between the German combat proficiency (that exceeded that of the Americans in the vast majority of encounters) and German cognitive and command flexibility (approximately 20 percent higher than that of the Americans). Samuels came to similar conclusions regarding German superiority against the British in WW I (Martin Samuels, *Command or Control? Command, Training and Tactics in the British and German Armies, 1888–1918* [London: Frank Cass, 1995], 282–285). Murray claims that the British navy's effectiveness in WW II was due to its superb flexibility, creativity, and adaptability in the underwater war in the Atlantic, as compared to the RAF's lower level

of adaptability. The British army, on the other hand, rarely managed to translate its combat experience into an improved warfighting doctrine. Its failures did not encourage the development of mental flexibility and initiative (Murray, "British Military Effectiveness in the Second World War," 112–122). Millett analyzed the effectiveness of the United States Army in WW II and found that whereas it developed physical mobility more than cognitive flexibility during the war, it also made great strides in adapting to new situations. The Americans compensated for their shortcomings at the tactical and operational levels by logistical flexibility that was expressed in the enormous flow of fighting materiel to combat units (Allan R. Millett, "The United States Armed Forces in the Second World War," in *Military Effectiveness*, vol. 3, 71–74).

6. Dixon studied the way inept command influences military confrontations. His research does not explicitly mention the lack of flexibility as a factor in command incompetence but does note that it is one of the elements that can be linked to mental dogmatism, such as fundamental conservatism, loyalty to a tradition that has outlived its practicality, inability to learn from mistakes and make a fresh start, tenacious attachment to a mission despite incontrovertible evidence of its futility, and failure to seize the opportunity of a fortuitous development (Dixon, *On the Psychology of Military Incompetence*, 152–153).

7. Gary D. Sheffield, "Introduction: Command, Leadership and the Anglo-American Experience," in *Leadership and Command: The Anglo-American Military Experience Since 1861*, ed. Gary D. Sheffield (London: Brassey's, 1997), 3. For an analysis of Field Marshal William Joseph Slim's leadership, see Robert Lyman, "The Art of Maneuver at the Operational Level of War: Lieutenant-General W. J. Slim and Fourteenth Army, 1944–45," in *The Challenges of High Command: The British Experience*, ed. Gary D. Sheffield and Geoffrey Till (New York: Palgrave Macmillan, 2003), 105. Many works on the art of command, such as *The Warlords* edited by Field Marshal Michael Carver (Michael Carver, ed., *The Warlords* [London: Weidenfeld & Nicolson, 1976]) and *Masters of the Art of Command* by Martin Blumenson and James L. Stokesbury (*Masters of the Art of Command* [Boston: Houghton Mifflin Company, 1975]) make no mention of command flexibility as an essential trait or ability that characterizes a good commander.

8. The Israeli military psychologist Reuven Gal has written that the commander of the future will have to learn quickly and adapt swiftly to new situations. He will need a high level of tolerance—especially because of battle fog and uncertainty—and the willingness and ability to constantly reassess rapidly changing conditions (Reuven Gal, "Military Leadership in the Light of Research" [lecture, November 18, 1991, Leadership Seminar at the Open University, Tel Aviv, Zichron Yakov: Israeli Institute for Military Research, 1991]).

Yogev, who also dealt with the battlefield of the future, claims that the primary required qualities of future military commanders will be open-mindedness and cognitive flexibility (Amnon Yogev, interviewed by Haim Lapid and Hagai Ben-Zvi, "When the

Eyes No Longer Surround the Fighting Theater: Preparation for the Battlefield of the Future Involves Changes in Weapons Systems and Thinking Patterns," *Skira Hodshit* 36, no. 12 [1990]: 37–38). Macgregor writes that the "Information Age" commander needs to value learning and to design organizations for change and decentralization (*Transformation Under Fire*, 192). Hosek asserts that one of the two most important traits of a combat commander of the 21st century is versatility in the sense of multitasking (James R. Hosek, "The Soldier of the 21st Century," *New Challenges, New Tools for Defense Decisionmaking*, ed. Stuart E. Johnson, Martin C. Libicki, and Gregory F. Treverton (RAND, MR 1576, 2003). Dickerson's research on adaptability found that the American army must develop manpower trained in adaptability if it hopes to cope effectively with future challenges (Dickerson, "Adaptability: A New Principle of War.").

In his book *Men Against Fire*, S. L. A. Marshal writes:

> It is not within the ingenuity of man ever to fully close the gap between training and combat. Once that fact is fully grasped, we have no choice but to incorporate its meaning into the working philosophy of training. So doing, we can arrive at a fresh application of what is really intended by the somewhat vague statement that "plasticity of mind" is the desirable mental attitude in the commander. Thus by a rough approximation: 60 percent of the art of command is the ability to anticipate; 40 percent of the art of command is the ability to improvise, to reject the preconceived idea that has been tested and proved wrong in the crucible of operations, and to rule by action instead of acting by rules (S. L. A. Marshall, *Men Against Fire* [New York: The Infantry Journal & William Morrow, 1947], 108).

> To square training with the reality of war, it becomes a necessary part of the young officer's mental equipment for training to install in him the full realization that in combat many things can and will go wrong without it being anyone's fault in particular . . . It therefore follows that the far object of a training system is to prepare the combat officer mentally so that he can cope with the unusual and the unexpected as if it were the altogether normal and give him poise in a situation where all else is in disequilibrium (ibid., 116).

9. Citino, *The German Way of War*, 302.

10. Hanan Shai (Shwartz), *Command and Control in the Modern Military Organization* [Hebrew] (I.D.F: The Command and Staff College, "Barak" Program, 1994), vii.

11. Helmuth von Moltke, *Moltke: On the Art of War—Selected Writings*, ed. Daniel J. Hughes, trans. Daniel J. Hughes and Harry Bell (New York: Ballantine, 1993), 132–133.

12. Ibid., 92.

13. Peter Paret, ed., *Makers of Modern Strategy: From Machiavelli to the Nuclear Age* (Princeton, N.J.: Princeton University Press, 1986), 301.

14. Shai, *Command and Control in the Modern Military Organization*, 157.

15. Paret, *Makers of Modern Strategy*, 304.

16. Condell and Zabecki, *On the German Art of War—Truppenführung*, 17.

17. Ibid., 9.
18. Shai, *Command and Control in the Modern Military Organization*, 130.
19. IDF Education and Training Department, "Command and Control by Employing 'Mission Commands,'" [Hebrew] (1987): 20.
20. Dupuy, *A Genius for War*, 305.
21. "German Training and Tactics: An Interview With Col. Pestke," *Marine Corps Gazette* 67, no. 10 (October 1983): 59.
22. Rommel, *The Rommel Papers*, 523.
23. U.S. Army, *German Military Improvisations During the Russian Campaign* (1971), originally published by the American Army as *MS T-21 German Military Improvisations*.
24. Bruce I. Gudmundsson, *Stormtroop Tactics: Innovation in the German Army, 1914–1918* (New York: Praeger, 1989), 172.
25. Hasso Von Manteuffel, "The Tank Battle of Târgul Frumos," *Military Review* 36, no. 6 (1956): 82.
26. Ibid., 82–83.
27. Steven J. Zaloga and James Grandsen, *The T-34 Tank* (London: Osprey, 1980), 29.
28. Bryan Perrett, *Knights of the Black Cross: Hitler's Panzerwaffe and Its Leaders* (London: Wordsworth, 1986), 191.
29. "Târgul Frumos and the Counter Stroke at Facuti. Operational and Tactical Lessons—War on the Eastern Front, 1941–45," Camberly Staff College (n.d.): 10-1–10-11.
30. David M. Glantz and Jonathan M. House, *When Titans Clashed: How the Red Army Stopped Hitler* (Lawrence, Kans.: University of Kansas, 1995), 63.
31. Ibid., 65.
32. S. A. Tyushkevich, *The Soviet Armed Forces: A History of Their Organizational Development: A Soviet View*, trans. CIS Multilingual Section, Translation Bureau, Secretary of State Department, Ottawa, Canada (Superintendent of documents, U.S. Government Printing Office, 1978), 277.
33. Glantz and House, *When Titans Clashed*, 65–66.
34. House, *Combined Arms Warfare in the Twentieth Century*, 131.
35. Tyushkevich, *The Soviet Armed Forces*, 278–281.
36. Shai, *Command and Control in the Modern Military Organization*, 279.
37. Leites, *Soviet Style in War*, 180–193, 340–341; see also Simpkin, *Race to the Swift*, 237–238.
38. Leonhard, *The Art of Maneuver*, 53–55.
39. Shai, *Command and Control in the Modern Military Organization*, 283. See also Col. Pestke's reply to a question on this matter ("German Training and Tactics," 64–65).
40. Condell and Zabecki, *On the German Art of War—Truppenführung*, 282–283.
41. Cushman, "Challenge and Response at the Operational and Tactical Levels, 1914–45," 329.

42. Van Creveld, *Fighting Power*, 40.
43. Shai, *Command and Control in the Modern Military Organization*, 298.
44. Rosen, *Winning the Next War*, 131–142.

Chapter 5

1. Karl E. Weick and Kathleen M. Sutcliffe, *Managing the Unexpected: Assuring High Performance in an Age of Complexity* (San Francisco: Jossey-Bass, 2001), 3–17.
2. Ibid., 70.
3. Rosen, *Winning the Next War*, 180–182.
4. Weick and Sutcliffe, *Managing the Unexpected*, 128–137.
5. Ibid., 56.
6. "Agranat Commission Report," 1413.
7. Gal Hirsh, "Operations at the Speed of Thought," *Ma'arachot* 380–381 (2001): 34.
8. Richard D. Downie, *Learning From Conflict: The U.S. Military in Vietnam, El Salvador, and the Drug War* (Westport, Conn.: Praeger, 1998).
9. Marshall, *Men Against Fire*, 113.
10. Ibid., 114.
11. Weick and Sutcliffe, *Managing the Unexpected*, 57–58.
12. Hirsh, "Operations at the Speed of Thought," 38.
13. "What was remarkable was that the 90th applied none of the lessons of the second attack." James Carafano, *GI Ingenuity: Improvisation, Technology and Winning WW II* (Mechanicsburg, Pa.: Stackpole Books, 2006), 29.
14. Leasor, *Singapore*, 161–163.
15. Allen, *Singapore 1941–1942*, 52.
16. Lowe, "Great Britain's Assessment of Japan Before the Outbreak of the Pacific War," 472–473.
17. Churchill, *The Second World War*, vol. 3, 619.
18. John Toland, *The Rising Sun: The Decline and Fall of the Japanese Empire, 1936–1945* (London: Penguin, 2001), 232–233.
19. Masatake Okumiya and Jiro Horikoshi, *Zero!* (New York: Ballantine, 1956), 57.
20. Toland, *The Rising Sun*, 234.
21. *Flying Tiger: Chennault of China*, explains how Chennault developed aerial combat techniques based on "pairs of planes" that compensated for the P-40's deficiencies. The American pilots took advantage of the Zero's shortcomings, such as a large blind spot in the front of the plane, the vulnerability of its oil radiator, and the Japanese pilot's exposure when separated from his formation. The Flying Tigers chalked up an impressive kill rate: between December 8, 1941 and July 1942, they shot down 299 Japanese planes against eight American pilots killed. Robert Lee Scott, *Flying Tiger: Chennault of China* (New York: Doubleday, 1959), 65–70.
22. Mahnken, *Uncovering Ways of War*, 79–81.

23. Gordon R. Sullivan and Michael V. Harper, *Hope Is Not a Method: What Business Can Learn From America's Army* (New York: Times Business, 1996), 204–210.

24. Ariely, "Operational Knowledge Management in the Military," 718.

25. Gil Ariely, *Learning to Digest During Fighting—Real Time Knowledge Management*, 2006 [accessed Aug. 5, 2009], http://www.ict.org.il/Articles/tabid/66/Articlsid/229/Default.aspx; Gil Ariely, "Learning While Fighting (during 2nd Lebanon War)," *Ma'arachot* 412 (2007): 4–13; Gil Ariely, "Learning While Fighting in 'Cast Lead' Operation," *Ma'arachot* 425 (2009): 12–21.

26. Heilmeier, "Guarding Against Technological Surprise."

27. "The Falklands Campaign: The Lessons" (London: Her Majesty's Stationery Office, December 1982), 24–25.

28. Cordesman, *The Iraq War*, 180.

Chapter 6

1. Alfred Price, *Luftwaffe Handbook 1939–1945* (New York: Charles Scribner's Sons, 1977b), 25.

2. Luttwak, *Strategy: The Logic of War and Peace*, 27–28.

3. Price, *Luftwaffe Handbook 1939–1945*, 25.

4. Jones, *The Wizard War*, 122.

5. Price, *Luftwaffe Handbook 1939–1945*, 25.

6. Price, *Instruments of Darkness*, 107–109.

7. Jones, *The Wizard War*, 290–293.

8. Louis Brown, *A Radar History of World War II: Technical and Military Imperatives* (Washington, D.C.: Carnegie Institute of Washington, 1999), 296.

9. Price, *Instruments of Darkness*, 142.

10. Jones, *The Wizard War*, 300.

11. Cajus Bekker, *The Luftwaffe War Diaries* (New York: Ballantine Books, 1964), 459–460.

12. According to Bekker, the first time the Germans encountered the use of chaff, "the inconceivable now took place. . . . The whole thing was a mystery. . . . The screens of the Würzburgs, operating on 53-cm became an indecipherable jumble of echo points resembling giant insects, from which nothing could be recognized at all" (ibid., 457–458). Weise, the head of German anti-aircraft defenses, testified at the end of the war that the use of Window immediately and completely paralyzed the use of radar for guiding interceptors and anti-aircraft batteries (Hubert Weise, "The Overall Defense of the Reich: 1940–1944 (January)," in *Fighting the Bombers: The Luftwaffe's Struggle Against the Allied Bomber Offensive*, ed. David C. Isby [London: Greenhill, 2003], 60). See also Luttwak, *Strategy: The Logic of War and Peace*, 27–28. See Jones for a description of the problem faced by German ground controllers on the second night of chaff when Essen was bombed (Jones, *The Wizard War*, 300–301).

13. Jones, *The Wizard War*, 305.
14. Price, *Instruments of Darkness*, 165.
15. Stanley Baldwin expressed this concept in no uncertain terms as far back as 1932: "The bomber will always get through." Alan J. Levine, *The Strategic Bombing of Germany, 1940–1945* (Westport, Conn.: Praeger, 1992), 7.
16. Martin Middlebrook, "Harris," in *The War Lords*, ed. Michael Carver (London: Weidenfeld & Nicolson, 1976), 324.
17. Churchill, *The Second World War* 5:519–520.
18. Ibid., 521.
19. Ibid., 4:288.
20. Jones, *The Wizard War*, 302.
21. Middlebrook, "Harris," 326.
22. Churchill, *The Second World War* 5:521–522.
23. Alfred Price, *Battle Over the Reich* (London: Ian Allan, 1973), 68.
24. Ibid.
25. Ibid.
26. Murray, 1985: 171–173.
27. Price, *Battle Over the Reich*, 70.
28. Bekker, *The Luftwaffe War Diaries*, 493.
29. Ibid., 461.
30. Price, *Battle Over the Reich*, 70.
31. Bekker, *The Luftwaffe War Diaries*, 461–462, 493.
32. Price, *Battle Over the Reich*, 71.
33. Williamson R. Murray, *Luftwaffe* (London: George Allen & Unwin, 1985), 173.
34. Price, *Battle Over the Reich*, 77–78.
35. Bekker, *The Luftwaffe War Diaries*, 496.
36. Brown, *A Radar History of World War II*, 298–299.
37. Price, *Battle Over the Reich*, 79–80.
38. Jones, *The Wizard War*, 466. Thompson, a former Pathfinder pilot, wrote that in January 1944 neither he nor his colleagues were aware of the deployment of the German Naxos radar and that all the lead planes, in the squadron he had recently left, were shot down that month, undoubtedly due to the Germans' use of this radar (Walter Thompson, *Lancaster to Berlin* [Guernsey, Channel Islands: Goodall, 1997], 192). For more details on German interception tactics, see Josef Scholls, "Night Fighter Tactics (NJ6)," in *Fighting the Bombers*, 233.
39. Thompson, *Lancaster to Berlin*, 193.
40. Murray, *Luftwaffe*, 203–205.
41. Bekker: *The Luftwaffe War Diaries*, 501.
42. Ibid., 500.
43. Jones, *The Wizard War*, 393–395.

44. Thompson, *Lancaster to Berlin*, 193.
45. Brown, *A Radar History of World War II*, 319.
46. Scholls, "Night Fighter Tactics (NJ6)," 234.
47. Murray, *Luftwaffe*, 203.
48. Price, *Battle Over the Reich*, 80–81.
49. Thompson, *Lancaster to Berlin*, 150–151.
50. Jennie Gray, *Fire by Night: The Dramatic Story of One Pathfinder Crew and Black Thursday, 16/17 December 1943* (London: Grub Street, 2000), 125.
51. Murray, *Luftwaffe*, 203.
52. Bekker, *The Luftwaffe War Diaries*, 501–502.
53. Ibid., 503–504.
54. Charles Webster and Noble Frankland, *The Strategic Air Offensive Against Germany 1939–1945*, vol. 2: *Endeavor* (London: Her Majesty's Stationery Office, 1961), 206.
55. Ibid., 193.
56. Murray, *Luftwaffe*, 210.
57. Brown, *A Radar History of World War II*, 309–310.
58. Manfred Messerschmidt, "German Military Effectiveness," in *Military Effectiveness*, vol. 3, 245.
59. James S. Corum and Richard R. Muller, *The Luftwaffe's Way of War: German Air Force Doctrine 1911–1945* (Baltimore: Nautical and Aviation, 1998), 119.
60. Ibid., 124.
61. Ibid.
62. Ibid., 149.
63. Ibid., 152.
64. Jeremy Howard-Williams, *Night Intruder: A Personal Account of the Radar War Between the RAF and Luftwaffe Night Fighter Force* (Vancouver: David & Charles, 1976), 93.
65. Brown, *A Radar History of World War II*, 299.
66. Ibid., 318.

Chapter 7

1. Steiger, *Armour Tactics in the Second World War*, 150. Forward armor plating 60 mm thick was installed at an angle that protected it from Panzer III 37-mm and 50-mm shells and Panzer IV short-barreled–75-mm shells. Its 76-mm gun easily penetrated the German tanks at long range, and its wide tracks were an advantage in moving through the mud and snow compared to the narrow tracks of the German tanks.

2. In 1939 Guderian visited a Soviet tank factory. Before the war he wrote that the Soviet Union's 10,000 tanks put the Red Army "at the head of the world's armies in respect of motorization" (Steiger, *Armour Tactics in the Second World War*, 78). In reality, historians estimate that the number of Soviet tanks was between 17,000 (David M. Glantz, ed. *The Initial Period of War on the Eastern Front: 22 June–August 1941* [London:

Frank Cass, 1993], 29) and 22,000 (Steven J. Zaloga and James Grandsen, *Soviet Heavy Tanks* [London: Osprey, 1981], 13) or even 24,000 (Kahn, *Hitler's Spies,* 457–458).

3. Glantz, *The Initial Period of War on the Eastern Front,* 336.
4. Steiger, *Armour Tactics in the Second World War,* 79–81.
5. Guderian, *Panzer Leader,* 235.
6. R. H. S. Stolfi, *Hitler's Panzers East: World War II Reinterpreted* (Norman, Okla.: University of Oklahoma Press, 1991), 164–165.
7. Zaloga and Grandsen, *The T-34 Tank,* 8–9.
8. Guderian, *Panzer Leader,* 237–238.
9. Zaloga and Grandsen, *The T-34 Tank,* 9.
10. F. W. von Mellenthin, *Panzer Battles* (Norman, Okla.: Oklahoma University Press, 1956), 153.
11. Liddell Hart, *The Other Side of the Hill,* 330, "Their T-34 tank was the finest in the world."
12. Steiger, *Armour Tactics in the Second World War,* 150.
13. Orgill, *The Tank,* 192.
14. Guderian, *Panzer Leader,* 237–238.
15. Glantz, *The Initial Period of War on the Eastern Front,* 451–452.
16. Kahn, *Hitler's Spies,* 458.
17. Betts, *Surprise Attack,* 131; Kahn, *Hitler's Spies,* 460.
18. Kahn, *Hitler's Spies,* 528.
19. Steiger, *Armour Tactics in the Second World War,* 81.
20. Charles C. Sharp, *German Panzer Tactics in World War II: Combat Tactics of German Armored Units From Section to Regiment* (West Chester, Ohio: Nafziger Collection, 1998), 42.
21. Cited in Hecht, *Military History,* 115.
22. U.S. Army, *German Military Improvisations During the Russian Campaign,* 48.
23. Steiger, *Armour Tactics in the Second World War,* 53.
24. Stolfi, *Hitler's Panzers East,* 158.
25. U.S. Army, *German Military Improvisations During the Russian Campaign,* 68.
26. Glantz, *The Initial Period of War on the Eastern Front,* 340.
27. U.S. Army, *German Military Improvisations During the Russian Campaign,* 68–69.
28. Steiger, *Armour Tactics in the Second World War,* 139.
29. Citino, *The Path to Blitzkrieg,* 226.
30. Ibid.; Condell and Zabecki, *On the German Art of War—Truppenführung,* 92–98 (paragraphs 339–346).
31. Sharp, *German Panzer Tactics in World War II,* 8. See "Training Directive for the Light and Medium Tank Companies in Combat," March 1, 1939, which states: "Support of the panzer attack by other arms (especially artillery, combat engineers, and motorized infantry) is indispensable."

32. Citino, *The Path to Blitzkrieg*, 231.

33. Robert M. Citino, *The German Way of War: From the Thirty Years War to the Third Reich* (Lawrence, Kans.: University Press of Kansas, 2005).

34. Deighton, *Blitzkrieg*, 164–168.

35. Citino, *The Path to Blitzkrieg*, 235.

36. Condell and Zabecki, *On the German Art of War—Truppenführung*, 197.

37. Sharp, *German Panzer Tactics in World War II*, 26.

38. Deighton, *Blitzkrieg*, 154.

39. Ibid.

40. R. M. Ogorkiewicz, *Armoured Forces: A History of Armoured Forces and Their Vehicles* (London: Arms & Armour Press, 1960), 216–217. The Panzer II tanks had a long-barrel (L/60 long-barrel = 50*60–3000 mm = 3 meters) instead of the 50-mm short-barrel (L/42). At almost the same time, the 75-mm–short-barreled (L/24) cannon was replaced by one almost double in length (the L/43, which was later replaced by a slightly longer barrel—the L/46). Lengthening the barrel increased muzzle velocity and the time the shell remained under pressure and enhanced the accuracy and power of impact. In these respects the Panzer IV was superior to the T-34. For details on the types and number of tanks of each model that entered service, see Walter J. Speilberger, *Panzer III & Its Variants* (Atglen, Pa.: Schiffer Military History, 1993), 53–58.

41. Ibid., 150–165.

42. Walter J. Speilberger, *Panzer IV & Its Variants* (Atglen, Pa.: Schiffer Military History, 1993b), 50.

43. Guderian, *Panzer Leader*, 238.

44. Steiger, *Armour Tactics in the Second World War*, 83–84.

45. Ogorkiewicz, *Armoured Forces*, 215–216.

46. Thomas L. Jentz, ed., *PanzerTruppen (pt. 1): The Complete Guide to the Creation & Combat Employment of Germany's Tank Force 1933–1942* (Atglen, Pa.: Schiffer Military History, 1996), 39.

47. Thomas L. Jentz, ed. *PanzerTruppen (pt. 2): The Complete Guide to the Creation & Combat Employment of Germany's Tank Force 1943–1945* (Atglen, Pa.: Schiffer Military History, 1996b), 96–97.

48. Gudmundsson, *Stormtroop Tactics*, 176.

49. Millett et al., *Military Effectiveness*, vol. 1, 205.

50. Förster, "The Dynamics of *Volkegemeinschaft*," 209; see also Williamson R. Murray, "Contingency and Fragility of the German RMA," in *The Dynamics of Military Revolution 1300–2050*, ed. MacGregor Knox and Williamson R. Murray (Cambridge, U.K.: Cambridge University Press, 2001), 162–166; Williamson R. Murray, "Innovation: Past and Future," in *Military Innovation in the Interwar Period*, ed. Williamson R. Murray and Allan R. Millett (Cambridge, U.K.: Cambridge University Press, 1996c), 314. Naveh tends to differ with these scholars, citing the Germans' failure to derive tactical

lessons in WW II as an example of the absence of system-wide work patterns in that period. Naveh claims that the diversity of problems that surfaced during the campaigns in Poland and France—including technical problems related to the tanks' guns and protective armor and to fighting techniques—was not dealt with even as Operation Barbarossa approached (Naveh, *In Pursuit of Military Excellence*, 131).

51. Murray, "Innovation: Past and Future," 314. German interest in the testing was paralleled by a desire for getting to the bare truth. Ludendorf accurately described the German military ethos in his tour of the Western Front in 1916: "[The staffs] knew I wanted to hear their real views and have a clear idea of the true situation, not a favorable report made to order" (Murray, "Contingency and Fragility of the German RMA," 158).

52. Förster, "The Dynamics of *Volkegemeinschaft*," 209. Another question is what happened to these reports when the German High Command was unable to replenish weapons or come forth with additional manpower.

53. Glantz, *The Initial Period of War on the Eastern Front*, 168.

54. Förster, "The Dynamics of *Volkegemeinschaft*," 209–210.

Chapter 8

1. Dani Asher, "The Syrian Plan for the Yom Kippur War and the War's Moves" [Hebrew], in *Yom Kippur War Studies*, ed. Hagai Golan and Shaul Shai (Tel Aviv: Ma'arachot, 2003), 299–301.

2. "Agranat Commission Report," 1415–1419.

3. Ibid., 1475–1477.

4. Ibid.

5. Yishai Wechsler and Yehuda Tal, *198 Tank Battalion at Yom Kippur War* [Hebrew] (Tel Aviv: Defense Ministry Publications, 2002), 88–89. See also Arad and Laskov, *Looking Back*, 18–19.

6. Dani Asher, *Breaking the Concept* [Hebrew] (Tel Aviv: Ma'arachot, 2003b), 222–223.

7. Lorber, *Science, Technology and the Battlefield*, 81.

8. Shimon Iftach, *Missiles and Rockets in Modern Warfare* [Hebrew] (Tel Aviv: Ma'arachot, 1979), 43.

9. Ibid., 38.

10. Edward Luttwak and Dan Horowitz, *The Israeli Army* (London: Allen Lane, 1975), 346.

11. Interview with Dani Asher, 2004.

12. Dani Asher, "From 'Order 41' to 'Tahrir 41': From the Egyptian Warfighting Doctrine to War," [Hebrew] *Ma'arachot* 332 (1993): 49.

13. Lorber, *Science, Technology and the Battlefield*, 81.

14. "Anti-Tank Fighting," *Ma'arachot* 209 (1970): 28.

15. Lanir, *Fundamental Surprise*, 49.
16. Haim Herzog, *The War of Atonement* [Hebrew] (Jerusalem: Yedioth Ahronot, 1975), 68.
17. Lanir, *Fundamental Surprise*, 49.
18. Arad and Laskov, *Looking Back*, 35.
19. Yehuda Wallach, "Requiem for the Tank," *Ma'arachot Shiryon* 23 (1971): 41.
20. Yehuda Wallach, "Is the Tank Dead? Another Nail in Armor's Coffin," *Ma'arachot Shiryon* 25 (1972): 4–9.
21. Ibid., 9.
22. Yehuda Wallach, "Can We Learn from History?" *Ma'arachot* 254 (1977): 32.
23. Uri Bar-Joseph, *The Watchman Fell Asleep: The Surprise of Yom Kippur and Its Sources* [Hebrew] (Lod: Zmora-Beitan, 2001), 419–436.
24. Azriel Lorber, *Misguided Weapons: Technological Failure and Surprise on the Battlefield* (Washington, D.C.: Brassey's, 2002), 76–80. See also the view of the Egyptian chief of staff, Saad el Din Shazly, *The Crossing of the Suez: The Egyptian Chief of Staff's Memoirs From the Yom Kippur War* [Hebrew] (Tel Aviv: Ma'arachot, 1987), 56.
25. Erez Wiener, "Formation and Crisis in the IDF's Warfighting Doctrine," in *Yom Kippur War Studies*, 83–84.
26. Arad and Laskov, *Looking Back*, 18–19, 35.
27. Ben-Israel, 1993, 12.
28. Lanir, *Fundamental Surprise*, 48.
29. "Agranat Commission Report," 1346. See also Israel Tal's testimony regarding the Germans' successful use of anti-tank guns as defensive weapons in the Western Desert in WW II (ibid., 1444–1445).
30. Ibid., 1350–1351.
31. Ibid., 1447.
32. Shimshi, *Storm in October*, 30–34; meaning an unknown weapon.
33. Arad and Laskov, *Looking Back*, 18–19.
34. Ibid., 36.
35. Binyamin Zeev Kedar, *October 1973—The Story of "Smash" Battalion* [Hebrew] (Tel Aviv: Tamuz, 1975), 48–49.
36. Avraham Adan, *On Both Banks of the Suez* [Hebrew] (Jerusalem: Edanim, 1979), 287.
37. Anthony H. Cordesman and Abraham R. Wagner, *The Lessons of Modern War, Vol.1: The Arab-Israeli Conflicts, 1973–1989* (London: Westview, 1990a), 61.
38. Millett et al., "The Effectiveness of Military Organizations," 26.
39. Haim Laskov, "The Lessons of the Sinai Campaign," *Ma'arachot* 21 (1970): 5–6. See also Menahem Einan, "The Need for Quality in Light of the Limitations of Quantity," in *Quantity and Quality in Force Planning*, 482.
40. Levita, *Offense and Defense in Israeli Military Doctrine*, 128–129.

41. Luttwak and Horowitz, *The Israeli Army*, 190–191.
42. Ibid., 370.
43. Luttwak, *Strategy: The Logic of War and Peace*, 38.
44. Luttwak and Horowitz, *The Israeli Army*, 363.
45. "Agranat Commission Report," 1437–1438.
46. Ibid., 1432.
47. Ibid., 1454.
48. Cordesman and Wagner, *The Lessons of Modern War*, vol. 1, 55. See also Uri Ben Ari's testimony on the mechanized infantry's readiness (equipment and transportation) in the war ("Agranat Commission Report," 1418).
49. Cordesman and Wagner, *The Lessons of Modern War*, vol. 1, 54, 66.
50. "The Israeli Artillery Corp in the Yom Kippur War" [Hebrew], part of a series of seminars on the Yom Kippur War (Ramat Efal: Israel Galili Foundation for Research on Defense Force, The Israeli Military History Foundation at Tel Aviv University, July 6, 2000), 39.
51. Cordesman and Wagner, *The Lessons of Modern War*, vol. 1, 60.
52. Dani Asher and Itai Asher, "The Israeli Artillery on the Southern Front in the Yom Kippur War," *Ma'arachot* 354 (1997): 12.
53. Ibid., 20.
54. Cordesman and Wagner, *The Lessons of Modern War*, vol. 1, 60.
55. Asher and Asher, "The Israeli Artillery on the Southern Front in the Yom Kippur War," 20.
56. Adan, *On Both Banks of the Suez*, 160.
57. Ibid., 203.
58. Shimshi, *Storm in October*, 28.
59. Adan, *On Both Banks of the Suez*, 160.
60. "Agranat Commission Report," 528–530.
61. Ibid., 1348–1349.

Chapter 9

1. Weizman, *On Eagles' Wings*, 329.
2. Elhanan Oren, *A History of the Yom Kippur War* [Hebrew] (IDF: History Department, 2004), 355.
3. Shimon Iftach, "Missiles in Egypt," *Ma'arachot* 217a–218 (1971): 18–20.
4. Dima Adamsky, *Operation Kavkaz: Soviet Intervention and Israeli Intelligence Failure in the War of Attrition* [Hebrew] (Tel Aviv: Ma'arachot, 2006).
5. Ehud Yonay, *No Margin for Error: The Story of the Israeli Air Force* [Hebrew] (Jerusalem: Keter, 1995), 219.
6. Eliezer Cohen (Cheetah) and Zvi Lavi, *The Sky Is Not the Limit: The Story of the Israeli Air Force* [Hebrew] (Jerusalem: Sifriyat Hapoalim, 1990), 403–408.

7. Ibid., 445.

8. Yossi Abudi, "The Israeli Air Force in War," part of a series of seminars on the Yom Kippur War, Study Day No. 7, *The Air Arena in the Yom Kippur War* (lecture, Ramat Efal: Israel Galili Foundation for Research on Defense Force, The Israeli Military History Foundation at Tel Aviv University, May 4, 2000).

9. Cohen and Lavi, *The Sky Is Not the Limit*, 428–430.

10. Yonay, *No Margin for Error*, 251.

11. Ibid., 251–252.

12. Bar-Joseph, *The Watchman Fell Asleep*, 96.

13. Shmuel Gordon, "Future Lessons From a Past War" [Hebrew], in *Yom Kippur War Studies*, 369.

14. "Agranat Commission Report," 215.

15. Gordon, "Future Lessons From a Past War," 369.

16. Cohen and Lavi, *The Sky Is Not the Limit*, 445.

17. "Agranat Commission Report," 216.

18. Ibid.

19. Yonay, *No Margin for Error*, 253.

20. Shimon Golan, "The Preemptive Air Strike at the Outbreak of the Yom Kippur War," *Ma'arachot* 373 (2000): 59. Shmuel Gordon, *Thirty Hours in October* [Hebrew] (Tel Aviv: Ma'ariv Book Guild, 2008), 226.

21. Gordon's *Thirty Hours in October* discusses in depth this decision (321–351).

22. Yonay, *No Margin for Error*, 256.

23. Bar-Joseph, *The Watchman Fell Asleep*, 378.

24. Ran Ronen (Peker), *Hawk in the Sky* [Hebrew] (Tel Aviv: Yedioth Ahronot, 2002), 342.

25. Cohen and Lavi, *The Sky Is Not the Limit*, 475–476.

26. Iftach Spector, *Dream in Blue and Black* [Hebrew] (Jerusalem: Keter, 1991), 42; see also 22. For details on this day, see Iftach Spector, *Loud and Clear* [Hebrew] (Tel Aviv: Miskal-Yedioth Ahronoth and Chemed Books, 2008), 44–50.

27. Gordon, "Future Lessons From a Past War," 374.

28. Herzog, *The War of Atonement*, 232.

29. Oren, *A History of the Yom Kippur War*, 315.

30. Ben-Israel, "Technological Lessons," 12.

31. Muhammad Fawzi, chapters in his book *The October '73 War* (published in the Lebanese newspaper *al Sharia* between August 8, 1988 and October 24, 1988), [Hebrew] Hatzav 843/013 (1989).

32. Gordon, "Future Lessons From a Past War," 385.

33. Shmuel Gordon, "The Paradox of October 7" [Hebrew] *Ma'arachot* 361 (1998b): 44–57, 49.

34. Clausewitz, *On War*, 120.

35. Moltke, *Moltke: On the Art of War—Selected Writings*, 92.

36. Giora Rom, "The Israeli Air Force in War," part of a series of seminars on the Yom Kippur War, Study Day No. 7, *The Air Arena in the Yom Kippur War* [Hebrew] (lecture, Ramat Efal: Israel Galili Foundation for Research on Defense Force, The Israeli Military History Foundation at Tel Aviv University, May 4, 2000). Further editing by Giora Rom in 2005, during an interview.

37. Hoshen, "Intelligence in Weapons Development," 528.

38. Price, *Instruments of Darkness*, 272.

39. Ben-Israel, "Technological Lessons," 11.

40. Hecht, "Technological Military Deception in the Twentieth Century," 36.

41. Ben-Israel, "Technological Lessons," 8.

42. Rom, "The Israeli Air Force in War," 107.

43. Interview with Giora Rom, 2005.

44. Meirav Halperin and Aharon Lapidot, *G Suit: Pages in the Log Book of the Israel Air Force* [Hebrew] (Tel Aviv: IAF Publications, 2000), 164.

45. Aharon Lapidot, ed., *101: The First Fighter Squadron of the Israel Air Force* [Hebrew], The Heritage Series of the Israel Air Force (IAF: Defense Ministry, 1998), 113–115.

46. Aharon Lapidot and Dan Sela, *Thunder of the Hammers: The 69th Squadron of the Israel Air Force* [Hebrew], The Heritage Series of the Israel Air Force (Tel Aviv: IAF Defense Ministry, 1998b), 115.

47. Yonay, *No Margin for Error*, 251–252. See also Spector, *Loud and Clear*, 236.

48. Spector, *Dream in Blue and Black*, 55–56.

49. Interview with Giora Rom, 2005.

50. Amos Amir, *Flames in the Sky* [Hebrew] (Tel Aviv: Defense Ministry, 2000), 231.

51. Amir Oren, "Where Did the Red Army Go?" *Ha'aretz* (October 7, 2005): 3b.

52. "*The 16th Century: The 16th Squadron Heritage Book*" [Hebrew] (Tel Aviv: IAF, 1987), 204.

53. Abudi, "The Israeli Air Force in War," 6.

54. Ibid.

55. Oren, *A History of the Yom Kippur War*, 145.

56. Yonay, *No Margin for Error*, 269.

57. Eli Zeira, *The October 1973 War: Myth Against Reality* [Hebrew] (Tel Aviv: Yedioth Ahronot, 1993), 177.

58. Yonay, *No Margin for Error*, 271.

59. Zeev Bonen, *RAFAEL*, 60–64. Another homing device, in better condition, was found on the 24th near the Suez Canal (Gordon, *Thirty Hours in October*, 427).

60. Ronen Bergman, "No Stopping on Red," *Yedioth Ahronot*, 7 Days [appendix], no. 2180 (November 4, 2005): 21.

61. Aharon Lapidot and Dan Arkin, *IAF Logistics* [Hebrew] (Tel Aviv: IAF, Defense

Ministry, 2001), 38. See also Aryeh Hillel, *The Human Hand Under Their Wings: The History of the Israeli Air Force's Technological Layout, 1948–1973* [Hebrew] (IAF: Logistics Wing/Ma'arachot/Defense Ministry, 1990), 157.

62. Lapidot and Arkin, *IAF Logistics*, 38.

63. Yair Rosenblum and Oded Feldman, "Air Force Ensemble," a poem inspired by Major General Benny Peled's speech.

64. Iftach Spector, "Squadron Command in Wartime—Personal Lessons," in *Air Force Headquarters—Air Division: Echoes From the War: Air Commanders and Airmen Speak About Combat* [Hebrew] (IAF, 1984), 136–141.

65. Gordon, "Future Lessons From a Past War," 150–153.

66. Interview with Giora Rom, 2005.

67. Weizman, *On Eagles' Wings*, 192.

68. Interview with Giora Rom, 2005.

69. Y and Y, "The Airplane in the Ground Battle: End of the Road or Crossroads?" *Ma'arachot* 266 (1978): 43–46.

Chapter 10

1. George Forty, *The Armies of Rommel* (London: Arms & Armour, 1997), 103.

2. Condell and Zabecki, *On the German Art of War—Truppenführung*, 10.

3. Handel, "Technological Surprise in War," 4.

4. Stephen Bungay, *Alamein* (London: Aurum, 2002), 34. See also James Lucas, *Panzer Army Africa* (London: Macdonald and Jane's, 1977), 26.

5. Eado Hecht, *Military History: A Summary of Lectures and Reading at the IDF Command and Staff College "Barak,"* pt. 2 (IDF Command and Staff College, 2004), 261.

6. Bungay, *Alamein*, 29.

7. Michael Carver, *Tobruk* (London: B.T. Batsford, 1964), 28.

8. Handel, "Technological Surprise in War," 4. Regarding surprise at the tactical level, see Cyril Joly's description in *Take These Men* (London: Constable, 1955), 135; and Robert Crisp's description in *Brazen Chariots* (London: Transworld, 1959), p. 51.

9. Churchill, *The Second World War*, 3:343.

10. Carver, *Tobruk*, 39.

11. B. H. Liddell Hart, *The Strategy of Indirect Approach* (London: Faber and Faber, 1954), 278.

12. Mellenthin, *Panzer Battles*, 55.

13. Ibid., 94.

14. Ibid., 55.

15. Heinz Werner Schmidt, *With Rommel in the Desert* (London: Panther, 1960), 51.

16. Wolf Heckmann, *Rommel's War in Africa*, trans. Stephan Seago (New York: Smithmark, 1995), 93. On the 3.7-inch gun's technical shortcomings as an anti-tank weapon, and the debate in England over its preferred deployment, see Shelford Bidwell

and Dominick Graham, *Fire-Power: The British Army Weapons & Theories of War 1904–1945* (S. Yorkshire: Pen & Sword Military Classics, 2004), 230–231.

17. Carver, *Tobruk*, 211–212.
18. Heckmann, *Rommel's War in Africa*, 94.
19. Louis Allen, *Singapore 1941–1942* (London: Frank Cass, 1977), 147.
20. Woodburn S. Kirby, *Singapore: The Chain of Disaster* (London: Camelot Press, 1971), 149–150.
21. Allen, *Singapore 1941–1942*, 204. John Ferris, "Worthy of Some Better Enemy?: The British Estimate of the Imperial Japanese Army, 1919–41, and the Fall of Singapore," *Canadian Journal of History* 28 (1993): 246–247.
22. Noel Barber, *Sinister Twilight: The Fall of Singapore* (London: Collins, 1973), 60.
23. Dixon, *On the Psychology of Military Incompetence*, 132.
24. Ferris, "Worthy of Some Better Enemy?" 250–251.
25. Ibid., 133.
26. Bungay, *Alamein*, 32.
27. Kier, *Imagining War*, 90.
28. Murray, "British Military Effectiveness in the Second World War," 125.
29. Kier, *Imagining War*, 130.
30. Murray, "British Military Effectiveness in the Second World War," 127.
31. Rommel, *The Rommel Papers*, 520.
32. Ibid., 184.
33. Mellenthin, *Panzer Battles*, 148.
34. Murray, "British Military Effectiveness in the Second World War," 127.
35. Millett et al., "The Effectiveness of Military Organizations," 22.
36. Murray, "British Military Effectiveness in the Second World War," 129.
37. Ibid., 112.
38. Brian Bond and Williamson R. Murray, "The British Armed Forces," in *Military Effectiveness: Vol. 2: The Interwar Period*, ed. Allan R. Millett and Williamson R. Murray (Boston: Allen & Unwin, 1988), 121.
39. William G. F. Jackson, *The North African Campaign 1940–43* (London: B.T. Batsford, 1975), 131.
40. Murray, "British Military Effectiveness in the Second World War," 128.
41. Carver, *Tobruk*, 255.
42. Bungay, *Alamein*, 45–46.
43. House, *Combined Arms Warfare in the Twentieth Century*, 124–125.
44. Hecht, *Military History*, 162.
45. Deighton, *Blitzkrieg*, 128–29.
46. House, *Combined Arms Warfare in the Twentieth Century*, 124–125.
47. Heckmann, *Rommel's War in Africa*, 39.
48. Bungay, *Alamein*, 43.

49. Carver, *Tobruk*, 28.
50. Orgill, *The Tank*, 170–171.
51. House, *Combined Arms Warfare in the Twentieth Century*, 125.
52. Shelford Bidwell, *Gunners at War* (Aberdeen: Arms and Armour, 1970), 172–173, 234–235.
53. Forty, *The Armies of Rommel*, 145; Lucas, *Panzer Army Africa*, 27.
54. House, *Combined Arms Warfare in the Twentieth Century*, 125.
55. Orgill, *The Tank*, 183–184.
56. Murray, "Innovation: Past and Future," 324.
57. Bond and Murray: "The British Armed Forces," 121.
58. Murray, "Contingency and Fragility of the German RMA," 157.
59. Murray, "Innovation: Past and Future," 314.
60. Murray, "British Military Effectiveness in the Second World War," 124–125.
61. Murray, "Innovation: Past and Future," 324.
62. Millett et al., "The Effectiveness of Military Organizations," 25. According to Timothy Place, who studied British military training during WW II, the first of the *Notes From Theaters of War* was published on February 19, 1942 and dealt with the lessons of the Crusader offensive (November 1941). This is one example of Britain's slow response to changing battlefield conditions (eight months passed between Operation Battleaxe and the publication of the first note). Timothy H. Place, *Military Training in the British Army 1940–1944* (London: Frank Cass, 2000), 12. Concerning the " . . . slowness of military thought in Middle East Headquarters in mid-1942 . . . ," see Bidwall and Graham, *Fire-Power*, 240.
63. Hecht, *Military History*, 162.

Chapter 11

1. Anthony H. Cordesman and Abraham R. Wagner, *The Lessons of Modern War, vol. 3: The Afghan and Falklands Conflicts* (London: Westview Press, 1990c), 1–2; Stephen Blank, "Soviet Forces in Afghanistan: Unlearning the Lessons of Vietnam," in *Responding to Low-Intensity Conflict Challenges*, ed. Stephen Blank and others (Maxwell Air Base, Ala.: Air University Press, 1990b), 53–126.
2. The Russian General Staff, *The Soviet-Afghan War: How a Superpower Fought and Lost*, trans. and ed. Lester W. Grau and Michael A. Gress (Lawrence, Kans.: University Press of Kansas, 2002), 305.
3. Blank, "Soviet Forces in Afghanistan," 84.
4. Mark Urban, *War in Afghanistan* (London: Macmillan, 1988), 65.
5. The Russian General Staff, *The Soviet-Afghan War*, 91.
6. Lester W. Grau, "The Soviet-Afghan War: A Superpower Mired in the Mountains," *Journal of Slavic Military Studies* 17, no. 1 (2004): 148.
7. Mohammad Yahya Nawroz and Lester W. Grau, "The Soviet Experience in Afghanistan," *Military Review* 75, no. 5 (1995): 22.

8. Grau, "The Soviet-Afghan War," 147. Blank states that the Soviet High Command prepared for combat against the Afghanis as though the latter were similar to European enemies. There are indications that the Soviet army was planning to use chemical weapons in Afghanistan to achieve a quick victory through a deep-penetrating strike (Stephen Blank, "Imaging Afghanistan: Lessons of a 'Small War,'" *Journal of Soviet Military Studies* 3, no. 3 [1990]: 469). Reports show that the Soviet army employed chemical weapons on at least thirty-six occasions between 1980 and 1982 (Joseph J. Collins, "The Soviet Military Experience in Afghanistan," *Military Review* 65 [May 1985]: 20). It was also preparing for enemy air strikes and for this reason, deployed air defense units (Blank, "Soviet Forces in Afghanistan," 80).

9. Ibid., 87.

10. Nawroz and Grau, "The Soviet Experience in Afghanistan," 24–25.

11. John E. Sray, *Mountain Warfare: The Russian Perspective* (FMSO: March 1994), 2–8, http://fmso.leavenworth.army.mil/documents/mountain.htm.

12. Stephen Blank, *Afghanistan and Beyond: Reflections on the Future of Warfare* (Carlisle, Pa.: Strategic Studies Institute, U.S. Army War College, 1993), 1, http://www.au.af.mil/au/awc/awcgate/ssi/afghan-blank.pdf.

13. Grau and Gress, *The Soviet-Afghan War*, 19.

14. Scott R. McMichael, "The Soviet Army, Counterinsurgency, and the Afghan War," *Parameters* 19 (1989): 23.

15. Grau and Gress, *The Soviet-Afghan War*, 18; see also Blank, "Soviet Forces in Afghanistan," 80; McMichael, "The Soviet Army, Counterinsurgency, and the Afghan War," 23.

16. Rafael Reuveny and Aseem Prakash, "The Afghanistan War and the Breakdown of the Soviet Union," *Review of International Studies* 25 (1999): 707.

17. Douglas A. Borer, *Superpowers Defeated: Vietnam and Afghanistan Compared* (London: Frank Cass, 1999), 214–215.

18. Grau and Gress, *The Soviet-Afghan War*, xix.

19. Blank, "Imaging Afghanistan," 469.

20. Cordesman and Wagner, *The Lessons of Modern War*, vol. 3, 3.

21. Robert S. Litwak, "The USSR in Afghanistan," in *Foreign Military Intervention*, ed. Ariel Levite, Bruce Jentelson, and Larry Berman (New York: Columbia University Press, 1992), 79; see also McMichael, "The Soviet Army, Counterinsurgency, and the Afghan War," 31.

22. Narwoz and Grau, "The Soviet Experience in Afghanistan," 20.

23. McMichael, "The Soviet Army, Counterinsurgency, and the Afghan War," 24–25.

24. Urban, *War in Afghanistan*, 150.

25. McMichael, "The Soviet Army, Counterinsurgency, and the Afghan War," 24–25.

26. Edward B. Westermann, "The Limits of Soviet Airpower: The Failure of Military Coercion in Afghanistan, 1979–89," *Journal of Conflict Studies* 19, no. 2 (1999): http://www.lib.unb.ca/Texts/JCS/bin/get5.cgi?directory=fall99/&filename=WESTERMA.htm 2.

27. Robert M. Cassidy, *Russia in Afghanistan and Chechnya: Military Strategic Culture and the Paradoxes of Asymmetric Conflict* (Carlisle, Pa.: Strategic Studies Institute, U.S. Army War College, 2003), 19, http://www.strategicstudiesinstitute.army.mil/pubs/display.cfm?pubID=125.

28. For more details on these issues, see Grau and Gress, *The Soviet-Afghan War*, 93–149; for an analysis of encounters at the tactical level and various methods of operation, see Lester W. Grau, ed., *The Bear Went Over the Mountain: Soviet Combat Tactics in Afghanistan* (London: Frank Cass, 1998).

29. Nawroz and Grau, "The Soviet Experience in Afghanistan," 23; Grau and Gress, *The Soviet-Afghan War*, 311.

30. Cordesman and Wagner, *The Lessons of Modern War*, vol. 3, 124–125.

31. McMichael, "The Soviet Army, Counterinsurgency, and the Afghan War," 26–27.

32. Sray, *Mountain Warfare: The Russian Perspective*, 17–18. Nawroz and Grau also claim that despite the establishment in many districts of instruction centers for mountain warfare, the Soviets were slow in adopting new tactics for fighting a stubborn enemy in harsh terrain. Nawroz and Grau, "The Soviet Experience in Afghanistan," 22).

33. Grau and Gress, *The Soviet-Afghan War*, 189.

34. Ibid., 20.

35. Cordesman and Wagner, *The Lessons of Modern War*, vol. 3, 4–8.

36. Urban, *War in Afghanistan*, 67–68.

37. Cordesman and Wagner, *The Lessons of Modern War*, vol. 3, 117.

38. These included: the establishment of infantry mountain battalions, rifle companies made up of four platoons instead of three, combat service support brigades and battalions, and equipping the special forces with the "roomier" BMPs and BTRs rather than the BMDs (Narwoz and Grau, "The Soviet Experience in Afghanistan," 25–26).

39. Urban, *War in Afghanistan*, 65.

40. Grau and Gress, *The Soviet-Afghan War*, 35–36.

41. This included replacing the BMP-1 with the BMP-2 in 1985 and the BTR-60 with the BTR-70 and later with the BTR-80 in 1985. The soldiers compensated for the latter's defense drawbacks by mounting a coiled netting on the vehicle's sides and revetting the broadsides of the vehicle with jerry cans of water, sand bags, and so forth (Grau and Gress, 2002, 37). One solution to the problem of the size of the AFVs on narrow mountainous roads in Afghanistan was the rather extensive use in combat of the relatively small, air-transportable BMDs. A relatively quick solution to the shortcomings of the 73-mm anti-tank cannon with its limited-range (1300 meters), limited elevation (33 degrees), and slow muzzle velocity was to replace it with a mounted 30-mm cannon of greater range and improved elevation (50 degrees) in mid-1981 (Cordesman and Wagner, *The Lessons of Modern War*, vol. 3, 150–151). Infantry weapons were also upgraded by equipping the troops with the AK-47 assault rifle, which used a 5.45-mm bullet instead of the 7.62, and other weapons, such as grenade launchers.

42. Collins, "The Soviet Military Experience in Afghanistan," 23.
43. McMichael, "The Soviet Army, Counterinsurgency, and the Afghan War," 28.
44. Cordesman and Wager, *The Lessons of Modern War*, vol. 3, 118.
45. McMichael, "The Soviet Army, Counterinsurgency, and the Afghan War," 28.
46. Blank, *Afghanistan and Beyond*, 7.
47. Blank, "Imaging Afghanistan," 479–481.
48. McMichael, "The Soviet Army, Counterinsurgency, and the Afghan War," 33.
49. Cassidy, *Russia in Afghanistan and Chechnya*, 36.
50. Blank, "Soviet Forces in Afghanistan," 58.
51. Theo Farrell, "World Culture and Military Power," *Security Studies* 14, no.3 (2005): 485.
52. Cassidy, *Russia in Afghanistan and Chechnya*, 11.
53. Grau and Gress, *The Soviet-Afghan War*, 310–311.
54. Deborah D. Avant, *Political Institutions and Military Change: Lessons From Peripheral Wars* (Ithaca, N.Y.: Cornell University Press, 1994), 29–36.
55. Grau and Gress, *The Soviet-Afghan War*, 310.

Chapter 12

1. May, *Strange Victory*, 8–10.
2. Ibid.
3. Ibid., 236.
4. Liddell Hart, *The Strategy of Indirect Approach*, 247.
5. Marc Bloch, *Strange Defeat: A Statement of Evidence Written in 1940*, trans. Gerard Hopkins (New York: W.W. Norton & Company, 1999), 36–37.
6. Ibid., 45.
7. Deighton, *Blitzkrieg*, 254.
8. J. F. C. Fuller, *The Second World War 1939–1945* (London: Eyre & Spottiswoode, 1954), 78. For the sake of argument, it makes no difference whether the method was a warfighting doctrine (Naveh, "Was the Blitzkrieg a Warfighting Doctrine? Part I."). The historical fact is that in the tactical and operational battle environment, the successful application of German forces, formerly known as *blitzkrieg*, caught the French, British, and Soviet armies by surprise in the 1940 invasion of France and in the 1941 invasion of the Soviet Union. In our discussion, this manner of force application is called the "warfighting doctrine" in the tactical-operational sense.
9. House, *Combined Arms Warfare in the Twentieth Century*, 117.
10. Fuller, *The Second World War 1939–1945*, 73–74.
11. Steiger, *Armour Tactics in the Second World War*, 19–20.
12. Alistair Horne, *To Lose a Battle: France 1940* (London: MacMillan, 1969), 357.
13. Gat, *British Armour Theory and the Rise of the Panzer Arm*, 81.
14. Deighton, *Blitzkrieg*, 123–124; Gat, "Ideology, Policy, Technology and Strategic

Doctrine Between Two World Wars"; John J. Mearsheimer, *Liddell Hart and the Weight of History* (Ithaca, N.Y.: Cornell University Press, 1988), 266–300.

15. Gat, *British Armour Theory and the Rise of the Panzer Arm.*

16. Beginning in 1925 German intelligence published a bimonthly summary of technological and doctrinal developments in armor throughout the world. Articles were translated from American, British, French, Austrian, Polish, and Russian military journals and books (Corum, *The Roots of Blitzkrieg*, 131). British field regulations published in 1927 served as the main reference book for German armor until 1933 (Hecht, The "Operational Breakthrough" in German Military Thought 1870–1945, 120–121). The German military attaché in England, who was present at maneuvers, transmitted up-to-the-minute reports to Germany on the development of the British armor branch between 1933 and 1937 (Gat, *British Armour Theory and the Rise of the Panzer Arm*, 63–66). Close cooperation went on between the German and Soviet armies (Hecht, The "Operational Breakthrough" in German Military Thought 1870–1945, 121; Erickson, *The Soviet High Command*, 258). Charles de Gaulle's book, *The Army of the Future*, published in 1934, developed his ideas regarding the role of the armored force in a future war and in the French army. The book was translated into German in 1935. In his memoirs, de Gaulle recalls that he was aware that Hitler and his colleagues had read the book and found it most interesting (de Gaulle, *The Complete War Memoirs of Charles de Gaulle*, 15–16).

Guderian's memoirs note that he read books and articles by Fuller and Mertal, a British military thinker (Liddell Hart's name was mentioned only in the English edition of Guderian's memoirs), and that they sparked his interest and gave him food for thought (Guderian, *Panzer Leader*, 20). In his book *Achtung Panzer!* published in 1937, Guderian stated: "After mature consideration it was decided that until we had accumulated sufficient experience on our own account, we should base ourselves principally on the British notions as expressed in *Provisional Instructions for Tank and Armoured Car Training II 1927*" (Guderian, *Achtung-Panzer!* 167). Red Army officers, too, studied British military thinkers (Erickson, "Threat Identification and Strategic Appraisal by the Soviet Union, 1930–1941," 399). Fuller's book *Tanks in the Great War* was published in Russia in 1923 (Messenger, *The Blitzkrieg Story*, 61), and his book *Armored Warfare* was made compulsory reading for Russian commanders (Yehuda Wallach, *Military Doctrines: Their Development in the 19th and 20th Centuries* [Hebrew] (Tel Aviv: Ma'arachot, 1977b), 203.

17. Gat, "Ideology, Policy, Technology and Strategic Doctrine Between Two World Wars."

18. Horne, *To Lose a Battle*, 33.

19. Robert A. Doughty, *The Seeds of Disaster: The Development of French Army Doctrine 1919–1939* (Hamden, Conn.: Archon, 1985), 164–166.

20. Young, "French Military Intelligence and Nazi Germany, 1938–1939," 302–303.

21. Horne, *To Lose a Battle*, 60.

22. Ibid., 63–64.

23. Young, "French Military Intelligence and Nazi Germany, 1938–1939," 302–303.
24. May, *Knowing One's Enemies*, 510–511.
25. May, *Strange Victory*, 290.
26. Strong, Kenneth, *Men of Intelligence* (London: Cassel, 1970), 55.
27. Doughty, *The Seeds of Disaster*, 177.
28. May, *Strange Victory*, 477. Horne gives the number 3000 against approximately 2400 German tanks (Horne, *To Lose a Battle*, 156–157); Doughty notes 2400 French tanks (Doughty, *The Seeds of Disaster*, 177) but adds that approximately 700 of them were spread out among light cavalry/mechanized divisions, 150 among infantry battalions, and the rest in three new mechanized divisions (the fourth, de Gaulle's division, was still being formed). Each division in the French layout had half the number of tanks as in the ten German mechanized divisions. De Gaulle states that "there were, however, 3000 up-to-date French tanks and 800 motorized machine-guns. The Germans had no more. But ours were, according to plan, distributed up and down the sectors of the front" (de Gaulle, *The Complete War Memoirs of Charles de Gaulle*, 36).
29. Steiger, *Armour Tactics in the Second World War*, 73–74.
30. Robert A. Doughty, "The French Armed Forces 1918–1940," in *Military Effectiveness*, vol. 2, 54.
31. Kier, *Imagining War*, 40.
32. Kiesling, *Arming Against Hitler*, 167–169.
33. Posen, *The Sources of Military Doctrine*, 235.
34. Kier, *Imagining War*, 144–146.
35. Horne, *To Lose a Battle*, 29.
36. Posen, *The Sources of Military Doctrine*, 128.
37. Ibid., 129–130.
38. Ibid., 131.
39. Ibid., 86, 132.
40. Kiesling, *Arming Against Hitler*, 137.
41. Doughty, "The French Armed Forces 1918–1940," 55–60.
42. Chene, "Adaptation," *Military Review* 36, no. 6 (1956): 85.
43. Doughty, "The French Armed Forces 1918–1940," 57.
44. Doughty, *The Seeds of Disaster*, 179.
45. Doughty, "The French Armed Forces 1918–1940," 57.
46. Liddell Hart, *The Strategy of Indirect Approach*, 245.
47. Horne, *To Lose a Battle*, 244.
48. Deighton, *Blitzkrieg*, 247.
49. Doughty, "The French Armed Forces 1918–1940," 55.
50. Ibid., 55.
51. Doughty, *The Seeds of Disaster*, 179.
52. Ibid., 152–153.

53. Ibid., 109–111.
54. Posen, *The Sources of Military Doctrine*, 131.
55. Doughty, *The Seeds of Disaster*, 11.
56. Murray, "Innovation: Past and Future," 324.
57. Murray, "Contingency and Fragility of the German RMA," 164.
58. Kier, *Imagining War*, 86.
59. Doughty, *The Seeds of Disaster*, 11.
60. Kier, *Imagining War*, 86.
61. De Gaulle, *The Complete War Memoirs of Charles de Gaulle*, 7.
62. May, *Strange Victory*, 283.
63. Deighton, *Blitzkrieg*, 116.
64. Doughty, "The French Armed Forces 1918–1940," 58.
65. Murray, "Contingency and Fragility of the German RMA," 160.
66. Deighton, *Blitzkrieg*, 175–176.
67. Doughty, *The Seeds of Disaster*, 12.
68. Chene, "Adaptation," 86.

Summary and Conclusions

1. Eado Hecht, "Combat Helicopters in Iraq: Failure and Success," *Ma'arachot* 395 (2004b), 12–21.
2. *Winograd Committee Final Report on Israel's 2006 War in Lebanon* [Hebrew], 2008, 253.
3. Ibid., 277.
4. Ibid., 424.
5. Matt M. Matthews, "*We Were Caught Unprepared: The 2006 Hezbollah-Israeli War*," The Long War Series Occasional Paper 26 (Fort Leavenworth, Kans.: U.S. Army Combined Arms Center, Combat Studies Institute Press, 2008), 63.
6. Ibid., 2, 62.
7. *Winograd Committee Final Report*, 273–274.
8. Ibid., 407.
9. Matt M. Matthews, "Hard Lessons Learned: A Comparison of the 2006 Hezbollah-Israeli War and Operation CAST LEAD: A Historical Overview," in *Back to Basics: A Study of the Second Lebanon War and Operation CAST LEAD*, ed. Scott C. Farquhar (Fort Leavenworth, Kans.: U.S. Army Combined Arms Center, Combat Studies Institute Press, May 2009), 5–44.
10. "Hybrid wars," defined by defense analyst Frank Hoffman as a "blend of the lethality of state conflict with the fanatical and protracted fervor of irregular war" (Frank G. Hoffman, *Conflict in the 21st Century: The Rise of Hybrid Wars* [Arlington, Va.: Potomac Institute for Policy Studies, 2007], 28).
11. David E. Johnson, *Military Capabilities for Hybrid War: Insights From the Is-*

rael Defense Forces in Lebanon and Gaza (RAND, op. 285, 2010), 7–8. See also Sean MacFarland, Michael Shields, and Jeffrey Snow, *The King and I: The Impending Crisis in Field Artillery's Ability to Provide Fire Support to Maneuver Commanders* (2008). See also Matthews: "The missteps committed by the IDF in this war provide the US Army with valuable examples of potential difficulties when counterinsurgency operations are abruptly changed to major combat operations. For the US Army, which has been almost exclusively involved in irregular warfare for years, this issue is of paramount importance. While the US Army must be proficient in conducting major combat operations around the world, it is possible that years of irregular operations have chipped away at this capability, not unlike the situation encountered by the IDF (Matthews, *We Were Caught Unprepared*, 65).

12. Posen, *The Sources of Military Doctrine*, 34–80.
13. Marten Zisk, *Engaging the Enemy*, 3–5.
14. Kier, *Imagining War*, 10–38.
15. Avant, *Political Institutions and Military Change*, 117–129.
16. Ibid., 29–36.
17. Murray, "Innovation: Past and Future," 312–318.
18. Williamson R. Murray, "Military Culture Does Matter," *Foreign Policy Research Institute* 7, no. 2 (January 1999): http://www.fpri.org/fpriwire/0702.199901.murray.militaryculturedoesmatter.html
19. Farrell and Terriff, "The Sources of Military Change," 7–10.
20. Ibid., 273.
21. Farrell, "World Culture and Military Power."
22. Cassidy, *Russia in Afghanistan and Chechnya*, 9–12.

Appendixes to Chapter 1

1. Harkabi, 1990, 337–338.
2. IDF Doctrine and Training Division, 1996, 18.
3. Harkabi, *War and Strategy*, 494.
4. Luttwak, *Strategy: The Logic of War and Peace*, 111.
5. Handel, "Technological Surprise in War," 42–43.
6. Lanir, *Fundamental Surprise*, 40–49.
7. Kam, *Surprise Attack*, 59.
8. Ben-Porat, "Intelligence Estimates—Why Do They Collapse?" 235–236.
9. Lanir, *Fundamental Surprise*, 152–156.
10. Uri Bar-Joseph, "Methodological Magic," *Intelligence and National Security* 3, no.4 (1988): 142–149.
11. Hecht, "Technological Military Deception in the Twentieth Century," 38.
12. Michael Howard, "The Forgotten Dimensions of Strategy," in *The Causes of War and Other Essays* (Cambridge, Mass.: Harvard University Press, 1984), 104–111.

13. "Tools, or weapons, if only the right ones can be discovered, form 99 percent of victory . . . Strategy, command, leadership, courage, discipline, supply, organization and all the moral and physical paraphernalia of war are nothing to a high superiority of weapons . . ." (J. F. C. Fuller, *Armament & History* [New York: De Capo, 1998], 31).

14. Isaac Ben-Israel, "Where Did Clausewitz Go Wrong?: Clausewitz and the War Principles in Light of Modern Technology," *Ma'arachot* 311 (1988): 24.

15. In his book *Military Power* (28–51), Biddle argues that technology's place as an explanation of military success is overestimated relative to the "modern system" of force employment. Modern system offensive includes the use of cover and concealment to reduce the attacker's exposure when advancing; the use of suppressive fire to keep the defender's head down while the attacker is exposed. The modern defense system requires the same exposure-reduction tactics of cover, concealment, dispersion, suppression, combined arms, and independent small-unit maneuver as the modern offense system—adapted to the problems of the defense.

16. Kober, *Decision*.

17. Kam, *Surprise Attack*; Betts, *Surprise Attack*.

18. Avi Kober, "Has Battlefield Decision Become Obsolete? The Commitment to the Achievement of Battlefield Decision Revisited," *Contemporary Security Policy* 22, no.2 (2001b): 96–97.

19. Michael I. Handel, *War, Strategy and Intelligence* (London: Frank Cass, 1989), 65–66.

20. Sherwin and Whaley, "Understanding Strategic Deception," 177–194.

21. Simpkin, *Race to the Swift*, 179.

22. Michael O'Hanlon, *Technological Change and the Future of Warfare* (Washington, D.C: Brookings Institution Press, 2000), 83–84.

23. Gordon, *The Bow of Paris*, 86.

24. O'Hanlon, *Technological Change and the Future of Warfare*, 34.

25. Gordon, *The Bow of Paris*, 86; Richard K. Betts, "Surprise Despite Warning: Why Sudden Attacks Succeed," *Political Science Quarterly* 95, no. 4 (1980–1981): 551–572.

26. Anthony H. Cordesman and Abraham R. Wagner, *The Lessons of Modern War, Vol. 2: The Iran-Iraq War* (London: Westview, 1990b), 81.

27. Gordon, *The Bow of Paris*, 105.

28. David Bnaya, "Technical Intelligence," in *Intelligence and National Security*, ed. Zvi Ofer and Avi Kober (Tel Aviv,: Ma'arachot, 1987), 521.

29. Lanir, *Fundamental Surprise*, 103–105.

30. Mahnken, *Uncovering Ways of War*, 178.

31. Dromi: "The Mutual Relationship Between Doctrine, Technology and Weapons Systems," 442.

32. Walter J. Boyne, *Clash of Titans: World War II at Sea* (New York: Simon & Schuster, 1995), 107.

33. Zeev Bonen, "Sophisticated Conventional War," in *Advanced Technology and Future War* (Ramat Gan: BESA Center, Bar-Ilan University, 1996), 20.

34. Bonen, *RAFAEL*, 80–99.

35. Mahnken, *Uncovering Ways of War*, 178.

36. Emily O. Goldman and Richard B. Andres, "Systematic Effects of Military Innovation and Diffusion," *Security Studies* 8, no. 4 (1999): 79–125.

37. Leonhard, *The Principles of War for the Information Age*, 187.

38. Hecht, "Technological Military Deception in the Twentieth Century," 40.

39. For a survey, see Hecht, "Technological Military Deception in the Twentieth Century."

40. Ben-Israel, "Where Did Clausewitz Go Wrong?" 16.

41. Mahnken, *Uncovering Ways of War*, 178.

42. Ariel Sobelman, "Information on the Modern Battlefield and Information Warfare," in *Israel's Security Web—Core Issues of Israel's National Security in Its Sixth Decade* [Hebrew], ed. Hagai Golan (Tel Aviv: Ma'arachot, 2001), 234.

BIBLIOGRAPHY

Abudi, Yossi. "The Israeli Air Force in War." Published lecture, part of a series of seminars on the Yom Kippur War. Study Day no. 7, *The Air Arena in the Yom Kippur War*. Ramat Efal: Israel Galili Foundation for Research on Defense Force, The Israeli Military History Foundation at Tel Aviv University, May 4, 2000.

Adamsky, Dima. *Operation Kavkaz: Soviet Intervention and Israeli Intelligence Failure in the War of Attrition* [Hebrew]. Tel Aviv: Ma'arachot, 2006.

Adan, Avraham. *On Both Banks of the Suez* [Hebrew]. Jerusalem: Edanim, 1979.

Agranat Commission Report: Investigating Commission on the Yom Kippur War. Third and Last Report. Vol. 4 [Hebrew]. Jerusalem, 1975.

Alger, John I. *The Quest for Victory: The History of the Principles of War*. London: Greenwood, 1982.

Allen, Louis. *Singapore 1941–1942*. London: Frank Cass, 1977.

Allon, Yigal. *Contriving Warfare* [Hebrew]. Tel Aviv: Hakibbutz Hame'uchad, 1990.

Almog, Doron. "Lessons From the Six-Day War as a Crisis in the Development of the Warfighting Doctrine." *Ma'arachot* 354 (1997): 2–9.

Amir, Amos. *Flames in the Sky* [Hebrew]. Tel Aviv: Defense Ministry, 2000.

"Anti-Tank Fighting." *Ma'arachot* 209 (1970): 27–30.

Arad, Aryeh, and Laskov, Haim. *Looking Back: Soldiers Discuss the Yom Kippur War Battles They Participated In* [Hebrew]. Israel: Arad Family, 2003.

Ariely, Gil. "Learning to Digest During Fighting—Real Time Knowledge Management." 2006. http://www.ict.org.il/Articles/tabid/66/Articlsid/229/Default.aspx (accessed Aug 5, 2009).

———. "Operational Knowledge Management in the Military." In *Encyclopedia of Knowledge Management*, edited by David Schwartz, 713–720. Hershey, Pa.: Idea Group, 2006.

———. "Learning While Fighting (During 2nd Lebanon War)." *Ma'arachot* 412 (2007): 4–13.

———. "Learning While Fighting in "Cast Lead" Operation." *Ma'arachot* 425 (2009): 12–21.

Asher, Dani. "From 'Order 41' to 'Tahrir 41': From the Egyptian Warfighting Doctrine to War." *Ma'arachot* 332 (1993): 46–53.

———. "The Syrian Plan for the Yom Kippur War and the War's Moves." In *Yom Kippur War Studies* [Hebrew], edited by Hagai Golan and Shaul Shai, 289–310. Tel Aviv: Ma'arachot, 2003.

———. *Breaking the Concept* [Hebrew]. Tel Aviv: Ma'arachot, 2003b.

Asher, Dani, and Asher, Itai. "The Israeli Artillery on the Southern Front in the Yom Kippur War." *Ma'arachot* 354 (1997): 10–20.

Atkinson, Rick. *An Army at Dawn: The War in North Africa, 1942–1943.* New York: Henry Holt and Company, 2002.

Avant, Deborah D. *Political Institutions and Military Change: Lessons From Peripheral Wars.* Ithaca, N.Y.: Cornell University Press, 1994.

Barber, Noel. *Sinister Twilight: The Fall and Rise Again of Singapore.* London: Collins, 1973.

Bar-Joseph, Uri. "Methodological Magic." *Intelligence and National Security* 3, no.4 (1988): 134–155.

———. *The Watchman Fell Asleep: The Surprise of Yom Kippur and Its Sources* [Hebrew]. Lod: Zmora-Beitan, 2001.

Bar-Joseph, Uri, and Arie W. Kruglanski. "Intelligence Failure and the Need for Cognitive Closure: On the Psychology of the Yom Kippur Surprise." *Political Psychology* 24, no.1 (2003): 75–99.

Bar-Lev, Haim. "A Lecture on War Principles at the IDF Command and Staff College." In *War Principles—Selected Readings.* IDF Command and Staff College Publications, October 1, 1982.

Bartov, Hanoch. *Daddo: 48 Years and 20 More Days.* Vol. 1 [Hebrew]. Tel Aviv: Ma'ariv, 1978.

Bekker, Cajus. *The Luftwaffe War Diaries.* New York: Ballantine, 1964.

Ben-Gurion, David. *Army and Security* [Hebrew]. Tel Aviv: Ma'arachot, 1955.

Ben-Israel, Isaac. "The Logic of Military Lesson Learning." *Ma'arachot* 305 (1986): 24–29.

———. "Where Did Clausewitz Go Wrong? Clausewitz and the War Principles in Light of Modern Technology." *Ma'arachot* 311 (1988): 16–25.

———. "Technological Lessons." *Ma'arachot* 332 (1993): 8–12.

———. "The Theory of Relativity in Force Planning." *Ma'arachot* 352–353 (1997): 33–42.

———. *The Philosophy of Military Intelligence* [Hebrew]. Tel Aviv: Defense Ministry, 1999.

Ben-Porat, Yoel. "Intelligence Estimates, Why Do They Collapse?" In *Intelligence and National Security* [Hebrew], edited by Zvi Ofer and Avi Kober, 223–250. Tel Aviv: Ma'arachot, 1987.

———. "Impossible Prediction, An Interview With Haim Lapid and Hagai Ben-Zvi." *Skira Hodshit* 36, no.12 (1990): 52–53.

Ben-Zvi, Abraham. "Hindsight and Foresight: A Conceptual Framework for the Analysis of Surprise Attack." *World Politics* 28 (1976): 381–397.

Bergman, Ronen. "No Stopping on Red." *Yedioth Ahronot*, 7 Days [weekend appendix], 2180 (November 4, 2005): 21.

Betts, Richard K. "Surprise Despite Warning: Why Sudden Attacks Succeed." *Political Science Quarterly* 95, no. 4 (1980–1981): 551–572.

———. *Surprise Attack: Lessons for Defense Planning*. Washington, D.C.: Brookings Institution, 1982.

Biddle, Stephan. *Military Power: Explaining Victory and Defeat in Modern Battle*. Princeton, N.J.: Princeton University Press, 2004.

Bidwell, Shelford. *Gunners at War*. Aberdeen: Arms and Armour, 1970.

Bidwell, Shelford, and Dominick Graham. *Fire-Power: The British Army Weapons and Theories of War 1904–1945*. S. Yorkshire: Pen and Sword Military Classics, 2004.

Blank, Stephen, "Imaging Afghanistan: Lessons of a 'Small War.'" *Journal of Soviet Military Studies* 3, no. 3 (1990): 468–490.

———. "Soviet Forces in Afghanistan: Unlearning the Lessons of Vietnam." In *Responding to Low-Intensity Conflict Challenges*, edited by Stephen Blank, Lawrence E. Grinter, Karl P. Magyer, Lewis B. Ware, and Bynum E. Weathers, 53–126. Maxwell Air Base, Ala.: Air University Press, 1990b).

———. *Afghanistan and Beyond: Reflections on the Future of Warfare*. Carlisle, Pa.: Strategic Studies Institute, U.S. Army War College, 1993. http://www.au.af.mil/au/awc/awcgate/ssi/afghan-blank.pdf.

Bloch, Marc. *Strange Defeat: A Statement of Evidence Written in 1940*. Translated by Gerard Hopkins. New York: W.W. Norton & Company, 1999.

Blumenson, Martin. *Kasserine Pass*. Boston: Riverside, 1967.

———. "Kasserine Pass 30 January–22 February 1943." In *America's First Battles 1776–1965*, edited by Charles E. Heller and William A. Stofft, 226–265. Lawrence, Kans.: University Press of Kansas, 1986.

Blumenson, Martin, and James L. Stokesbury. *Masters of the Art of Command*. Boston: Houghton Mifflin Company, 1975.

Bnaya, David. "Technical Intelligence." In *Intelligence and National Security* [Hebrew], edited by Zvi Ofer and Avi Kober, 519–525. Tel Aviv: Ma'arachot, 1987.

Bond, Brian. *British Military Policy Between the Two World Wars*. New York: Oxford University press, 1980.

Bond, Brian, and Williamson R. Murray. "The British Armed Forces." In *Military Effectiveness: Vol. 2—The Interwar Period*, edited by Allan R. Millett and Williamson R. Murray, 98–130. Boston: Allen & Unwin, 1988.

Bonen, Zeev. "Problems in Developing Military Layouts." *Ma'arachot* 245 (1975): 2–6.

———. "Tactical Testing and Weapon Development." *Ma'arachot* 272 (1980): 50–52.

———. "Sophisticated Conventional War." In *Advanced Technology and Future War*, 19–30. Ramat Gan: BESA Center, Bar-Ilan University, 1996.

———. *RAFAEL* [Hebrew]. Tel Aviv: N.N.D. Media, 2003.

Borer, Douglas A. *Superpowers Defeated: Vietnam and Afghanistan Compared*. London: Frank Cass, 1999.

Boyne, Walter J. *Clash of Titans: World War II at Sea*. New York: Simon and Schuster, 1995.

Brown, Louis. *A Radar History of World War II: Technical and Military Imperatives*. Washington, D.C.: Carnegie Institute of Washington, 1999.

Bungay, Stephan. *Alamein*. London: Aurum, 2002.

Bush, Vannevar. *Modern Arms and Free Man: A Discussion of the Role of Science in Preserving Democracy*. New York: Simon and Schuster, 1949.

Carafano, James J. *GI Ingenuity: Improvisation, Technology and Winning WW II*. Mechanicsburg, Pa.: Stackpole Books, 2006.

Carver, Michael. *Tobruk*. London: B.T. Batsford, 1964.

———, ed. *The Warlords*. London: Weidenfeld and Nicolson, 1976.

Cary, James. *Tanks and Armor in Modern Warfare*. New York: Franklin Watts, 1966.

Cassidy, Robert M. *Russia in Afghanistan and Chechnya: Military Strategic Culture and the Paradoxes of Asymmetric Conflict*. Carlisle, Pa.: Strategic Studies Institute, U.S. Army War College, 2003. http://www.strategicstudiesinstitute.army.mil/pubs/display.cfm?pubID=125.

Chene. "Adaptation." *Military Review* 36, no. 6 (1956): 84–89.

Churchill, Winston. *The Second World War*. 6 vol. Boston: Houghton Mifflin Company, 1951.

Citino, Robert M. *The Path to Blitzkrieg: Doctrine and Training in the German Army, 1920–1939*. Boulder, Colo.: Lynne Rienner, 1999.

———. *The German Way of War: From the Thirty Years War to the Third Reich*. Lawrence, Kans.: Kansas University Press, 2005.

Clausewitz, Carl von. *On War*. Edited and translated by Michael Howard and Peter Paret. Princeton, N.J.: Princeton University Press, 1976.

Cohen (Cheetah), Eliezer, and Zvi Lavi. *The Sky Is Not the Limit: The Story of the Israeli Air Force* [Hebrew]. Jerusalem: Sifriyat Hapoalim, 1990.

Cohen, Eliot A. "American Views of the Revolution in Military Affairs." In *Advanced Technology and Future War*, 3–18. Ramat Gan: BESA Center, Bar-Ilan University, 1996.

Cohen, Eliot A., and John Gooch. *Military Misfortunes: The Anatomy of Failure in War*. New York: Vintage Books, 1991.

Collins, Joseph J. "The Soviet Military Experience in Afghanistan." *Military Review* 65 (May 1985): 16–28.

Condell, Bruce, and David T. Zabecki. *On the German Art of War—Truppenführung*. Boulder, Colo.: Lynne Rienner, 2001.

Constant, James N. *Fundamentals of Strategic Weapons: Offense and Defense Systems.* The Hague: Martinus Nijhoff, 1981.
Cooper, Matthew. *The German Army 1933–1945: Its Political and Military Failure.* New York: Stein and Day, 1978.
Cordesman, Anthony H. *The Iraq War: Strategy, Tactics, and Military Lessons.* London: Praeger, 2003.
Cordesman, Anthony H., and Abraham R. Wagner. *The Lessons of Modern War. Vol. 1: The Arab-Israeli Conflicts, 1973–1989.* London: Westview, 1990a.
———. *The Lessons of Modern War. Vol. 2: The Iran-Iraq War.* London: Westview, 1990b.
———. *The Lessons of Modern War. Vol. 3: The Afghan and Falklands Conflicts.* London: Westview, 1990c.
Corum, James S. *The Roots of Blitzkrieg: Hans Von Seeckt and German Military Reform.* Lawrence, Kans.: University Press of Kansas, 1992.
———. "A Comprehensive Approach to Change: Reform in the German Army in the Interwar Period." In *The Challenge of Change: Military Institutions and New Realities, 1918–1941*, edited by Harold R. Winton and David R. Mets, 35–73. Lincoln, Neb.: University Press of Nebraska, 2000.
Corum, James S., and Richard R. Muller. *The Luftwaffe's Way of War: German Air Force Doctrine 1911–1945.* Baltimore, Md.: Nautical and Aviation, 1998.
Courtney, Hugh, Jane Kirkland, and Patrick Viguerie. "Strategy for Uncertainty." *Harvard Business Review* 75, no.6 (1997): 66–80.
Crisp, Robert. *Brazen Chariots.* London: Transworld, 1959.
Cushman, John H. "Challenge and Response at the Operational and Tactical Levels, 1914–45." In *Military Effectiveness: Vol. 3—The Second World War*, edited by Allan R. Millett and Williamson R. Murray, 320–340. Boston: Allen & Unwin, 1988.
Davis, Paul K. "Uncertainty-Sensitive Planning." In *New Challenges, New Tools for Defense Decisionmaking*, edited by Stuart E. Johnson, Martin C. Libicki, and Gregory F. Treverton. RAND, MR 1576, 2003.
De Gaulle, Charles. *The Complete War Memoirs of Charles de Gaulle* Vol. 1, *The Call of Honor.* Translated by Jonathan Griffin and Richard Howard. New York: Simon and Schuster, 1964.
Deighton, Len. *Blitzkrieg: From the Rise of Hitler to the Fall of Dunkirk.* London: Jonathan Cape, 1979.
Dewar, James A. *Assumption-Based Planning: A Tool for Reducing Avoidable Surprises.* RAND Studies in Policy Analysis. Cambridge: Cambridge University Press, 2002.
Dickerson, Brian. "Adaptability: A New Principle of War." In *National Security Challenges for the 21st Century*, edited by Williamson R. Murray. Carlisle, Pa.: Strategic Studies Institute, U.S. Army War College, 2003.
Dixon, Norman. *On the Psychology of Military Incompetence.* New York: Basic, 1976.
———. *Our Worst Enemy.* London: Jonathan Cape, 1987.

Doughty, Robert A. *The Seeds of Disaster: The Development of French Army Doctrine 1919–1939*. Hamden, Conn.: Archon, 1985.

———. "The French Armed Forces 1918–1940." In *Military Effectiveness: Vol. 2—The Interwar Period*, edited by Allen R. Millett and Williamson R. Murray, 70–97. Boston: Allen & Unwin, 1988.

Downie, Richard D., *Learning From Conflict: The U.S. Military in Vietnam, El Salvador, and the Drug War*. Westport, Conn.: Praeger, 1998.

Dromi, Uri. "The Mutual Relationship Between Doctrine, Technology and Weapons Systems." In *Quantity and Quality in Force Planning* [Hebrew], edited by Zvi Ofer and Avi Kober, 433–446. Tel Aviv: Ma'arachot, 1985.

Dror, Yehezkel. "The Quality of the Decision-Making Process as a Critical Factor in Force Design and Development." In *Quantity and Quality in Force Planning* [Hebrew], edited by Zvi Ofer and Avi Kober, 457–466. Tel Aviv: Ma'arachot, 1985.

Dupuy, T. N. *A Genius for War: The German Army and General Staff 1807–1945*. London: Macdonald and Jane's, 1977.

Einan, Menahem. "The Need for Quality in Light of the Limitations of Quantity." In *Quantity and Quality in Force Planning* [Hebrew], edited by Zvi Ofer and Avi Kober, 477–484. Tel Aviv, Ma'arachot, 1985.

Erickson, John. *The Soviet High Command*. London: MacMillan, 1967.

———. "Threat Identification and Strategic Appraisal by the Soviet Union, 1930–1941." In *Knowing One's Enemies: Intelligence Assessment Before the Two World Wars*, edited by Ernest R. May, 375–423. Princeton, N.J: Princeton University Press, 1984.

Eyal, Reuven. "The Intelligence Consumer's Four Paradoxes." In *Intelligence and National Security* [Hebrew], edited by Zvi Ofer and Avi Kober, 475–485. Tel Aviv, Ma'arachot, 1987.

"*Falklands Campaign: The Lessons*." London: Her Majesty's Stationery Office, December 1982.

Farrell, Theo. "World Culture and Military Power." *Security Studies* 14, no. 3 (2005): 448–488.

Farrell, Theo, and Terry Terriff. "The Sources of Military Change." In *The Sources of Military Change: Culture, Politics, Technology*, edited by Theo Farrell and Terry Terriff. London: Lynne Rienner, 2002.

Fawzi, Muhammad. Chapters in his book *The October '73 War*, originally published in *al Sharia* (August 8, 1988 through October 24, 1988) [Hebrew]. Hatzav 843/013 (1989).

Ferris, John. "Worthy of Some Better Enemy? The British Estimate of the Imperial Japanese Army, 1919–41, and the Fall of Singapore." *Canadian Journal of History* 28 (1993): 223–256.

Ferris, John, and Michael Handel. "Clausewitz, Intelligence, Uncertainty and the Art of Command and Military Operations." *Intelligence and National Security* 10 (1995): 1–58.

Finkel, Meir. "Where Did Flexibility Go?" *Ma'arachot* 393 (2004): 54–55.
Förster, Jürgen. "The Dynamics of *Volkegemeinschaft*: The Effectiveness of the German Military Establishment in the Second World War." In *Military Effectiveness: Vol. 3— The Second World War*, edited by Allen R. Millett and Williamson R. Murray, 180–220. Boston: Allen & Unwin, 1988.
Forty, George. *The Armies of Rommel*. London: Arms and Armour, 1997.
Frost, Robert S. *The Growing Imperative to Adopt "Flexibility" as an American Principle of War*. Carlisle, Pa.: Strategic Studies Institute, U.S. Army War College, 1999.
Fuller, J. F. C. *The Second World War 1939–1945*. London: Eyre and Spottiswoode, 1954.
———. *Armament and History*. New York: De Capo, 1998.
Gal, Reuven. "Military Leadership in the Light of Research." Lecture, leadership seminar at the Open University, Tel Aviv, Zichron Yakov: Israeli Institute for Military Research, November 18, 1991.
Gat, Azar. "Weapons, Warfighting Doctrine and Basic Organization" [Hebrew]. *Ma'arachot* 278 (1981): 50–52.
———. *The Development of Military Thought: The Nineteenth Century*. Oxford: Clarendon Press, 1992.
———. *British Armour Theory and the Rise of the Panzer Arm: Revising the Revisionists*. London: Macmillan, 2000.
———. "Ideology, Policy, Technology and Strategic Doctrine Between Two World Wars." *Ma'arachot* 390 (2003): 4–15.
Gazit, Shlomo. "Intelligence Estimates and the Decision Maker." In *Intelligence and National Security* [Hebrew], edited by Zvi Ofer and Avi Kober, 455–473. Tel Aviv: Ma'arachot, 1987.
———. *Between Deterrence and Surprise, Responsibility for Formulating Israel's National Intelligence Assessment* [Hebrew]. Tel Aviv: Jaffee Center, University Tel Aviv, 2003.
Gazit, Uri. "The 'Crocodile' Marches Past." *Shiryon* 11 (2001): 30–31.
Gerard, Beth M. "Mistakes in Force Structure and Strategy on the Eve of the Great Patriotic War." *Journal of Soviet Military Studies* 4, no. 3 (1991): 471–486.
"German Training and Tactics: An Interview With Col Pestke." *Marine Corps Gazette* 67, no. 10 (October 1983): 58–65.
Geyer, Michael. "National Socialist Germany: The Politics of Information." In *Knowing One's Enemies: Intelligence Assessment Before the Two World Wars*, edited by Ernest R. May), 310–346. Princeton, N.J: Princeton University Press, 1984.
Glantz, David M., ed. *The Initial Period of War on the Eastern Front: 22 June–August 1941*. London: Frank Cass, 1993.
Glantz, David M., and Jonathan M. House. *When Titans Clashed: How the Red Army Stopped Hitler*. Lawrence, Kans.: University of Kansas, 1995.
Golan, Shimon. "The Preemptive Air Strike at the Outbreak of the Yom Kippur War." *Ma'arachot* 373 (2000): 52–61.

Goldman, Emily O., and Richard B. Andres. "Systematic Effects of Military Innovation and Diffusion." *Security Studies* 8, no. 4 (1999): 79–125.

Gordon, Shmuel. *The Bow of Paris* [Hebrew]. Tel Aviv: Sifriyat Hapoalim, 1997.

———. *The Last Order of Knights: Modern Air Strategy* [Hebrew]. Tel Aviv: Ramot-Tel Aviv University Publications, 1998.

———. "The Paradox of October 7." *Ma'arachot* 361 (1998b): 44–57.

———. "Future Lessons From a Past War." In *Yom Kippur War Studies* [Hebrew], edited by Hagai Golan and Shaul Shai, 359–388. Tel Aviv: Ma'arachot, 2003.

———. *Thirty Hours in October* [Hebrew]. Tel Aviv: Ma'ariv Book Guild, 2008.

Grau, Lester W., ed. *The Bear Went Over the Mountain: Soviet Combat Tactics in Afghanistan.* London: Frank Cass, 1998.

———. "The Soviet-Afghan War: A Superpower Mired in the Mountains." *Journal of Slavic Military Studies* 17, no. 1 (2004): 129–151.

Graves, Donald E., "*War is an Art.*" Review of *On the German Art of War—Truppenführung*, by Bruce Condell and David T. Zabecki. *The Army Training and Doctrine Bulletin* 5, no. 2 (2002): 71–74.

Gray, Jennie. *Fire by Night: The Dramatic Story of One Pathfinder Crew and Black Thursday, 16/17 December 1943.* London: Grub Street, 2000.

Guderian, Heinz. *Achtung-Panzer! The Development of Tank Warfare.* Translated by Christopher Duffy. London: Cassel, 1992.

———. *Panzer Leader.* Translated by Constantine Fitzgibbon. London: Penguin Books, 1996.

Gudmundsson, Bruce. I. *Stormtroop Tactics: Innovation in the German Army, 1914–1918.* New York: Praeger, 1989.

Haber, Eitan, and Zeev Shif. *Yom Kippur War Lexicon* [Hebrew]. Tel Aviv: Zmora-Beitan-Dvir, 2003.

Halperin, Meirav, and Aharon Lapidot. *G Suit: Pages in the Log Book of the Israel Air Force* [Hebrew]. Tel Aviv: IAF Publications, 2000.

Handel, Michael I. *Perception, Deception and Surprise: The Case of the Yom Kippur War.* The Leonard Davis Institute for International Relations, Jerusalem Papers on Peace Problems 19. Jerusalem: The Magnes Press, The Hebrew University, 1976.

———. *Military Deception in Peace and War.* The Leonard Davis Institute for International Relations, Jerusalem Papers on Peace Problems 38. Jerusalem: Magnes, The Hebrew University, 1985.

———. "Clausewitz in the Age of Technology." *The Journal of Strategic Studies* 9, nos. 2–3 (1986): 51–92.

———. "Technological Surprise in War." *Intelligence and National Security* 2, no.1 (1987): 1–53.

———. *War, Strategy and Intelligence.* London: Frank Cass, 1989.

———. "Intelligence in Historical Perspective." In *Go Spy the Land: Military Intelligence in History*, edited by Keith Neilson and B. J. C. McKercher. Westport, Conn.: Praeger, 1992.

———. *Masters of War: Classical Strategic Thought*. London: Frank Cass, 2001.

Harkabi, Yehoshafat. "Complications Between Military Intelligence and the Leader." In *Intelligence and National Security* [Hebrew], edited by Zvi Ofer and Avi Kober, 439–453. Tel Aviv: Ma'arachot, 1987.

———. *War and Strategy* [Hebrew]. Tel Aviv: Ma'arachot, 1990.

Hecht, Eado. *The "Operational Breakthrough" in German Military Thought 1870–1945* [Hebrew]. Tel Aviv: Ma'arachot, 1999.

———. "Technological Military Deception in the Twentieth Century." *Ma'arachot* 364 (1999b): 30–43.

———. *Military History: Pt. 2*. A Summary of Lectures and Reading at the IDF Command and Staff College Barak Program, 2004.

———. "Combat Helicopters in Iraq: Failure and Success." *Ma'arachot* 395 (2004b): 12–21.

Heckmann, Wolf. *Rommel's War in Africa*. Translated by Stephan Seago. New York: Smithmark, 1995.

Heilmeier, George H. "Guarding Against Technological Surprise." *Air University Review* 27, no. 6 (1976): 2–7.

Herwig, Holger H. "Innovation Ignored: The Submarine Problem—Germany, Britain and the United States, 1919–1939." In *Military Innovation in the Interwar Period*, edited by Williamson R. Murray and Allan R. Millett, 227–264. Cambridge, U.K.: Cambridge University Press, 1996.

Herzog, Haim. *The War of Atonement* [Hebrew]. Jerusalem: Yedioth Ahronot, 1975.

Heuer, Richards J., Jr. "Cognitive Factors in Deception and Counter-Deception." In *Strategic Military, Deception*, edited by Donald C. Daniel and Katherine L. Herbig, 31–69. New York: Pergamon, 1982.

———. *Psychology of Intelligence Analysis*. C.I.A: Center for the Study of Intelligence, 1999.

Hillel, Aryeh. *The Human Hand Under Their Wings: The History of the Israeli Air Force's Technological Layout, 1948–1973* [Hebrew]. IAF: Logistics Wing/Ma'arachot/Defense Ministry, 1990.

Hirsh, Gal. "Operations at the Speed of Thought." *Ma'arachot* 380–381 (2001): 34–39.

Hoffman, Frank G. *Conflict in the 21st Century: The Rise of Hybrid Wars*. Arlington, Va.: Potomac Institute for Policy Studies, 2007.

Horne, Alistair. *To Lose a Battle: France 1940*. London: MacMillan, 1969.

Hosek, James R. "The Soldier of the 21st Century." In *New Challenges, New Tools for Decision Making*, edited by Stuart E. Johnson, Martin C. Libicki, and Gregory F. Treverton. RAND, MR 1576, 2003.

Hoshen, Gideon. "Intelligence for Weapons Development." In *Quantity and Quality in Force Planning* [Hebrew], edited by Zvi Ofer and Avi Kober, 527–533. Tel Aviv: Ma'arachot, 1987.

House, Jonathan M. *Combined Arms Warfare in the Twentieth Century.* Lawrence, Kans.: University Press of Kansas, 2001.

Howard, Michael. "Military Science in an Age of Peace." *Journal of Royal United Services Institute for Defence Studies* 119, no.1 (1974): 3–11.

———. "The Forgotten Dimensions of Strategy." In *The Causes of War and Other Essays*, 101–115. Cambridge, Mass: Harvard University Press, 1984.

Howard-Williams, Jeremy. *Night Intruder: A Personal Account of the Radar War Between the RAF and Luftwaffe Night Fighter Force.* (Vancouver: David and Charles, 1976).

Huntington, Samuel. *The Soldier and the State.* New York: Random House, 1957.

Iftach, Shimon. "Missiles in Egypt." *Ma'arachot* 217a–218 (1971): 18–25.

———. *Missiles and Rockets in Modern Warfare* [Hebrew]. Tel Aviv: Ma'arachot, 1979.

"Israeli Artillery Corp in the Yom Kippur War." Part of a series of seminars on the Yom Kippur War. Ramat Efal: Israel Galili Foundation for Research on Defense Force, The Israeli Military History Foundation at Tel Aviv University, July 6, 2000.

Jackson, William G. F. *The North African Campaign 1940–43.* London: B.T. Batsford, 1975.

Janis, Irving L. *Victims of Groupthink.* Boston: Houghton Mifflin, 1972.

Janis, Irving L., and Leon Mann. *Decision Making: A Psychological Analysis of Conflict, Choice, and Commitment.* New York: Free Press, 1977.

Jentz, Thomas L., ed. *PanzerTruppen (Pt. 1): The Complete Guide to the Creation and Combat Employment of Germany's Tank Force 1933–1942.* Atglen, Pa.: Schiffer Military History, 1996a.

———. *PanzerTruppen (Pt. 2): The Complete Guide to the Creation and Combat Employment of Germany's Tank Force 1943–1945.* Atglen, Pa.: Schiffer Military History, 1996b.

Jervis, Robert. "Hypothesis and Misperception." *World Politics* 20, no. 3 (1968): 454–479.

———. *Perception and Misperception in International Politics.* Princeton, N.J.: Princeton University Press, 1976.

Johnson, David E. *Fast Tanks and Heavy Bombers: Innovation in the U.S. Army 1917–1945.* Ithaca, N.Y.: Cornell University Press, 1998.

———. *Military Capabilities for Hybrid War: Insights From the Israel Defense Forces in Lebanon and Gaza.* RAND, OP.285, 2010.

Joly, Cyril. *Take These Men.* London: Constable, 1955.

Jones, R. V. *The Wizard War: British Scientific Intelligence 1939–1946.* New York: Coward, McCann and Geoghegan, 1978.

———. *Future Conflict and New Technology.* Washington, D.C.: The Washington Papers, The Center for Strategic and International Studies, Georgetown University, 1981.

———. *Reflections on Intelligence.* London: Mandarin, 1989.

Kahn, David. *Hitler's Spies: German Military Intelligence in World War II*. New York: Macmillan, 1978.

Kam, Ephraim. *Surprise Attack: The Victim's Perspective*. Cambridge, Mass.: Harvard University Press, 2004.

Karnow, Stanley. *Vietnam: A History*. New York: Penguin, 1984.

Kedar, Binyamin Zeev. *October 1973—The Story of "Smash" Battalion* [Hebrew]. Tel Aviv: Tamuz, 1975.

Kier, Elizabeth. *Imagining War: French and British Military Doctrine Between the Wars*. Princeton, N.J.: Princeton University Press, 1997.

Kiesling, Eugenia C. *Arming Against Hitler: France and the Limits of Military Planning*. Lawrence, Kans.: University Press of Kansas, 1996.

Kirby, Woodburn S. *Singapore: The Chain of Disaster*. London: Camelot Press, 1971.

Knorr, Klaus. *Military Power and Potential*. Lexington, Mass.: Heath Lexington, 1970.

Kober, Avi. "Theory, Doctrine and Planning in Military Force Design and Development." In *Quantity and Quality in Force Planning* [Hebrew], edited by Zvi Ofer and Avi Kober, 79–95. Tel Aviv: Ma'arachot, 1985.

———. *Decision: Military Decision in Arab-Israeli Wars 1948–1982* [Hebrew]. Tel Aviv, Ma'arachot, 2001.

———. "Has Battlefield Decision Become Obsolete? The Commitment to the Achievement of Battlefield Decision Revisited." *Contemporary Security Policy* 22, no.2 (2001b): 96–120.

Kress, Moshe. *Operational Logistics* [Hebrew]. Tel Aviv: Ma'arachot, 2002.

Kriebel, Rainer. *Inside the Afrika Korps—The Crusader Battles, 1941–1942*. Edited by Bruce Gudmundsson. London: Greenhill Books, 1999.

Lanir, Zvi. *Fundamental Surprise: The National Intelligence Crisis* [Hebrew]. Tel Aviv: Center of Strategic Studies, 1983.

Lapidot, Aharon, ed. *101: The First Fighter Squadron of Israel Air Force* [Hebrew]. The Heritage Series of the Israel Air Force, IAF: Defense Ministry, 1998a.

Lapidot, Aharon, and Dan Arkin. *IAF Logistics* [Hebrew]. IAF: Defense Ministry, 2001.

Lapidot, Aharon, and Dan Sela. *Thunder of the Hammers: The 69th Squadron of the Israel Air Force* [Hebrew]. The Heritage Series of the Israel Air Force, IAF: Defense Ministry, 1998b.

Laskov, Haim. "The Lessons of the Sinai Campaign." *Ma'arachot* 21 (1970): 5–7.

———. *Military Leadership* [Hebrew]. Tel Aviv, Ma'arachot, 1985.

Leasor, James. *Singapore: The Battle That Changed the World*. London: Hodder and Stoughton, 1968.

Leeb, Wilhelm Ritter von. "Defense." In *Roots of Strategy Vol. 3: 3 Military Classics*. Harrisburg, Pa.: Stackpole Books, 1991.

Leites, Nathan. *Soviet Style in War*. New York: Crane Russak, 1982.

Lempert, Robert J., Steven W. Popper, and Steven C. Bankes. *Shaping the Next One*

Hundred Years: New Methods for Quantitative Long-Term Policy Analysis. RAND, MR-1626, 2003.

Leonhard, Robert R. *The Art of Maneuver: Maneuver-Warfare Theory and AirLand Battle.* New York: Ballantine, 1991.

———. *The Principles of War for the Information Age.* California: Presidio, 2000.

Levine, Alan J. *The Strategic Bombing of Germany, 1940–1945.* Westport, Conn.: Praeger, 1992.

Levita, Ariel. *Offense and Defense in Israeli Military Doctrine* [Hebrew]. Tel Aviv: Center of Strategic Studies, 1988.

Levran, Aharon. "Surprise and Warning: Reflections on Basic Questions." *Ma'arachot* 276–277 (1980): 17–21.

Liddell Hart, B. H. *Thoughts on War.* London: Faber and Faber, 1944.

———. *The Strategy of Indirect Approach.* London: Faber and Faber, 1954.

———. *The Other Side of the Hill* (enlarged and revised edition). London: Cassell, 1956.

Litwak, Robert S. "The USSR in Afghanistan." In *Foreign Military Intervention*, edited by Ariel Levite, Bruce Jentelson, and Larry Berman, 65–95. New York: Columbia University Press, 1992.

Lorber, Azriel. *Science, Technology and the Battlefield.* [Hebrew]. Tel Aviv: Kronenberg, 1997.

———. *Misguided Weapons: Technological Failure and Surprise on the Battlefield.* Washington, D.C.: Brassey's, 2002.

Lowe, Peter. "Great Britain's Assessment of Japan Before the Outbreak of the Pacific War." In *Knowing One's Enemies: Intelligence Assessment Before the Two World Wars*, edited by Ernest R. May, 456–475. Princeton, N.J.: Princeton University Press, 1984.

Lucas, James. *Panzer Army Africa.* London: Macdonald and Jane's, 1977.

Luck, Hans von. *Panzer Commander: The Memoirs of Colonel Hans von Luck.* New York: Dell, 1989.

Luttwak, Edward, and Dan Horowitz. *The Israeli Army.* London: Allen Lane, 1975.

———. *Strategy: The Logic of War and Peace* (revised and enlarged edition). Cambridge, Mass.: Harvard University Press, 2001.

Lyman, Robert. "The Art of Maneuver at the Operational Level of War: Lieutenant-General W.J. Slim and Fourteenth Army, 1944–45." In *The Challenges of High Command: The British Experience*, edited by Gary Sheffield and Geoffrey Till), 88–112. New York: Palgrave Macmillan, 2003.

Machiavelli, Niccolo. *The Prince and Discourses.* New York: The Modern Library, 1950.

MacFarland, Sean, Michael Shields, and Jeffrey Snow. *The King and I: The Impending Crisis in Field Artillery's Ability to Provide Fire Support to Maneuver Commanders.* 2008. http://www.npr.org/documents/2008/may/artillerywhitepaper.pdf

Macgregor, Douglas A. *Transformation Under Fire: Revolutionizing How America Fights.* Westport, Conn.: Praeger, 2003.

Mahan, Alfred Thayer. *The Influence of Sea Power Upon History, 1660–1783*. London: Methuen, 1965.
Mahnken, Thomas G. "Uncovering Foreign Military Innovation." *The Journal of Strategic Studies* 22, no. 4 (1999): 26–54.
———. *Uncovering Ways of War: U.S. Intelligence and Foreign Military Innovation, 1918–1941*. Ithaca, N.Y.: Cornell University Press, 2002.
Manteuffel, Hasso von. "The Tank Battle of Târgul Frumos." *Military Review* 36, no. 6 (1956): 78–84.
Mao Tse-Tung. *Six Essays on Military Affairs*. Peking: Foreign Languages, 1985.
Marshall, A. W. *Problems of Estimating Military Power*. RAND, P-3417, 1966.
Marshall, S. L. A. *Men Against Fire*. New York: The Infantry Journal and William Morrow, 1947.
Matthews, Matt M. *We Were Caught Unprepared: The 2006 Hezbollah-Israeli War*. The Long War Series Occasional Paper 26. Fort Leavenworth, Kans.: U.S. Army Combined Arms Center, Combat Studies Institute Press, 2008.
———. "Hard Lessons Learned: A Comparison of the 2006 Hezbollah-Israeli War and Operation CAST LEAD: A Historical Overview." In *Back to Basics: A Study of the Second Lebanon War and Operation CAST LEAD*, edited by Scott C. Farquhar, 5–44. Fort Leavenworth, Kans.: U.S. Army Combined Arms Center, Combat Studies Institute Press, May 2009.
May, Ernest R. *Strange Victory: Hitler's Conquest of France*. New York: Hill and Wang, 2000.
May, Ernest R., ed. *Knowing One's Enemy: Intelligence Assessment Before the Two World Wars*. Princeton, N.J.: Princeton University Press, 1984.
Mayr, Ernst. *What Evolution Is*. New York: Basic, 2001.
McMichael, Scott R. "The Soviet Army, Counterinsurgency, and the Afghan War." *Parameters* 19 (1989): 21–35.
McNerney, Michael. "Military Innovation During War: Paradox or Paradigm?" *Defense and Security Analysis* 21, no. 2 (2005): 201–212.
Mearsheimer, John J. *Liddell Hart and the Weight of History*. Ithaca, N.Y.: Cornell University Press, 1988.
———. "Hitler and the Blitzkrieg Strategy." In *The Use of Force: Military Power and International Politics*, edited by Robert J. Art and Kenneth N. Waltz, 130–144. Lanham, Md.: Rowman & Littlefield, 1999.
Mellenthin, F. W. von. *Panzer Battles*. Translated by H. Betzler. Norman, Okla.: Oklahoma University Press, 1956.
Messenger, Charles. *The Blitzkrieg Story*. New York: Scribners, 1976.
Messerschmidt, Manfred. "German Military Effectiveness." In *Military Effectiveness: Vol. 3—The Second World War*, edited by Allan R. Millett and Williamson R. Murray, 218–255. Boston: Allen & Unwin, 1988.

Middlebrook, Martin. "Harris." In *The War Lords*, edited by Michael Carver. London: Weidenfeld & Nicolson, 1976.

Millett, Allan R. "The United States Armed Forces in the Second World War." In *Military Effectiveness: Vol. 3—The Second World War*, edited by Allan R. Millett and Williamson R. Murray, 45–89. Boston: Allen & Unwin, 1988b.

Millett, Allan R., Williamson R. Murray, and Kenneth H. Watman. "The Effectiveness of Military Organizations." In *Military Effectiveness: Vol. 1—The First World War*, edited by Allan R. Millett and Williamson R. Murray, 1–30. Boston: Allen & Unwin, 1988.

Mintzberg, Henry. "The Fall and Rise of Strategic Planning." *Harvard Business Review* 72, no.1 (1994): 107–114.

———. *The Rise and Fall of Strategic Planning*. London: Prentice Hall, 2000.

Moltke, Helmuth Graf von. *Moltke: On the Art of War—Selected Writings*. Edited by Daniel J. Hughes. Translated by Daniel J. Hughes and Harry Bell. New York: Ballantine, 1993.

Murray, Williamson R. *Luftwaffe*. London: George Allen & Unwin, 1985.

———. "British Military Effectiveness in the Second World War." In *Military Effectiveness: Vol. 3—The Second World War*. Edited by Allan R. Millett and Williamson R. Murray, 90–135. Boston: Allen & Unwin, 1988.

———. "Armored Warfare: The British, French and German Experiences." In *Military Innovation in the Interwar Period*, edited by Williamson R. Murray and Allan R. Millett, 6–49. Cambridge, U.K.: Cambridge University Press, 1996a.

———. "Strategic Bombing: The British, American and German Experiences." In *Military Innovation in the Interwar Period*, edited by Williamson R. Murray and Allan R. Millett, 96–143. Cambridge, U.K.: Cambridge University Press, 1996b.

———. "Innovation: Past and Future." In *Military Innovation in the Interwar Period*, edited by Williamson R. Murray and Allan R. Millett), 300–328. Cambridge, U.K.: Cambridge University Press, 1996c.

———. "Military Culture Does Matter." *Foreign Policy Research Institute* 7, no. 2 (January 1999). http://www.fpri.org/fpriwire/0702.199901.murray.militaryculturedoesmatter.html

———. "Contingency and Fragility of the German RMA." In *The Dynamics of Military Revolution 1300–2050*, edited by Macgregor Knox and Williamson R. Murray. Cambridge, U.K.: Cambridge University Press, 2001.

Na'aman, Shlomo, and Roni Cohen. "Tank Armament—Past, Present and Future." *Ma'arachot* 247–248 (1975): 22–56.

Nagar, Shaul. "Our Water Crossing, Our Victory." *Shiryon* 19 (2003): 136–140.

Naveh, Shimon. "Was the Blitzkrieg a Warfighting Doctrine? Pt. 1." *Ma'arachot* 330 (1993): 28–43.

———. "The Cult of the Offensive Preemption and Future Challenges for Israeli Operational Thought." In *Between War and Peace: Dilemmas of Israeli Security*, edited by Efraim Karsh, 168–187. London: Frank Cass, 1996.

———. *In Pursuit of Military Excellence: The Evolution of Operational Theory*. London: Frank Cass, 1997.

Nawroz, Mohammad Yahya, and Lester W. Grau. "The Soviet Experience in Afghanistan." *Military Review* 75, no. 5 (1995): 17–27.

Nazareth, J. *Dynamic Thinking for Effective Military Command*. New Delhi: Tata McGraw-Hill, 1977.

Ogorkiewicz, R. M. *Armoured Forces: A History of Armoured Forces and Their Vehicles*. London: Arms and Armour, 1960.

O'Hanlon, Michael. *Technological Change and the Future of Warfare*. Washington, D.C.: Brookings Institution, 2000.

Okumiya, Masatake, and Jiro Horikoshi. *Zero!* New York: Ballantine, 1956.

Oren, Amir. "Where Did the Red Army Go?" *Ha'aretz*, October 7, 2005, 3b.

Oren, Elhanan. *A History of the Yom Kippur War* [Hebrew]. IDF: History Unit, 2004.

Orgill, Douglas. *The Tank: Studies in the Development and Use of a Weapon*. London: Heinemann, 1970.

Overy, Richard. *Russia's War*. London: Penguin, 1997.

Owens, Bill. *Lifting the Fog of War*. Baltimore: Johns Hopkins University Press, 2001.

Paret, Peter, ed. *Makers of Modern Strategy: From Machiavelli to the Nuclear Age*. Princeton, N.J.: Princeton University Press, 1986.

Perrett, Bryan. *Knights of the Black Cross: Hitler's Panzerwaffe and Its Leaders*. London: Wordsworth, 1986.

Place, Timothy H. *Military Training in the British Army 1940–1944*. London: Frank Cass, 2000.

Posen, Barry R. *The Sources of Military Doctrine: France, Britain, and Germany Between the World Wars*. Ithaca, N.Y.: Cornell University Press, 1984.

———. "Explaining Military Doctrine." In *The Use of Force: Military Power and International Politics*, edited by Robert J. Art and Kenneth N. Waltz, 23–43. Lanham, Md.: Rowman and Littlefield, 1999.

Price, Alfred. *Battle Over the Reich*. London: Ian Allan, 1973.

———. *Instruments of Darkness: The History of Electronic Warfare*. London: Macdonald and Jane's, 1977a.

———. *Luftwaffe Handbook 1939–1945*. New York: Charles Scribner's Sons, 1977b.

Pridor, Adir. "The Ability to Measure Military Power." In *Quantity and Quality in Force Planning* [Hebrew], edited by Zvi Ofer and Avi Kober, 55–69. Tel Aviv: Ma'arachot, 1985.

Reuveny, Rafael, and Aseem Prakash. "The Afghanistan War and the Breakdown of the Soviet Union." *Review of International Studies* 25 (1999): 693–708.

Rom, Giora. "The Israeli Air Force in War." Lecture, part of a series of seminars on the

Yom Kippur War. Study Day no. 7. *The Air Arena in the Yom Kippur War*. Ramat Efal: Israel Galili Foundation for Research on Defense Force, The Israeli Military History Foundation at Tel Aviv University, May 4, 2000.

Rommel, Ervin. *The Rommel Papers*. Edited by B. H. Liddell Hart. Translated by Paul Findlay. London: Collins, 1953.

Ronen (Peker), Ran. *Hawk in the Sky* [Hebrew]. Tel Aviv: Yedioth Ahronot, 2002.

Rosen, Stephen P. *Winning the Next War*. Ithaca, N.Y.: Cornell University Press, 1991.

Russian General Staff. *The Soviet-Afghan War: How a Superpower Fought and Lost*. Translated and edited by Lester W. Grau and Michael A. Gress. Lawrence, Kans.: University Press of Kansas, 2002.

Samuel, Yitzhak. *Organizations: Features, Structures, Processes* [Hebrew]. Haifa: Haifa University Press, 1996.

Samuels, Martin. *Command or Control?: Command, Training and Tactics in the British and German Armies 1888–1918*. London: Frank Cass, 1995.

Sanders, Charles W., Jr. *No Other Law: The French Army and the Doctrine of the Offensive*. RAND, P-7331, 1987.

Schelling, Thomas C. *Arms and Influence*. New Haven: Yale University Press, 1966.

Schmidt, Heinz Werner. *With Rommel in the Desert*. London: Panther, 1960.

Scholls, Josef. "Night Fighter Tactics (NJ6)." In *Fighting the Bombers: The Luftwaffe's Struggle Against the Allied Bomber Offensive*, edited by David C. Isby, 230–236. London: Greenhill, 2003.

Schwartz, Peter. *The Art of the Long View: Planning for the Future in an Uncertain World*. Chichester, U.K.: John Wiley, 1998.

Scott, Robert Lee. *Flying Tiger: Chennault of China*. New York: Doubleday, 1959.

Senger Und Etterlin, F. M. von. *Neither Fear Nor Hope*. Translated by George Malcolm. London: MacDonald, 1963.

Shai (Shwartz), Hanan. *Command and Control in the Modern Military Organization*. IDF: The Command and Staff College, Barak Program, 1994.

Shamir, Ami. "The Combat Engineers Corps" [Hebrew]. In *IDF—A Military and Security Encyclopedia*. Tel Aviv: Revivim and Sifriyat Ma'ariv, 1981.

Sharon, Dan. "Surprise, Problem Solving and Commanders Education." *Ma'arachot* 278 (1981): 30–34.

Sharp, Charles C. *German Panzer Tactics in World War II: Combat Tactics of German Armored Units From Section to Regiment*. West Chester, Ohio: Nafziger Collection, 1998.

Shazli, Saad el Din. *The Crossing of the Suez: The Egyptian Chief of Staff's Memoirs From the Yom Kippur War* [Hebrew]. Tel Aviv: Ma'arachot, 1987.

Sheffield Gary D. "Introduction: Command, Leadership and the Anglo-American Experience." In *Leadership and Command: The Anglo-American Military Experience Since 1861*, edited by Gary D. Sheffield, 1–16. London: Brassey's, 1997.

Sherwin, Ronald G., and Barton Whaley. "Understanding Strategic Deception: An Analysis of 93 Cases." In *Strategic Military Deception*, edited by Donald C. Daniel and Katherine L. Herbig, 177–194. New York: Pergamon, 1982.

Shimshi, Eliyashiv. *Storm in October* [Hebrew]. Tel Aviv: Ma'arachot, 1987.

Shor, Zvi. "Military Strength in the Budget Clamp." In *Quantity and Quality in Force Planning* [Hebrew], edited by Zvi Ofer and Avi Kober, 305–311. Tel Aviv: Ma'arachot, 1985.

Shvili, Eli. "Thinking About Countermeasures." *Ma'arachot* 355 (1998): 38–45.

Shy, John. "First Battles in Retrospect." In *America's First Battles 1776–1965*, edited by Charles E. Heller and William A. Stofft, 327–352. Lawrence, Kans.: University Press of Kansas, 1986.

Simpkin, Richard E. *Race to the Swift: Thoughts on Twenty-First Century Warfare*. London: Brassey's, 1985.

16th Century: The 16th Squadron Heritage Book [Hebrew]. IAF, 1987.

Snyder, Jack. "The Cult of the Offensive." In *The Use of Force: Military Power and International Politics*, edited by Robert J. Art and Kenneth N. Waltz, 113–129. Lanham, Md.: Rowman and Littlefield, 1999.

Sobelman, Ariel. "Information on the Modern Battlefield and Information Warfare." In *Israel's Security Web—Core Issues of Israel's National Security in Its Sixth Decade* [Hebrew], edited by Hagai Golan, 218–236. Tel Aviv: Ma'arachot, 2001.

Spector, Iftach. "Squadron Command in Wartime—Personal Lessons." In *Air Force Headquarters—Air Division. Echoes from the War: Air Commanders and Airmen Speak about Combat*. IAF, 1984.

———. *Dream in Blue and Black* [Hebrew]. Jerusalem: Keter, 1991.

———. *Loud and Clear* [Hebrew]. Tel Aviv: Miskal-Yedioth Ahronoth and Chemed Books, 2008.

Speilberger, Walter J. *Panzer III and Its Variants*. Atglen, Pa.: Schiffer Military History, 1993a.

———. *Panzer IV and Its Variants*. Atglen, Pa.: Schiffer Military History, 1993b.

Sray, John E. *Mountain Warfare: The Russian Perspective* FMSO, March 1994. http://fmso.leavenworth.army.mil/documents/mountain.htm.

Steiger, Rudolf. *Armour Tactics in the Second World War: Panzer Army Campaigns of 1939–1941 in German War Diaries*. Translated by Martin Fry. New York: BERG, 191.

Stolfi, R. H. S. *Hitler's Panzers East: World War II Reinterpreted*. Norman, Okla.: University of Oklahoma Press, 1991.

Strategy Survival Guide. British Cabinet Office. Prime Minister's Strategy Unit, Ver. 2.1., May 2004.

Strong, Kenneth. *Men of Intelligence*. London: Cassel, 1970.

Sullivan, Gordon R., and Michael V. Harper. *Hope Is Not a Method: What Business Can Learn From America's Army*. New York: Times Business, 1996.

Sun Tzu. *Roots of Strategy*. Edited by Thomas R. Phillips. Harrisburg, Pa.: Stackpole Books, 1985.

Suvorov, Victor. *Icebreaker: Who Started the Second World War?* Translated by Thomas B. Beattie. London: Hamish Hamilton, 1990.

Tal, Israel. *National Security: The Few Against the Many* [Hebrew]. Tel Aviv: Dvir, 1996.

Tamari, Dov. "The Yom Kippur War—A Question of Ignorance." *Zmanim* 84 (2003): 18–31.

"Târgul Frumos and the Counter Stroke at Facuti. Operational and Tactical Lessons—War on the Eastern Front 1941–45." Camberly Staff College, no date.

Tellis, Ashley J., Janice Bially, Christopher Layne, and Melissa McPherson. *Measuring National Power in the Postindustrial Age.* RAND, MR-1110-A, 2000.

Thompson, Walter. *Lancaster to Berlin.* Guernsey, Channel Islands: Goodall, 1997.

Toland, John. *The Rising Sun: The Decline and Fall of the Japanese Empire 1936–1945.* London: Penguin, 2001.

Tversky, A., and D. Kahenman. "Judgment Under Uncertainty: Heuristics and Biases." *Science* 185 (1974): 1124–1131.

Tyushkevich, S. A. *The Soviet Armed Forces: A History of Their Organizational Development: A Soviet View.* Translated by the CIS Multilingual Section, Translation Bureau, Secretary of State Department, Ottawa, Canada. Superintendent of Documents U.S. Government Printing Office, 1978.

Urban, Mark. *War in Afghanistan.* London: Macmillan, 1988.

"Use of Scenarios in Long Term Defence Planning." 10.4.2007. http://www.plausible futures.com.cparticle55074-6691a.html

Van Creveld, Martin. *Supplying War: Logistics From Wallenstein to Patton.* Cambridge, U.K.: Cambridge University Press, 1977.

———. *Fighting Power: German and U.S. Army Performance, 1939–1945.* Westport, Conn.: Greenwood, 1982.

———. *Command in War.* Cambridge, Mass.: Harvard University Press, 1985.

Van Evera, Stephen. *Causes of War: Power and the Roots of Conflict.* Ithaca, N.Y.: Cornell University Press, 1999.

Wallach, Yehuda. "Requiem for the Tank." *Ma'arachot Shiryon* 23 (1971): 40–41.

———. "Is the Tank Dead? Another Nail in Armor's Coffin." *Ma'arachot Shiryon* 25 (1972): 4–9.

———. "Can We Learn From History?" *Ma'arachot* 254 (1977): 30–33.

———. *Military Doctrines, Their Development in the 19th and 20th Centuries* [Hebrew]. Tel Aviv: Ma'arachot, 1977b.

Wark, Wesley K. *The Ultimate Enemy: British Intelligence and Nazi Germany, 1933–1939.* Ithaca, N.Y.: Cornell University Press, 1985.

Watt, Donald C. "British Intelligence and the Coming of the Second World War in Europe." In *Knowing One's Enemies: Intelligence Assessment Before the Two World Wars,* edited by Ernest R. May, 237–270. Princeton, N. J.: Princeton University Press, 1984.

Wavell, A. P. *The Good Soldier.* London: MacMillan, 1948.

Webster, Charles, and Noble Frankland. *The Strategic Air Offensive Against Germany 1939–1945, Vol. 2: Endeavor.* London: Her Majesty's Stationery Office, 1961.
Wechsler, Yishai, and Yehuda Tal. *198 Tank Battalion at Yom Kippur War* [Hebrew]. Tel Aviv: Defense Ministry Publications, 2002.
Weick, Karl E., and Kathleen M. Sutcliffe. *Managing the Unexpected: Assuring High Performance in an Age of Complexity.* San Francisco: Jossey-Bass, 2001.
Weise, Hubert. "The Overall Defense of the Reich: 1940–1944 (January)." In *Fighting the Bombers: The Luftwaffe's Struggle Against the Allied Bomber Offensive*, edited by David C. Isby, 50–61. London: Greenhill, 2003.
Weizman, Ezer. *On Eagles' Wings: The Personal Story of the Commanding Officer of the Israeli Air Force* [Hebrew]. Tel Aviv: Ma'ariv, 1975.
Westermann, Edward B. "The Limits of Soviet Airpower: The Failure of Military Coercion in Afghanistan, 1979–89." *Journal of Conflict Studies* 19 (1999), no.2. http://www.lib.unb.ca/Texts/JCS/bin/get5.cgi?directory=fall99/andfilename=WESTERMA.htm 2.
Whaley, Barton. *Stratagem: Deception and Surprise in War.* Norwood, Mass.: Artech House, 2007.
Wiener, Erez. "Formation and Crisis in the IDF's Warfighting Doctrine" [Hebrew]. In *Yom Kippur War Studies*, edited by Hagai Golan and Shaul Shai, 83–84. Tel Aviv: Ma'arachot, 2003.
Winograd Committee Final Report on Israel's 2006 War in Lebanon. 2008.
Winton, Harold R. "Tanks, Votes and Budgets: The Politics of Mechanization and Armored Warfare in Britain, 1919–1939." In *The Challenge of Change: Military Institutions and New Realities, 1918–1941*, edited by Harold R. Winton and David R. Mets, 74–107. Lincoln, Neb.: University Press of Nebraska, 2000.
Wohlstetter, Roberta. *Pearl Harbor—Warning and Decision.* Stanford, Calif.: Stanford University Press, 1962.
Y and Y. "The Airplane in the Ground Battle: End of the Road or Crossroads?" *Ma'arachot* 266 (1978): 43–46.
Yoffe, Eli. "China—Awakening Giant" [Hebrew]. In *Quantity and Quality in Force Planning*, edited by Zvi Ofer and Avi Kober, 139–145. Tel Aviv: Ma'arachot, 1985.
Yogev, Amnon. "When the Eyes No Longer Surround the Fighting Theater: Preparation for the Battlefield of the Future Involves Changes in Weapons Systems and Thinking Patterns." An Interview with Haim Lapid and Hagai Ben-Zvi. *Skira Hodshit* 36, no. 12 (1990): 39–33.
Yonay, Ehud. *No Margin for Error: The Story of the Israeli Air Force* [Hebrew]. Jerusalem: Keter, 1995.
Young, Robert, J. "French Military Intelligence and Nazi Germany, 1938–1939." In *Knowing One's Enemies: Intelligence Assessment Before the Two World Wars*, edited by Ernest R. May, 271–309. Princeton, N.J.: Princeton University Press, 1984.
Zaloga, Steven J., and James Grandsen. *The T-34 Tank.* London: Osprey, 1980.

———. *Soviet Heavy Tanks*. London: Osprey, 1981.

Zeira, Eli. *The October 1973 War: Myth Against Reality* [Hebrew]. Tel Aviv: Yedioth Ahronot, 1993.

Zigdon, Ya'akov, and Nisim Ohana. "Force Preparation—The Bridge That Links Force Design and Development to Its Application." *Ma'arachot* 372 (2000): 50–53.

Zisk, Kimberly Marten. *Engaging the Enemy: Organizational Theory and Soviet Military Innovation, 1955–1991*. Princeton, N.J.: Princeton University Press, 1993.

Military Documents

British Army. *The Application of Force: An Introduction to British Army Doctrine and to the Conduct of Military Operations*, DGDandD/18/34/66. The British Army: Army Code no. 71622, 1998.

British Army. *Land Operations*, AC 71819. Directorate General Development and Doctrine, May 2005.

British Army. *United Kingdom Glossary of Joint and Multinational Terms and Definitions*. Joint Doctrine Publication 0–01.1 (JDP 0–01.1). Shrivenham: The Development, Concepts and Doctrine Centre, Ministry of Defence, Edition 7, June 2006, F-8.

German Army. *Command and Staff College Dictionary*. Hamburg, 1983.

IDF Doctrine and Training Division. "IDF Dictionary" [Hebrew]. Publication 1–10. 1998.

IDF Education and Training Department. "Command and Control by Employing 'Mission Commands'" [Hebrew]. 1987.

U.S. Army. *German Military Improvisations During the Russian Campaign*. 1971. Originally published by the American Army as MS T-21 "German Military Improvisations."

U.S. Army. *How the Army Runs: A Senior Leader Reference Handbook 2007–2008*. Carlisle, Pa.: U.S. Army War College, December 13, 2007.

U.S. Army. FM 3–0. Operations (2008).

U.S. Department of Defense. *Department of Defense Dictionary of Military and Associated Terms*. Joint Publication 1–02. U.S Department of Defense. 2001. Updated 2007.

Interviews

Finkel, Meir. *The Sagger Missile Surprise in the Yom Kippur War*. Interview with Dani Asher, director of Israel's Southern Command Intelligence Branch for studying the Egyptian army, 1970–1971. Glilot: December 29, 2003.

Finkel, Meir. *The Surface-to-Air Missile Surprise in the Yom Kippur War*. Interview with Giora Rom, commander of the 115th Skyhawk Squadron in the Yom Kippur War. Nahalat Yehuda: July 21, 2005.

INDEX

Adan, Avraham (Bren), 35, 65–66, 161–162
Aircraft:
  American:
    B-17, 38;
    B-52, 94, 234;
    Brewster-Buffalo, 116;
    Mustang, 93;
    P-40 (Curtiss), 117;
  British:
    Hurricane, 116;
    Lancaster, 126;
    Mosquito, 131, 136;
    Spitfire, 116;
  German:
    Junkers 88, 128;
    Messerschmitt 110, 128;
    Stuka (Junkers 87), 145;
  Israeli:
    Skyhawk (A-4), 169, 171, 175, 177;
    Phantom (F-4E), 94, 165–166, 168, 173–174;
  Japanese:
    Zero, 4, 115–117, 261n78, 263n99, 278n21;
  Soviet:
    Hind (Mi-24)
Air force:
  American, 38, 117, 164;
  British (RAF), 116, 123, 125–126, 130, 132–134, 254n27, 263n98, 274n5;
  German (Luftwaffe), 10–11, 35–36, 43, 81, 123, 125–128, 131, 135–137, 148, 183, 263n98;
  Israeli (IAF), 12, 16, 94, 164–178, 238, 260n65;
  Japanese, 4, 115–17, 261nn71, 78, 263n99, 278n21
Alam el-Halfa, 77
Anti-aircraft: *See* Surface-to-air missiles
Anti-tank:
  Guns:
    American:
      37 mm, 87;
    British:
      3.7-inch AA, 13, 105, 179, 182, 188, 289n16;
      40-mm, 187;
      Six-pound, 188;
    German:
      50-mm, 13, 181, 188;
      88-mm AA, 3–4, 13, 93–94, 105, 145, 179–183, 187–189
  Missiles:
    Arab (Soviet):
      Sagger (AT-3), 12, 24, 115, 150–163, 171, 224, 233, 235–236, 240, 244;
      Shmel, 151–153;
    Israeli (French):
      SS-10, 152;
      SS-11, 152
Arras, 4, 105
Armor warfare and formations (*See also* tanks):
  American: 87;
  British: 12, 59–60, 62, 77, 88, 181, 186–188, 295n16;

322 INDEX

French: 14, 43, 62, 209–211, 218–220, 259n60, 295n16;
German: 11, 38, 41, 44, 58, 60–62, 69–71, 77, 81, 88, 93, 108, 111, 114, 138, 140–143, 145, 148, 180, 187–188, 207–208, 211, 229, 234, 258n58, 261nn74, 76, 295n16;
Iraqi: 77–78;
Israeli: 12, 28, 35, 65–66, 80, 115, 150, 152–163, 167, 173, 178, 233, 236, 251n6, 254n27;
Russian (and Soviet): 11, 13–14, 62, 78, 105, 107, 138, 140, 142, 192–194, 197, 199–202
Army (excluding air forces and navies):
American: 33, 83, 87, 93, 101, 105, 108–109, 115, 117, 227, 251n5, 253n18, 273–274n4, 274n5, 276n8;
British: 55, 58–59, 101, 114, 179–180, 185–186, 189, 273n4, 275n5;
Chinese: 34, 261;
Egyptian: 12, 15, 24, 41, 43, 67, 150–153, 157–158, 161–162, 251n6;
German (Wehrmacht): 3, 8, 11, 32, 60, 62–64, 68–72, 76, 81, 101, 103–5, 114, 127, 135, 138, 141–142, 144–148, 180, 195, 205, 212, 225, 231, 256n53, 258n58, 266n18, 273n4, 274n5;
French: 14, 62, 66, 72, 111, 205–207, 209–211, 213–215, 218–219, 273n4, 295n16;
Israeli (IDF): 1, 12, 16, 31, 35, 43, 63–69, 72, 76, 79–80, 82, 86, 90–92, 101, 105, 114–115, 118, 150–168, 174–175, 178, 224–227, 231, 235–236, 240, 246, 250n2, 251nn5, 6, 253n18, 254n27, 274n4, 298n11;
Japanese: 115, 183–184, 261n76, 263n99;
Russian (Red Army, Soviet army): 10, 32, 44, 62, 78–79, 106–108, 141, 191, 193, 195–196, 202–204, 138, 141, 144, 203, 219, 231, 247, 253n18, 254n23, 273n4, 281n2, 292n8, 293n32, 295n16;
Syrian: 1, 43, 154, 167, 171

Battleaxe, 181, 291
Barbarossa, 6, 11, 31, 33, 75, 76, 79–81, 95, 99, 106, 138, 140, 148, 202, 234–236
Beck, Ludwig von, 60
Berlin raid, 10, 126, 130, 131, 133–34

Cambrai, 26, 252n10, 261n76

Cast Lead, 118, 127
Chennault, Claire, 117, 278n21
Churchill, Winston, 59, 116, 126, 181
Clausewitz, Carl von, 29, 31–33, 38, 49, 68, 101, 141, 169, 231, 240, 272n2
Crusader 181, 291n62

De Gaulle, Charles, 26, 62, 209, 219, 231, 295n16, 296n28
Doctrine:
American: 28, 32, 38, 86–88, 109, 110, 135, 274–275n5;
British: 3, 32, 43, 59, 179, 184–186, 273n4, 274–275n5;
Egyptian: 43, 235, 251n6;
French: 14, 16, 64–65, 206, 210, 212–22, 260n73;
German (Including *Truppenführung-HD-300*): 3, 10, 28–29, 32, 60–61, 64, 67–72, 80, 86, 101–102, 103–104, 119, 135–137, 144–145, -146, 148, 179–180, 206–208, 210–11, 216, 225, 234, 273n4, 274n5, 294n8;
Israeli: 3, 12, 28, 35, 64–69, 76, 79, 91, 150–151, 154–163, 166, 169–72, 174, 177–78, 250–251n2;
Japanese: 261n75;
Mujahideen: 13, 192, 193, 246;
NATO: 28;
Russian (and Soviet): 32, 79, 109, 142, 192, 195, 199, 203;
Syrian, 43
Dugman 5, 164, 166–69, 174

Flying Tigers, 4, 117, 278n21
Fording devices in the Yom Kippur War, 90–92, 145, 168
Focus (Moked), 24, 170, 243
Fuller, Frederick Charles (as an armor theorist before WWII), 58, 62, 185, 206–208, 295n16

Gamelin, Maurice, 210, 218–220, 259n61, 262–263n94
Goering, Hermann, 62, 128–129, 136
Golan Height, 12, 67, 150, 154, 158, 164–167, 171–172, 174–175
Guderian, Heinz, 59–63, 70–71, 103, 139–140, 146, 216–217, 221, 258n58, 266n18, 281n2, 295n16

Harris, Arthur, 126–127, 130–131, 134

Herrmann, Hajo, 128–29, 135
Hitler, Adolph, 25, 39, 41, 58, 61–62, 64, 68, 70–72, 141, 146, 205–6, 210, 229, 295n16
Hobart, Percy, 59, 185–186, 226n9
Hamburg raid, 10, 125–29, 131–32, 134

Intelligence:
    American: 43–44, 48, 115, 244, 261n75;
    British: 43, 125, 132–133, 184;
    French: 43, 209–10;
    German: 44, 62, 141, 236, 263n98, 295n16;
    Israeli: 12, 45, 153–56, 170, 235;
    Russian: 195

Jones, R.V., (as head of the scientific intelligence of Britain's air staff), 125, 132, 238, 254n24

Kammhuber, Josef, 123–24, 127–129, 135
Kasserine Pass, 86–88

Leeb, Wilhelm von, 69
Lesson Learning from and during Wars and Operations:
    Afghanistan: 202;
    Falkland: 118, 203;
    Iraq war: 78, 119;
    Six-Day War: 65–66, 153, 155–157, 159, 160, 164, 175, 259n60;
    Spanish Civil War: 111, 209–210, 219, 262n94;
    Vietnam: 164–65, 195, 198, 203, 229;
    World War I: 14, 58, 76, 189, 208, 216;
    World War II:
        France (1940): 148, 190, 259n61, 283–284n50;
        Eastern Front: 144, 148;
        Poland: 60, 147–148, 210, 283–284n50;
        North Africa: 190, 291n62;
        Yom Kippur War: 94, 151, 154, 162–163, 172–174, 176
Liddell Hart, Basil (as a military theoretician and advisor before WWII), 208, 266n10, 295n16
Lossberg, Victor von, 128

Maginot Line, 58, 213
Mahle, Paul, 133–35
Manstein, Erich von, 60–61, 103
Manteuffel, Hasso von, 4, 105–106
Methodical Battle: *See* French doctrine

Meuse river crossing, 216–217, 221
Milch, Erhard, 126, 128–29, 131
Moltke, Helmuth von, 32, 100, 102, 104, 169, 273n4
Montgomery, Bernard, 77, 273n4
Montgomery-Massingberd, Archibald, 59, 189, 266n12
Moscow, 139

Navy:
    American: 43, 110, 164, 261n75, 262n94;
    British: 116, 259n59, 274n5;
    Egyptian: 245, 260n65;
    Israeli: 236, 245, 260n65;
    Japanese: 110, 115, 261n83;
    Syrian: 245

Peled, Benny, 167–68, 176
Pearl Harbor, 6, 24, 40, 115, 116, 117, 234–235, 237, 259n65, 261n74, 262n94

Radar and Counter Radar:
    Radar:
        German:
            Corfu, 131;
            Flensburg, 131, 254n27;
            Freya, 123, 124;
            Lichtenstein SN-2, 15, 124–25, 127, 131, 132, 136;
            Monica, 131–132, 254n27;
            Naxos, 11, 131, 132, 280;
            Würzburg, 124, 125, 136, 277n12;
        British:
            G-H, 132;
            H2S, 131–132;
    Counter Radar:
        Window (Chaff), 9–10, 15, 89, 121, 123–28, 130–32, 134–36, 166, 229, 237, 254n24, 259–60n65, 279n12;
        Tame Boar, 11, 127–130;
        Wild Boar, 11, 127–130;
    German Radar navigation aids:
        Knickbein, 93;
        X-Gerät, 92;
        Y-Gerät, 92
Rom, Giora, 169, 171, 177
Rommel, Erwin, 4, 59, 71, 88, 94, 104–5, 114, 181–82, 185, 188, 207, 229
Rundstedt, Gerd von, 61, 258n58

Seeckt, Hans von, 102
Siegfried Line, 214

Singapore (and Malaya), 4, 115–116, 183–184, 228, 263n99
Spector, Iftach, 166, 168, 173–74, 176, 218
Stalin, Joseph, 78–79, 108
Stalin Line, 78
Stulpnagel, Otto von, 60, 258n58
Suez Canal, 3, 65, 67, 90–91, 150–51, 153, 165, 174, 236, 288n59
Surface-to-air missiles (SAMs), 12, 24, 164–65, 167–168, 170–172, 174, 178, 239, 246, 259–60n65:
    SA-2, 164–166;
    SA-3, 165–166, 168, 170;
    SA-6, 15, 118, 165–170, 174–175, 245;
    SA-7, 165, 175;
    Stinger, 96, 196, 239, 246

Tagar 4, 153, 166–169
Tal, Israel, 158–60
Tanks:
    American:
        Grant, 13, 189;
        Sherman, 87, 179, 189, 241;
    British:
        Centurion, 95;
        Crusader, 187–188;
        Matilda, 105, 187;
    German:
        Panzer II, 283n40;
        Panzer III, 11, 138, 143, 145–146, 179, 187, 211, 281n1;
        Panzer IV, 106, 138, 140, 143, 145–147, 179, 187–188, 211, 281n1, 283n40;
        Panzer V (Panther), 11, 15, 146, 147, 241;
        Panzer VI (Tiger), 11, 15, 105–106, 140, 146–147;
    French:
        Somua, 209, 211;
        Model B, 209, 211;
    Israeli:
        Merkava, 95, 253n12;
        Patton, 153, 160;
        Sho't (IDF codename for Centurion), 153;
    Japanese: 183–84;
    Russian:
        KV-1, 138, 140–41, 143, 147;
        T-34, 9, 11, 15, 63, 86, 95, 105, 118, 121, 138–141, 143, 146–148, 241, 283n40;
        Stalin, 4, 105–106
Târgul-Frumos, 4, 105, 277

Weise, Hubert, 128–129, 279n12

Ypres, 26, 252n10

The authorized representative in the EU for product safety and compliance is:
Mare Nostrum Group
B.V Doelen 72
4831 GR Breda
The Netherlands

www.ingramcontent.com/pod-product-compliance
Lightning Source LLC
Chambersburg PA
CBHW032101230426
43662CB00034B/151